LIBRARY
LYNDON STATE COLLEGE
LYNDONVILLE, VT 05851

# The Space Shuttle Decision,
## 1965-1972

# History of the Space Shuttle
Volume 1

# The Space Shuttle Decision, 1965–1972

T. A. HEPPENHEIMER

SMITHSONIAN INSTITUTION PRESS
WASHINGTON AND LONDON

By special arrangement with the National Aeronautics and
Space Administration, History Office, this publication is being offered for sale
by the Smithsonian Institution Press, Washington, D.C. 20560-0950.

Production editor: Ruth W. Spiegel

Library of Congress Cataloging-in-Publication Data
Heppenheimer, T. A., 1947–
History of the space shuttle/ T. A. Heppenheimer
    p.   cm.
Includes bibliographical references and index.
Contents: v. 1. Development of the space shuttle, 1965–1972–
v. 2. Development of the space shuttle, 1972–1981.
    ISBN 1-58834-014-7 (v. 1 : alk. paper)
    ISBN 1-58834-009-0 (v. 2 : alk. paper)
1. Space shuttles–United States–History. I. Title.
TL795.5.H4697   2002
629.44'1'0973—dc21                                                   2001049233

British Library Cataloguing-in-Publication available

Manufactured in the United States of America
09  08  07  06  05  04  03  02    5  4  3  2  1

∞ The paper used in this publication meets the minimum requirements of the
American National Standard for Information Sciences–Permanence of Paper
for Printed Materials ANSI Z39.48-1984.

For permission to reproduce illustrations appearing in this book, please correspond
directly with the owners of the works as listed in the individual captions.
The Smithsonian Institution Press does not retain rights
for these illustrations individually or maintain
a file of addresses for photo sources.

Para Beverley: mi vida, mi amor, la esposa de mi corazon.

# Contents

Acknowledgments . . . . . . . . . . . . . . . . . . . . . . . . . . . . . . . . . . . . . . . *v*
Introduction . . . . . . . . . . . . . . . . . . . . . . . . . . . . . . . . . . . . . . . . . . . *ix*
Abbreviations and Acronyms . . . . . . . . . . . . . . . . . . . . . . . . . . . . *xi*

**Chapter 1. Space Stations and Winged Rockets** . . . . . . . . . . . . . . . . . . . 1
   The *Collier's* Series . . . . . . . . . . . . . . . . . . . . . . . . . . . . . . . . . . . 1
   Background to the Space Station . . . . . . . . . . . . . . . . . . . . . . . 6
   Winged Rockets: The Work of Eugen Sänger . . . . . . . . . . . . . . . 12
   The Navaho and the Main Line of American Liquid Rocketry . . . . . 14
   The X-15: An Airplane for Hypersonic Research . . . . . . . . . . . . 25
   Lifting Bodies: Wingless Winged Rockets . . . . . . . . . . . . . . . . . . 35
   Solid-Propellant Rockets: Inexpensive Boosters . . . . . . . . . . . . . . 42
   Dyna-Soar: A Failure in Evolution . . . . . . . . . . . . . . . . . . . . . . . 49

**Chapter 2. NASA's Uncertain Future** . . . . . . . . . . . . . . . . . . . . . . . . 55
   Technology Bypasses the Space Station . . . . . . . . . . . . . . . . . . 55
   Apollo Applications: Prelude to a Space Station . . . . . . . . . . . . 60
   Space Station Concepts of the 1960s . . . . . . . . . . . . . . . . . . . . 66
   Early Studies of Low-Cost Reusable Space Flight . . . . . . . . . . . . 73
   Two Leaders Emerge: Max Hunter and George Mueller . . . . . . . . 84
   NASA and the Post-Apollo Future . . . . . . . . . . . . . . . . . . . . . . . 94

**Chapter 3. Mars and Other Dream Worlds** . . . . . . . . . . . . . . . . . . . 105
   Nuclear Rocket Engines . . . . . . . . . . . . . . . . . . . . . . . . . . . . . . 105
   A New Administrator: Thomas Paine . . . . . . . . . . . . . . . . . . . . 110
   Space Shuttle Studies Continue . . . . . . . . . . . . . . . . . . . . . . . . 116
   Space Shuttle Policy: Opening Gambits . . . . . . . . . . . . . . . . . . 119
   Paine Seeks a Space Station . . . . . . . . . . . . . . . . . . . . . . . . . . 126
   Space Shuttles Receive New Attention . . . . . . . . . . . . . . . . . . 131
   Space Task Group Members Prepare Plans . . . . . . . . . . . . . . . 136
   Agnew Leads a Push Toward Mars . . . . . . . . . . . . . . . . . . . . . 144

# THE SPACE SHUTTLE DECISION

**Chapter 4. Winter of Discontent** .............................. 151
    The Sixties ............................................. 152
    Mars: The Advance ...................................... 159
    Mars: The Retreat ...................................... 170
    The Turn of Congress ................................... 177
    Paine Leaves NASA ...................................... 186

**Chapter 5. Shuttle to the Forefront** ......................... 191
    The Air Force in Space ................................. 191
    The Air Force and NASA ................................. 198
    A New Shuttle Configuration ............................ 206
    Station Fades; Shuttle Advances ........................ 223
    The Space Shuttle Main Engine .......................... 235

**Chapter 6. Economics and the Shuttle** ....................... 245
    Why People Believed in Low-Cost Space Flight ........... 246
    The Shuttle Faces Questions ............................ 254
    Change at NASA and the Bureau of the Budget ............ 267
    The Fall of the Two-Stage Fully-Reusable Shuttle ....... 274

**Chapter 7. Aerospace Recession** .............................. 291
    The Boeing 747 ......................................... 293
    The Supersonic Transport (SST) ......................... 304
    The Lockheed L-1011 .................................... 318
    Aftermaths ............................................. 327

**Chapter 8. A Shuttle to Fit the Budget** ..................... 331
    The Orbiter: Convergence to a Good Solution ............ 331
    The Booster: Confusion and Doubt ....................... 346
    End Game in the Shuttle Debate ......................... 362
    TAOS: A New Alternative ................................ 372
    A Time to Decide ....................................... 380

*Contents*

**Chapter 9. Nixon's Decision** .................................. **389**
    Nixon and Technology ................................. 390
    Space Shuttle: The Last Moves ......................... 396
    The Hinge of Decision ................................ 408
    Loose Ends I: A Final Configuration ................... 415
    Loose Ends II: NERVA and Cape Canaveral ............... 423
    Awarding the Contracts ............................... 427

Bibliography ............................................. 437
Index .................................................... 447
The NASA History Series .................................. 467

# Acknowledgments

Some projects begin by happenstance, and in such a fashion, the present book grew out of my commercial work, *Countdown: A History of Space Flight* (John Wiley, 1997). In researching its source material, I made good use of the NASA SP series of books—and noticed that there was a significant gap in their coverage. The series included works on most of the principal piloted programs: Mercury, Gemini, Apollo, Skylab, Apollo-Soyuz. There was nothing, however, on the Space Shuttle.

In June 1996, I visited the NASA History Office to conduct research on an article for the magazine *Air & Space*. I heard someone say the name, "Roger," and quickly met the chief historian, Roger D. Launius. When I asked him about the lack of Space Shuttle coverage, he replied with a tale of woe. Launius had tried to interest aerospace historians in the project, but to no avail.

Seeing an opportunity, I suggested that I might write NASA's Space Shuttle book. Dr. Launius expressed interest and later, early in July, I submitted a formal proposal. Four months later, at the beginning of November, I was informed that I was to receive the assignment. The present book is the result of that effort.

It is a pleasure to note the people who have helped me in this work. Dr. Launius has been in the forefront, taking an active interest and steering me to other archives while offering full use of his own NASA holdings. His archivist Lee D. Saegesser, now retired, gave valuable help in finding specific documents. Other members of his staff have helped as well: Nadine Andreassen, Colin Fries, Mark Kahn, Stephen Garber, and Louise Alstork, who copyedited the typescript.

The correspondence, memos, and project documents that served as source material, exclusive of published books, filled three suitcases. Much of this material came from other NASA center archives, and I received valuable help from their own staff members. At Marshall Space Flight Center, key people include Mike Wright, Alan Grady, and Laura Ballentine. At Johnson Space Center, I received much help from Janine Bolton and Sharon Halprin. At Kennedy Space Center, I relied on Donna Atkins and Elaine Liston.

The Air Force maintains an extensive archive at Maxwell AFB, and I received good help from the librarian Ann Webb. A security officer, Archie DiFante, worked with classified materials and performed the highly valuable service of releasing some of their unclassified sections to me, on the spot. At

## THE SPACE SHUTTLE DECISION

the University of Michigan's Department of Aerospace Engineering, the librarian Kenna Gaynor helped as well. During my days as a graduate student in that department, early in the 1970s, I had compiled a collection of contractors' space shuttle documents, and had donated them to the department library. A quarter-century later, with help from my former professor Harm Buning, Ms. Gaynor found some of this material and sent it on to me.

I also learned much through interviews with key individuals: J. Leland Atwood, former chairman of North American Aviation (NAA); Robert Biggs, the corporate memory at Rocketdyne; Paul Castenholz, the man who made the Space Shuttle Main Engine; Maxime Faget, a principal Shuttle designer; and Dale Myers, NASA's Associate Administrator for Space Flight. This list of interviewees is somewhat limited, for a good reason: I was able to use transcripts from NASA's own program of oral history interviews.

In providing help, Professor John M. Logsdon stands in a class by himself. He personally conducted many of these interviews, and provided me with transcripts. He wrote a book-length monograph on the rise of the Space Shuttle. He also gave me the free use of his own archive at George Washington University, which contains thousands of neatly filed and readily accessible letters and memos. My thanks also go to Tammy Golbert of the Folsom Library at Rensselaer Polytechnic Institute. She gave me access to the papers of George Low that are on deposit within that library's special collections.

This project has been something of a family affair, and I note the help of my son Alex Heppenheimer, who used the Internet to secure a valuable guide to published source material. My former wife, Phyllis LaVietes, served as my secretary and took care of my word processing. I note as well the contributions of the artists Don Davis and Dan Gauthier, who provided me with illustrations. In addition to this, much of this book's line art and photography comes courtesy of the author Dennis Jenkins. His book *Space Shuttle* (Walsworth Publishing Company, 1996) contains much interior art, and he has generously made it available for use by NASA.

The NASA Headquarters Printing and Design Office developed the layout and handled printing for this volume. Specifically, I wish to acknowledge the work of Geoff Hartman, Janie Penn, Chris Pysz, and Kelly Rindfusz for their expert editorial and design work. In addition, Stanley Artis, Michael Crnkovic, and Jeffrey Thompson saw the book through the publication process. Thanks are due to all of them.

*Acknowledgements*

With considerable joy, I dedicate this book to my wife, the former Beverley Brownlee. We were married in West Palm Beach, Florida, on June 8, 1998, as I was completing the preparation of this work for the press.

**T.A. Heppenheimer**
Fountain Valley, California
June 28, 1998

# Introduction

The Space Shuttle took shape and won support, and criticism, as part of NASA's search for a post-Apollo future. As with the Army and Navy in World War II, NASA had grown rapidly during the 1960s. Similarly, just as those military services saw a sharp falloff in funding in the wake of victory, the success of the piloted Moon landings brought insistent demands that NASA should shrink considerably. In facing those demands, and in overcoming them to a degree, NASA established itself as a permanent player in Washington.

In civics books, we learn that the three branches of government include the White House, Congress, and the Supreme Court. In making policy and in carrying it out, however, the judiciary rarely plays a significant role. One may speak of a tripartite government with a different set of participants: the White House, Congress, and the Office of Management and Budget (OMB). Though the OMB is part of the Executive Branch and responds to the wishes of the President, its officials have considerable leeway to shape policy in their own right, by cutting budgets. In seeking its post-Apollo future, NASA repeatedly had to accept such cuts, as its senior officials struggled to win support within the White House.

During 1969, with Nixon newly elected and the first astronauts setting foot on the Moon, NASA Administrator Thomas Paine led a push for a future in space that promised to be expansive. He aimed at nothing less than a piloted expedition to Mars, propelled by nuclear rocket engines that were already in development. En route to Mars, he expected to build space stations and large space bases. Almost as an afterthought, he expected to build a space shuttle as well, to provide low-cost flight to these orbiting facilities.

Soon after Neil Armstrong made his one small step in the lunar Sea of Tranquillity, Paine received a cold bath in the Sea of Reality. Nixon's budget director, Robert Mayo, chopped a billion dollars from Paine's request. This brought an end to NASA's hopes for a space base and for flight to Mars. It appeared possible, however, to proceed with the space station and the shuttle, as a joint project. The shuttle drew particular interest within the Air Force, which saw it as a means to accomplish low-cost launches of reconnaissance satellites and other military spacecraft.

## WAY STATION TO SPACE

Congress, however, was deeply skeptical toward the proposed shuttle/station, as both the House and Senate came close to killing it in 1970. NASA responded to this near-death experience by placing the station on the shelf and bringing the Shuttle to the forefront. Its officials needed political support that could win over doubters in Congress, and they found this support within the Department of Defense.

The Air Force now found itself in a most unusual position. Its generals had worked through the 1960s to pursue programs that could put military astronauts in space. These programs had faltered, with the main ones, the Dyna-Soar and the Manned Orbiting Laboratory, being canceled in 1963 and 1969 respectively. Yet here was NASA offering the Pentagon a piloted Space Shuttle, and promising to design it to meet Air Force needs. Indeed, the Air Force would receive the Shuttle on a silver platter, for NASA alone would fund its development and construction. It is a measure of NASA's desperation that it accepted the Shuttle project on those terms. The ploy, however, worked. The Air Force gave its political support to the Shuttle, and NASA went on to quell the opposition on Capitol Hill.

The OMB was a tougher opponent. NASA tried to win it over by commissioning cost-benefit studies that sought to support the Shuttle on economic grounds. These studies, however, merely provided more ammunition for the OMB's critics. In mid-1971, these critics forced NASA to abandon plans for a shuttle with two fully reusable liquid-fueled stages, and to set out on a search for a shuttle design that would cost half as much to develop. Then, when the resulting design exercises promised success in meeting this goal, the OMB responded by arguing that this success showed that NASA could do still more to cut costs. Budget officials demanded a design that would be smaller and less costly, even though such a shuttle would have significantly less capability than the Air Force wanted.

By shrinking the Shuttle, however, NASA won support where it counted. Caspar Weinberger, the OMB's deputy director, gave his endorsement late in 1971. Nixon also decided that the nation should have the Shuttle. On the eve of decision, the key player proved to be OMB Director George Shultz. He decided that since the Shuttle was to serve the entire nation, it should have the full capability for which NASA hoped and the Air Force demanded. Shultz's decision reinforced Nixon's, putting an end to the OMB's continuing demands to downsize the design. The consequence was the Space Shuttle as we know it today.

# Abbreviations and Acronyms

| | |
|---|---|
| AACB | Aeronautics and Astronautics Coordinating Board (USAF-NASA) |
| AAF | Army Air Forces |
| AAS | American Astronautical Society |
| AB, A/B | airbreathing |
| ABES | Air Breathing Engine System (jet engine) |
| ACPS | attitude control propulsion system |
| AEC | Atomic Energy Commission |
| AFB | Air Force Base |
| AFL-CIO | American Federation of Labor-Congress of Industrial Organizations |
| AFSC | Air Force Systems Command (USAF) |
| AIAA | American Institute of Aeronautics and Astronautics |
| APU | auxiliary propulsion unit |
| ARDC | Air Research and Development Command (USAF) |
| ASSET | Aerothermodynamic/elastic Structural Systems Environmental Tests |
| ATM | Apollo Telescope Mount (Skylab) |
| ATSC | Air Technical Service Command (USAF) |
| AT&T | American Telephone and Telegraph Company |
| BoB | Bureau of the Budget |
| BTU | British thermal unit |
| CASI | Center for Aerospace Information |
| CD | certificate of deposit |
| CIA | Central Intelligence Agency |
| deg | degree |
| DoD | Department of Defense |
| EC/LS | environmental control/life support |
| ENG | engine |
| ESTP | Economics, Science, and Technology Programs (OMB) |
| °F | Fahrenheit degrees |
| F-1 | rocket engine designation |
| FAA | Federal Aviation Agency; after 1967, Federal Aviation Administration |
| FDL | Flight Dynamics Laboratory (Wright-Patterson AFB) |
| ft | foot |
| FWD | forward |
| FY | Fiscal Year |
| GD | General Dynamics Corporation |
| GE | General Electric Company |
| GLOW | gross liftoff weight |
| He | helium |
| HO | hydrogen-oxygen |
| HQ | headquarters |
| HS | Haynes Stellite (a class of superalloys) |
| IBM | International Business Machines Corporation |
| ICBM | intercontinental ballistic missile |
| IDA | Institute for Defense Analyses |

| | | |
|---|---|---|
| J-2 | rocket engine designation | |
| JFK | John F. Kennedy | |
| JP | Jet Propellant (grade of jet fuel) | |
| JPL | Jet Propulsion Laboratory | |
| K | thousand | |
| KISS | Keep It Simple, Stupid | |
| LACE | liquid air cycle engine | |
| LBJ | Lyndon Baines Johnson | |
| lb | pound | |
| LC | launch complex (Cape Canaveral) | |
| L/D | lift-to-drag ratio | |
| LH, $LH_2$ | liquid hydrogen | |
| LM | Lunar Module (Apollo) | |
| $LN_2$ | liquid nitrogen | |
| $LO_2$, LOX | liquid oxygen | |
| MIT | Massachusetts Institute of Technology | |
| MOL | Manned Orbiting Laboratory | |
| MOM | Manned Orbiting Module | |
| MORL | Manned Orbiting Research Laboratory | |
| M.P. | Member of Parliament | |
| MR | mixture ratio (rocket propellants) | |
| MSC | Manned Spacecraft Center (NASA) | |
| MSFC | Marshall Space Flight Center (NASA) | |
| MW | megawatts | |
| NAA | North American Aviation | |
| NACA | National Advisory Committee for Aeronautics | |
| NASA | National Aeronautics and Space Administration | |
| NASC | National Aeronautics and Space Council | |
| NERVA | Nuclear Engine for Rocket Vehicle Application | |
| NMI | NASA Management Instruction | |
| n.mi. | nautical miles | |
| NRO | National Reconnaissance Office | |
| NSC | National Security Council | |
| NSF | National Science Foundation | |
| NTOP | New Technology Opportunities Program | |
| OEO | Office of Economic Opportunity | |
| OMB | Office of Management and Budget | |
| OMS | orbital maneuvering system | |
| OMSF | Office of Manned Space Flight (NASA) | |
| OSSA | Office of Space Science and Applications (NASA) | |
| OST | Office of Science and Technology (White House) | |
| OWS | orbital workshop (Skylab) | |
| PARD | Pilotless Aircraft Research Division (NACA-Langley) | |
| PCG | Planning Coordination Group (NASA HQ) | |
| PRIME | Precision Recovery Including Maneuvering Entry | |
| PSAC | President's Science Advisory Committee | |
| PSG | Planning Steering Group (NASA HQ) | |
| psi | pounds per square inch | |
| PTA | Parent-Teacher Association | |
| R and D | research and development | |

## Abbreviations and Acronyms

| | |
|---|---|
| RATO | Rocket-Assisted Take-Off |
| RDT&E | research, development, test and engineering |
| RFP | Request for Proposal |
| RL-10 | rocket engine designation |
| ROMBUS | Reusable Orbital Module, Booster, and Utility Shuttle |
| RP | Rocket Propellant (rocket-grade kerosene) |
| rpm | revolutions per minute |
| SAB | Scientific Advisory Board (USAF) |
| SCLC | Southern Christian Leadership Conference |
| sec | second |
| SHHDC | Shuttle History Historical Documents Collection (NASA-MSFC) |
| S-IC | first stage of the Saturn V |
| S-II | second stage of the Saturn V |
| S-IVB | third stage of the Saturn V |
| SNECMA | Societe National d'Etude et de Construction de la Moteurs d'Aviation (France) |
| SP | Special Publication (NASA) |
| SRM | solid rocket motor |
| SSD | Space Systems Division (USAF) |
| SSME | Space Shuttle Main Engine |
| SST | supersonic transport |
| STAC | Science and Technology Advisory Committee (NASA HQ) |
| STG | Space Task Group |
| TAHO | Thrust Assisted Hydrogen-Oxygen |
| TAOS | Thrust Assisted Orbiter Shuttle |
| TEMPO | Technical Military Planning Operation (GE) |
| TPS | thermal protection system |
| TRW | Thompson Ramo Wooldridge, Incorporated |
| TVC | thrust vector control |
| TWA | Trans World Airlines |
| U.S. | United States |
| USAF | United States Air Force |
| USS | United States Ship (Navy) |
| VAB | Vehicle Assembly Building (Kennedy Space Center) |
| v.p. | vice president |
| wt | weight |
| XLR | Experimental Liquid Rocket |

# Space Stations and Winged Rockets

Before anyone could speak seriously of a space shuttle, there had to be widespread awareness that such a craft would be useful and perhaps even worth building. A shuttle would necessarily find its role within an ambitious space program. While science fiction writers had been prophesying such wonders since the days of Jules Verne, it was another matter to present such predictions in ways that smacked of realism. After World War II, however, the time became ripe. Everyone knew of the dramatic progress in aviation, which had advanced from biplanes to jet planes in less than a quarter-century. Everyone also recalled the sudden and stunning advent of the atomic bomb. Rocketry had brought further surprises as, late in the war, the Germans bombarded London with long-range V-2 missiles. Then, in 1952, a group of specialists brought space flight clearly into public view.

## *The Collier's Series*

One of these specialists, the German expatriate Willy Ley, had worked with some of the builders of the V-2 personally and had described his experiences, and their hopes, in his book *Rockets, Missiles, and Space Travel*.[1] The first version, titled *Rockets*, appeared in May 1944, just months before the first

---

1. Citation in bibliography.

## THE SPACE SHUTTLE DECISION

firings of the V-2 as a weapon. Hence, this book proved to be very timely. His publisher, Viking Press, issued new printings repeatedly, while Ley revised it every few years, expanding both the text and the title to keep up with fast-breaking developments.[2]

One day in the spring of 1951, Ley had lunch with Robert Coles, chairman of the Hayden Planetarium in Manhattan. He remarked that interest in astronautics was burgeoning in Europe. An international conference, held in Paris the previous October, had attracted over a thousand people. None had come from the U.S., however, and this suggested to Ley that Americans should organize a similar congress. Coles replied, "Go ahead, the planetarium is yours."

Ley proceeded to set up a symposium that took place on Columbus Day. Admission was by invitation only. Some invitations, however, went to members of the press. Among the attendees were a few staffers from *Collier's*, a magazine with a readership of ten million. Two weeks later, the managing editor, Gordon Manning, read a brief news item about an upcoming Air Force conference, in San Antonio, on medical aspects of space flight. He sent an associate editor, Cornelius Ryan, to cover this meeting and to see if it could be turned into a story.[3]

While no space enthusiast, Ryan was a meticulous reporter, as he would show in such books as *The Longest Day* and *A Bridge Too Far*. At the meeting, he fell in with Wernher von Braun, who had been the technical director of the V-2 project. Von Braun, a consummate salesman, had swayed even Hitler.[4] Over cocktails, dinner, and still more cocktails, Von Braun proceeded to deliver his pitch. It focused on a space station with an onboard crew living and working in space. Von Braun declared that it could be up and operating in orbit by 1967. It would have the shape of a ring, 250 feet in diameter, and would rotate to provide centrifugal force that could substitute for gravity in weightless space. The onboard staff of 80 people would include astronomers operating a major telescope. Meteorologists, looking earthward, would study cloud patterns and predict the weather.[5]

To serve the needs of the Cold War, von Braun emphasized the use a space station could have for military reconnaissance. He also declared that it

---

2. Expanded versions appeared in 1945, 1948, and 1952.
3. Ley, *Rockets*, pp. 330-331; AAS History Series, vol. 15, pp. 235-242.
4. Dornberger, *V-2*, pp. 103-111.
5. AAS History Series, vol. 15, pp. 235-242.

could operate as a high-flying bomber, dropping nuclear weapons with great accuracy. To build it, he called for a fleet of immense piloted cargo rockets (space shuttles, though the term had not yet entered use) each weighing 7,000 tons, 500 times the weight of the V-2. Yet the whole program—rockets, station and all—would cost only $4 billion, twice the budget of the wartime Manhattan Project that had built the atomic bomb.[6]

With its completion, the space station could serve as an assembly point for a far-reaching program of exploration. An initial mission would send a crew on a looping flight around the Moon, to photograph its unseen far side. Later, perhaps by 1977, a fleet of three rockets would carry as many as 50 people to the Moon's Bay of Dew for a six-week period of wide-ranging exploration using mobile vehicles.[7] Eventually, perhaps a century in the future, an even bolder expedition would carry astronauts to Mars.[8]

By the end of that evening, von Braun had converted Ryan, who now believed that piloted space flight was not only possible but imminent. Returning to New York, Ryan persuaded Manning that this topic merited an extensive series of articles that eventually would span eight issues of the magazine.[9] Manning then invited von Braun, together with several other specialists, to Manhattan for a series of interviews and discussions. These specialists included Willy Ley; the astronomer Fred Whipple of Harvard, a moon and Mars specialist; and Heinz Haber, an Air Force expert in the nascent field of space medicine.[10]

In preparing the articles, *Collier's* placed heavy emphasis on getting the best possible color illustrations. Artists included Chesley Bonestell, who had founded the genre of space art by presenting imagined views of planets such as Saturn, as seen closeup from such nearby satellites as its large moon Titan. Von Braun's engineering drawings and sketches of his rockets and spaceships were used by Bonestell and the other artists to create working drawings for Von Braun's review. They would execute the finished paintings only after receiving Von Braun's corrections and comments.[11]

---

6. *Ibid.*; *Time*, December 8, 1952, pp. 67, 71; *Collier's*, March 22, 1952, pp. 27-28.
7. *Collier's*, October 18, 1952, pp. 51-59; October 25, 1952, pp. 38-48.
8. *Ibid.*, April 30, 1954, pp. 22-29.
9. *Ibid.*, March 22, October 18 and October 25, 1952; February 28, March 7, March 14, and June 27, 1953; April 30, 1954. Reprinted in part in NASA SP-4407, vol. I, pp. 176-200.
10. *Collier's*, March 22, 1952, p. 23.
11. AAS History Series, vol. 15, p. 237; vol. 17, pp. 35-39.

THE SPACE SHUTTLE DECISION

Collier's, *March 22, 1952, spurred a surge of interest in space flight.* (Courtesy of Ron Miller)

The first set of articles appeared in March 1952, with the cover illustration of a space shuttle at the moment of staging, high above the Pacific. "Man Will Conquer Space *Soon*," blared the cover. "Top Scientists Tell How in 15 Startling Pages." Inside, an editorial noted "the inevitability of man's conquest of space" and presented "an urgent warning that the U.S. must immediately embark on a long-range development program to secure for the West 'space superiority.'"[12]

The series appeared while Willy Ley was bringing out new and updated editions of his own book. It followed closely *The Exploration of Space* by Arthur C. Clarke, published in 1951 and offered by the Book-of-the-Month Club.[13] The *Collier's* articles, however, set the pace. Late in 1952, *Time* magazine ran its own cover story on von Braun's ideas.[14] In Hollywood, producer

---

12. *Collier's*, March 22, 1953, p. 23.
13. Citation in bibliography.
14. *Time*, December 8, 1952.

*Space Stations and Winged Rockets*

*Cargo rocket of the* Collier's *series, with winged upper stage. (Art by Rolf Klep; courtesy of Ron Miller)*

George Pal was working already with Bonestell, and had brought out such science fiction movies as *Destination Moon* (1950) and *When Worlds Collide* (1951). In 1953, they drew on von Braun's work and filmed *The Conquest of Space*, in color. Presenting the space station and Mars expedition, the film proposed that the Martian climate and atmosphere would permit seeds to sprout in that planet's red soil.[15]

Walt Disney also got into the act, phoning Ley from his office in Burbank, California. He was building Disneyland, his theme park in nearby Anaheim, and expected to advertise it by showing a weekly TV program of that name over the ABC television network. With von Braun's help, Disney went on to produce an hour-long feature, *Man in Space*. It ran in October 1954, with subsequent reruns, and emphasized the piloted lunar mission. Audience-rating organizations estimated that 42 million people had watched the program.[16]

---

15. Miller and Durant, *Worlds Beyond*, pp. 100-102.
16. Ley, *Rockets*, p. 331.

## THE SPACE SHUTTLE DECISION

In its 1952 article, *Time* referred to von Braun's cargo rockets as "shuttles" and "shuttle rockets," and described the reusable third stage as "a winged vehicle rather like an airplane." His payload weight of 72,000 pounds proved to be very close to the planned capacity of 65,000 pounds for NASA's space shuttle.[17] He expected to fuel his rockets with the propellants nitric acid and hydrazine, which have less energy than the liquid hydrogen in use during the 1960s. Hence, his rockets would have to be very large. While his loaded weight of 7,000 tons would compare with the 2,900 tons of America's biggest rocket, the Saturn V,[18] his program cost of $4 billion was wildly optimistic.

Still, the influence of the *Collier's* series echoed powerfully throughout subsequent decades. It was this eight-part series that would define nothing less than NASA's eventual agenda for piloted space flight. Cargo rockets such as the Saturn V and the space shuttle, astronaut Moon landings, a space station, the eventual flight of people to Mars—all these concepts would dominate NASA's projects and plans. It was with good reason that, in the original *Collier's* series, the space station and cargo rocket stood at the forefront. By 1952, the concept of a space station had been in the literature for nearly 30 years, while large winged rockets were being developed as well.

## *Background to the Space Station*

The concept of a space station took root during the 1920s, in an earlier era of technical change that focused on engines. As recently as 1885, the only important prime mover had been the reciprocating steam engine. The advent of the steam turbine yielded dramatic increases in the speed and power of both warships and ocean liners. Internal-combustion engines, powered by gasoline, led to automobiles, trucks, airships, and airplanes. Submarines powered by diesel engines showed their effectiveness during World War I.[19]

After that war, two original thinkers envisioned that another new engine, the liquid-fuel rocket, would permit aviation to advance beyond the Earth's atmosphere and allow the exploration and use of outer space. These inventors were Robert Goddard, a physicist at Clark University in Worcester, Massachusetts, and Hermann Oberth, a teacher of mathematics in a *gymnasium*

---

17. *Time*, December 8, 1952, pp. 67, 68.
18. NASA SP-4012, vol. III, p. 27.
19. *Scientific American,* May 1972, pp. 102-111; April 1985, pp. 132-139.

in a German-speaking community in Romania.[20] Goddard experimented much, wrote little, and was known primarily for his substantial number of patents.[21] Oberth contented himself with mathematical studies and writings. His 1923 book, *Die Rakete zu den Planetenraumen* (The Rocket into Interplanetary Space), laid much of the foundation for the field of astronautics.

Both Goddard and Oberth were well aware of the ordinary fireworks rocket (a pasteboard tube filled with blackpowder propellant). They realized that modern technology could improve on this centuries-old design in two critical respects. First, a steel combustion chamber and nozzle in a rocket engine could perform much better than pasteboard. Second, the use of propellants such as gasoline and liquid oxygen would produce far more energy than blackpowder. Oberth produced two conceptual designs: the Model B, an instrument-carrying rocket for upper-atmosphere research, and the Model E, a spaceship.[22]

Having demonstrated to his satisfaction that space flight indeed was achievable, Oberth then considered its useful purposes. While he was not imaginative enough to foresee the advent of automated spacecraft (still well in the future), the recent war had shown that, using life support systems, submarines could support sizable crews underwater for hours at a time. Accordingly, he envisioned that similar crews, with oxygen provided through similar means, would live and carry out a variety of tasks in a space station as it orbited the Earth.

Without describing the station in any detail, he wrote that it could develop out of a plans for a large orbiting rocket with a mass of "at least 400,000 kg":

> *But if we should let a rocket of this size travel around the earth, it would constitute a sort of miniature moon. It would then no longer need to be designed or equipped for descent and landing. Traffic between this satellite and earth could be carried out with smaller vehicles and these large rockets (let us call them observation stations) could be built to further dimensions for their particular purpose. If ill effects result from experiencing weightlessness over long periods of time (which I doubt), two such rockets could be connected with a cable and caused to rotate about each other.*

---

20. Ley, *Rockets*, pp. 107, 116.
21. Lehman, *High Man*, pp. 360-363.
22. Ley, *Rockets*, pp. 108-112; NASA TT-F-9227, p. 98.

The station could serve as an astronomical observatory:

*In space, telescopes of any size could be used, for the stars would not flicker.... Sufficient for an objective glass would be a large, lightly shaded, concave mirror made of sheet metal. If this were mounted by means of three metal rods at a distance of several kilometers from the rocket, we would have a telescope which, for most purposes, would be one hundred times superior to the best instruments on earth.*

The station could also carry out Earth observations, while serving as a communications relay:

*With their sharp instruments they could recognize every detail on the earth and could give light signals to earth through the use of appropriate mirrors. They would enable telegraphic connections with places to which neither cables nor electrical waves can reach.... Their value to military operations would be obvious, be it that they are controlled by one of the belligerents or be it that high fees could be charged for the reports they could render. The station could observe every iceberg and could warn shipping, either directly or indirectly. The disaster of the Titanic of 1912, for example, could have been prevented in this way.*

Oberth also considered the building of immense orbiting mirrors, with diameters as large as 1,000 kilometers:

*For example, routes to Spitzbergen or to the northern Siberian ports could be kept free of ice. If the mirror had a diameter of only 100 km, it could make broad areas in the northern regions of the earth inhabitable through diffused light, and in our latitude it could prevent the fearful spring freezes and protect fruit crops from damage by night frosts in both spring and winter.*

He recommended sodium as a lightweight construction material. While it reacts strongly with oxygen, sodium would remain inert in airless space. He also described how the observation station also could serve as a fuel station:

*... if the hydrogen and oxygen are shielded from the sun's rays, they could be stored here for as long as desired in a solid state. A rocket which is filled here and launched from the observation station has no air resistance to overcome.... If we couple a large sphere of sodium sheet which is produced and*

## Space Stations and Winged Rockets

*filled with fuel on location with a small, stoutly built rocket which pushes its fuel supply ahead of it and is continually supplied by it, then we have a very powerful and long-range vehicle which is easily capable of making the trip to other bodies of the universe.*[23]

Although Oberth was shy and retiring by nature, the impact of his ideas, during subsequent decades, would rival that of von Braun's a generation later. *Die Rakete* spurred the founding of rocket-research groups in Germany, the U.S., and the Soviet Union. As early as 1898, Russia's Konstantin Tsiolkovsky, a provincial math teacher like Oberth, had developed ideas similar to those of Oberth's. Officials of the new Bolshevik government then dusted off Tsiolkovsky papers, showing that he had been ahead of the Germans. As his writings won new attention, the Soviet Union emerged as another center of interest in rocketry.[24]

Fritz Lang, a leading German film producer, then became interested. More than a filmmaker, Lang was a leader in his country's art and culture. Later, Willy Ley noted that at one of his premieres, "The audience comprised literally everyone of importance in the realm of arts and letters, with a heavy sprinkling of high government officials."[25] In 1926, Lang released the classic film *Metropolis*, with a robot in the leading role. Two years later, he set out to do the same for space flight with *Frau im Mond* (The Girl in the Moon).

Drawing heavily on Oberth's writings, Lang's wife, actress Thea von Harbou, wrote the script for *Frau im Mond*. Fritz Lang hired Oberth as a technical consultant. Oberth then convinced Lang to underwrite the building of a real rocket. After all, it would be great publicity for the movie were such a rocket to fly on the day of the premiere. The project attracted a number of skilled workers who went on to build Germany's first liquid-fuel rockets. Among them, a youthful Wernher von Braun went on to develop the V-2 with support from the German army.[26]

Even during the 1920s, Oberth's ideas drew enough attention to encourage other theorists and designers to pursue similar thoughts and to write their own books. Herman Potočnik, an engineer and former captain in the Austrian

---

23. NASA TT F-9227, pp. 92-97.
24. Ley, *Rockets*, pp. 100-104.
25. *Ibid.*, p. 124.
26. *Ibid.*, pp. 124-130; Neufeld, *Rocket and Reich*, pp. 11-23.

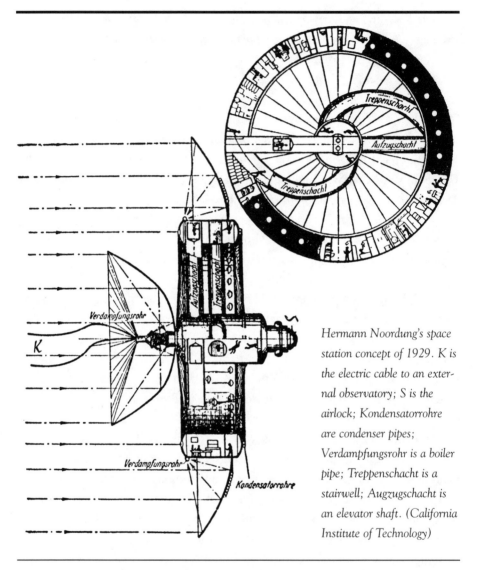

Hermann Noordung's space station concept of 1929. K is the electric cable to an external observatory; S is the airlock; Kondensatorrohre are condenser pipes; Verdampfungsrohr is a boiler pipe; Treppenschacht is a stairwell; Augzugschacht is an elevator shaft. (California Institute of Technology)

army, wrote under the pen name of Hermann Noordung. In 1929, he published *The Problem of Space Travel*, a book that addressed the issue of space station design. It was to be his last publication, however, for later that same year, he died of tuberculosis at the age of 36.[27]

Potočnik introduced the classic rotating wheel shape, proposing a diameter of 100 feet with an airlock at its hub. The sun would provide electric power,

---

27. NASA SP-4026, pp. xv-xvi.

though not with solar cells; these, too, lay beyond the imagination of that generation. Instead, a large parabolic mirror would focus sunlight onto boiler pipes in a type of steam engine. For more power, a trough of mirrors would run around the station's periphery concentrating solar energy on another system of pipes. Like a flower, the station would face the sun.[28]

Except for being two and a half times larger, von Braun's *Collier's* space station closely resembled that of Potočnik, and it is tempting to view von Braun as the latter's apt pupil. He certainly had the opportunity to read Potocnik's book (though published initially in its author's native language of Slovenian, it appeared quickly in German translation).[29] Moreover, von Braun's concept included a circumferential trough of solar mirrors for power. This, however, came not from Potočnik but rather from a suggestion of Fred Whipple (who had not read Potočnik's book), and thus represented an independent invention.[30] The influence of Potocnik on von Braun may have been only indirect.

The historian J.D. Hunley, who has prepared an English translation of Potočnik's book, describes its influence on von Braun as "probable but speculative." Nevertheless, he states unequivocally that "Potočnik's book was widely known even to people who may have seen only photographs of sections from the book in translation."[31] His concept of a large rotating wheel was sufficiently simple to permit von Braun and others to carry it in their heads for decades, developing this concept with fresh details when using it as the point of reference for an original design.

In the popular mind, if not for aerospace professionals, the *Collier's* series introduced the shape of a space station in definitive form. It carried over to Disney's *Man in Space*, and to George Pal's *Conquest of Space*. Fifteen years later, when producer Stanley Kubrick filmed Arthur C. Clarke's *2001: A Space Odyssey*, he too used the rotating-wheel shape, enlarging it anew to a diameter of a thousand feet.[32]

---

28. *Ibid.*, pp. 101-113.
29. *Ibid.*, pp. ix, xii.
30. Ley, *Rockets*, pp. 372-373.
31. NASA SP-4026, pp. xxii-xxiii.
32. Clarke, *2001*, photo facing p. 112.

## Winged Rockets: The Work of Eugen Sänger

While space stations came quickly to the forefront in public attention, it was another matter to build them, even in versions much smaller than von Braun's 250-foot wheel. Between 1960 and 1980 the concept flourished only briefly, in the short-lived Skylab program. The second major element of the *Collier's* scenario, the winged rocket, enjoyed considerably better prospects. At first merely topics for calculation and speculation, the development of long-range winged rockets during World War II was the departure point for a number of serious postwar projects.

In the 1930s, work on winged rockets foreshadowed the development of a high-speed airplane able to land on a runway for repeated flights. The first important treatment came from Eugen Sänger, a specialist in aeronautics and propulsion who received a doctorate at the *Technische Hochschule*[33] in Vienna and stayed on to pursue research on rocket engines. In 1933, he published *Raketenflugtechnik* (Rocket Flight Engineering). The first text in this field, it included a discussion of rocket-powered aircraft performance and a set of drawings. Sänger proposed achieving velocities as high as Mach 10, along with altitudes of up to 70 kilometers.[34]

While the turbojet engine was unknown at that time, it was this engine, rather than the rocket, that would offer the true path to routine high performance. Because a turbojet uses air from the atmosphere, a jet plane needs to carry fuel only, while its wings reduce the thrust and fuel consumption. Hence, it can maintain longer flight times. By contrast, a rocket must carry oxygen as well as fuel, and thus, while capable of high speeds, it lacks endurance. After World War II, rocket airplanes as experimental aircraft went on to reach speeds and altitudes far exceeding those of jets. Jet planes, however, took over the military and later the commercial realms.

During World War II, Sänger made a further contribution, showing how the addition of wings could greatly extend a rocket's range. Initially, a winged rocket would fly to modest range, along an arcing trajectory like that of an artillery shell. Upon reentering the atmosphere, however, the lift generated by

---

33. A technical institute that does not qualify as a university but that offers advanced academic studies, particularly in engineering.
34. AAS History Series, vol. 7, Part I, pp. 195, 203-206; vol. 10, pp. 228-230; Ley, *Rockets*, pp. 408-410.

The A-4b, a winged V-2 of 1945. (Smithsonian Institution Photo No. 76-7772)

the rocket's wings would carry it upward, causing it to skip off the atmosphere like a flat stone skipping over water. Sänger calculated that with a launch speed considerably less than orbital velocity, such a craft could circle the globe and return to its launch site.[35] After World War II, this concept drew high-level attention in Moscow, where, for a time, Stalin sought to use it as a basis for a serious weapon project.[36]

---

35. Ley, *Rockets*, pp. 428-434.
36. Zaloga, *Target*, pp. 121-124.

## The Navaho and the Main Line of American Liquid Rocketry

In haste and desperation, winged rockets entered the realm of hardware late in the war, as an offshoot of the V-2 program. The standard V-2 had a range of 270 kilometers. Following the Normandy invasion in 1944, as the Allies surged into France and the Nazi position collapsed, a group of rocket engineers led by Ludwig Roth sought to stretch this range to 500 kilometers by adding swept wings to allow the missile to execute a supersonic glide.

The venture was ill-starred from the outset. When winds blew on the wings during liftoff, the marginal guidance system could not prevent the vehicle from rolling and going out of control. In this fashion, the first winged V-2 crashed within seconds of its December 1944 launch. A month later, a second attempt was launched successfully and had transitioned to gliding flight at Mach 4. Then a wing broke off, causing the missile to break up high in the air.[37]

Nevertheless, this abortive effort provided an early point of departure for America's first serious long-range missile effort. In the Army Air Forces (AAF), the Air Technical Service Command (ATSC; renamed Air Materiel Command in March 1946) began by defining four categories of missiles: air-to-air, air-to-surface, surface-to-air, and surface-to-surface. The last of these included the V-2 and its potential successors.[38]

The program began with a set of military characteristics, outlined in August 1945, that defined requirements for missiles in these categories. AAF Headquarters published these requirements as a classified document. In November 1946, ATSC invited 17 contractors, most of them aircraft manufacturers, to submit proposals for design studies of specific weapons. One of these firms was North American Aviation (NAA) in Los Angeles.[39]

NAA had been a mainstay in wartime aircraft production. At the end of World War II, amid sweeping contract cancellations, the company dropped from 100,000 to 6,500 employees in about two months.[40] The few remaining contracts were largely in the area of jet-powered bombers and fighters. To NAA's president, James "Dutch" Kindelberger, these bombers repre-

---

37. Neufeld, *Rocket and Reich*, pp. 248-251, 281.
38. Neufeld, *Ballistic Missiles*, p. 26.
39. Fahrney, *History*, p. 1291; Neal, *Navaho*, pp. 1-2.
40. AAS History Series, vol. 20, pp. 121-132.

## Space Stations and Winged Rockets

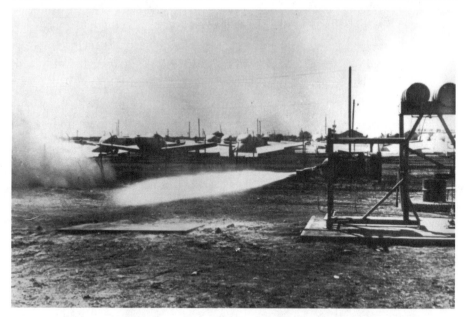

*Test of a small rocket engine in a parking lot at North American Aviation. (Rocketdyne)*

sented the way into the future. He decided to bring in the best scientist he could find and have him build a new research lab, staffed with experts in such fields as jet propulsion, rockets, gyros, electronics, and automatic control. The lab's purview, which would go well beyond the AAF study, was to work toward bringing in new business by extending the reach of the firm's technical qualifications.[41]

An executive recruiter, working in Washington, D.C., recommended William Bollay to head this lab. Bollay, who held a Ph.D. in aeronautical engineering from Caltech, had been a branch chief in the Navy's Bureau of Aeronautics, with responsibility for the development of turbojet engines. He came to NAA by November 1946, in time to deal with the AAF request for proposals. Working with the company's chief engineer, Raymond Rice, Bollay decided to pursue the winged V-2, which the Germans had designated as the A-9. During World War II, the Germans had regarded this missile as the next step beyond the standard V-2, hoping that its wings would offer a simple way to increase its range. The V-2's overriding priority had prevented serious

---

41. Author interview, J. Leland Atwood, Los Angeles, July 18, 1988.

work on its winged version. Late in 1945, however, the NAA proposal offered to "essentially add wings to the V-2 and design a missile fundamentally the same as the A-9."[42]

A letter contract, issued to the firm in April 1946, called for the study and design of a supersonic guided missile designated MX-770, with a range of 175 to 500 miles.[43] Meanwhile, rocket research was under way in an NAA company parking lot, with parked cars only a few yards away. A boxlike steel frame held a rocket motor; a wooden shack housed instruments. The steel blade of a bulldozer's scraper was used as a shield to protect test engineers in the event of an explosion.[44] A surplus liquid-fueled engine from Aerojet General, with a 1,000 pounds of thrust, served as the first test motor. The rocket researchers also built and tested home-brewed engines, initially with 50 to 300 pounds of thrust.[45] Some of these engines were so small that they seemed to whistle rather than roar. In the words of J. Leland Atwood, who became company president in 1948, "We had rockets whistling day and night for a couple of years."[46]

In June 1946, the first step toward a coordinated plan came in the form of a new company proposal. In the realm of large rocket-engine development, Bollay and his associates proposed a two-part program:

> *Phase I:* Refurbishment and testing of a complete V-2 propulsion system, to be provided as government-furnished equipment.
>
> *Phase II:* Redesign of this engine to American engineering standards and methods of manufacture, along with construction and testing.

In the spring of 1947, the company added a further step:

> *Phase III:* Design, construction and testing of a new engine, drawing on V-2 design but incorporating a number of improvements.[47]

Bollay and his colleagues also launched an extensive program of consultation with Wernher von Braun and his wartime veterans. These included

---

42. *Ibid.*; author interview, Jeanne Bollay, Santa Barbara, California, January 24, 1989; Report AL-1347 (North American), pp. 1-4; Neufeld, *Rocket and Reich*, p. 249.
43. Report AL-1347 (North American), pp. 5-6.
44. *Threshold*, Summer 1993, pp. 40-47.
45. Report AL-1347 (North American), p. 37.
46. Author interview, J. Leland Atwood, Los Angeles, July 18, 1988.
47. Report AL-1347 (North American), pp. 9-10, 34.

Walther Riedel, Hans Huter, Rudi Beichel, and Konrad Dannenberg. In addition, Dieter Huzel, a close associate of von Braun, went on to join NAA as a full-time employee.[48]

Bollay wanted to test-fire V-2 engines. Because their thrust of 56,000 pounds was far too great for the company's parking lot test center, Bollay needed a major set of test facilities. Atwood was ready to help. "We scoured the country," Atwood recalls. "It wasn't so densely settled then—and we located this land."[49] It was in the Santa Susana Mountains, at the western end of the San Fernando Valley. The landscape—stark, sere, and full of rounded reddish boulders—offered spectacular views. In March 1947, NAA leased the land and built a rocket test center on it as part of a buildup of facilities costing upwards of $1 million in company money and $1.5 million from the Air Force.[50]

In 1946, two government-furnished V-2 engines arrived at the site. Detailed designing of the Phase II engine began in June 1947; the end of September brought the first release of drawings and of the first fabricated parts. Early in 1949, the first such engine was completed. Two others followed shortly thereafter.[51]

Still very much a V-2 engine, it had plenty of room for improvement. Lieutenant Colonel Edward Hall, who was funding the work, declared that "it wasn't really a very good engine. It didn't have a proper injector, and that wasn't all. When we took it apart, we decided that was no way to go."[52] By fixing the deficiencies during Phase III, NAA expected to lay a solid foundation for future rocket engine development.

A particular point of contention involved this engine's arrangements for injecting propellants into its combustion chamber. Early in the German rocket program, Walter Riedel, von Braun's chief engine designer, had built a rocket motor with 3,300 pounds of thrust with a cup-shaped injector at the top of the thrust chamber. For the V-2, a new chief of engine design, Walter Thiel, grouped 18 such cups to yield its 56,000 pounds. Unfortunately, this arrangement did not lend itself to a simple design wherein a single liquid-oxygen line

---

48. *Threshold*, Summer 1991, pp. 52-63, Huzel, *Peenemünde*, pp. 226-228.
49. Author interview, J. Leland Atwood, Los Angeles, July 18, 1988.
50. Report AL-1347 (North American), pp. 23-26; Neal, *Navaho*, p. 29.
51. Report AL-1347 (North American), pp. 36-37; Fahrney, *History*, p. 1292; AAS History Series, vol. 20, pp. 133-144.
52. Author interview, Edward Hall, Los Angeles, January 25, 1989.

could supply all the cups. Instead, his "18-pot engine" required a separate oxygen line for each individual cup.[53]

Thiel had pursued a simpler approach by constructing an injector plate, resembling a showerhead, pierced with numerous holes to permit the rapid inflow and mixing of the rocket propellants. By the end of World War II, Thiel's associates had tested a version of the V-2 engine successfully that incorporated this feature, though it never reached production.[54] Bollay's rocket researchers, still working within the company parking lot, were upping their engines' thrust to 3000 pounds, and were using them to test various types of injector plates.[55] The best injector designs would be incorporated into the Phase III engine, bringing a welcome simplification and introducing an important feature that could carry over to larger engines with greater thrust. In September 1947, preliminary design of Phase III began, aiming at the thrust of the V-2 engine but with a weight reduction of 15 percent.[56]

Bollay had initially expected to design the 500-mile missile as a V-2 with swept wings and large control surfaces near the tail, closely resembling the A-9. Work in a supersonic wind tunnel built by Bollay's staff showed that this design would encounter severe stability problems at high speed. Thus, by early 1948, a new configuration emerged. With small forward-mounted wings (known as canards) that could readily control such instability, the new design moved the large wings well aft, replacing the V-2's horizontal fins. In January 1948, four promising configurations were tested in the Ordnance Aerophysics Laboratory wind tunnel in Daingerfield, Texas. By March, a workable preliminary design of the best of these four configurations was largely in hand.[57]

When it won independence from the Army, the U.S. Air Force received authority over programs for missiles with a range of 1,000 miles or more. Shorter-range missiles remained the exclusive domain of the Army. Accordingly, at a conference in February 1948, Air Force officials instructed NAA to stretch the range of their missile to 1000 miles.[58]

The 500-mile missile had featured a boost-glide trajectory. It used rocket power to arc high above the atmosphere and then its range was extended with

---

53. Ley, *Rockets*, pp. 204, 212, 215; Neufeld, *Rocket and Reich*, pp. 74-79, 84.
54. Neufeld, *Rocket and Reich*, p. 251.
55. Report AL-1347 (North American), p. 37; *Threshold*, Summer 1993, pp. 40-47.
56. Report AL-1347 (North American), p. 36.
57. *Ibid.*, pp. 30-33, 38-39.
58. Fahrney, *History*, pp. 1293-1294; Neal, *Navaho*, pp. 6-7.

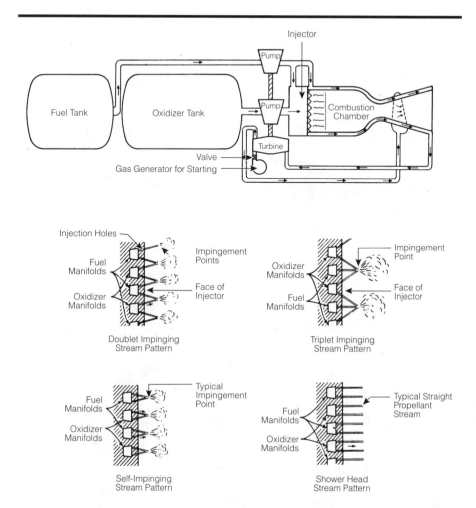

*Top, liquid-fuel rocket engine showing location of injector. Bottom, representative types of injector. (Cornelisse et al., p. 209; Sutton, p. 208)*

a supersonic glide. This approach was not well suited when the range was doubled. At the Air Force developmental center of Wright Field, near Dayton, Ohio, Colonel M. S. Roth proposed to increase the missile range anew by adding ramjets.[59] Unlike the turbojet engines of the day, the ramjet—which worked by ramming air into the engine at high speed—could

---

59. Letter, Colonel M. S. Roth to Power Plant Lab, 11 February 1948 (cited in Fahrney, *History*, p. 1294).

# THE SPACE SHUTTLE DECISION

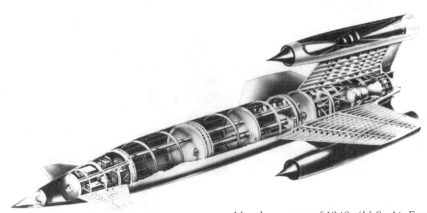

Navaho concept of 1948. (U.S. Air Force)

fly supersonically. A turbojet, however, could take off from a standing start whereas a ramjet needed a rocket boost to reach the speed at which this air-ramming effect would come into play.

A Navy effort, Project Bumblebee, had been under way in this area since World War II and NAA had done several relevant aerodynamic studies. In addition, at Wright Field, the Power Plant Laboratory included a Nonrotating Engine Branch that was funding the development of ramjets as well as rocket motors. Its director, Weldon Worth, dealt specifically with ramjets; Lieut. Col. Hall, who dealt with rockets, served as his deputy.[60]

Though designed for boost-glide flight, the new missile configuration readily accommodated ramjets and their fuel tanks for supersonic cruise. The original boost-glide missile thus evolved into a cruise missile when a modification of the design added two ramjet engines, mounting one at the tip of each of two vertical fins. These engines and their fuel added weight, which necessitated an increase in the planned thrust of the Phase III rocket motor. Originally it had been planned to match the 56,000 pound thrust of the V-2. In March 1948, however, the thrust of this design went up to 75,000 pounds. The missile was named the Navaho, reflecting a penchant at NAA for names beginning with "NA."[61]

---

60. Report AL-1347 (North American), p. 6; *Jet Propulsion*, vol. 25 (1955), pp. 604-614; author interview, Edward Hall, Los Angeles, August 29, 1996.
61. Report AL-1347 (North American), pp. 39, 42-43.

## Space Stations and Winged Rockets

By late November of 1949, the first version of this engine was ready for testing at the new Santa Susana facility. Because it lacked turbopumps, propellants were pressure-fed from heavy-walled tanks. Thus, this version of the engine was much simpler than its later operational type, which would rely on turbopumps to force propellants into the engine. Proceeding cautiously, the rocket crew began with an engine-start test at 10 percent of maximum propellant flow for 11 seconds. It was successful and led to somewhat longer starting tests in December. Then, as the engineers grew bolder, they hiked up the thrust. In March 1950, this simplified engine first topped its rated level of 75,000 pounds—for four and a half seconds. During May and June, the full-thrust runs went well, exceeding a minute in duration.

Meanwhile, a separate developmental effort was building the turbopumps. Late in March 1950, the first complete engine, turbopumps included, was assembled. In August, this engine fired successfully for a full minute—at 12.3 percent of rated thrust. Late in October, the first full-thrust firing reached 70,000 pounds—for less than five seconds. In seven subsequent tests during 1950, however, only one, in mid-November, topped its rated thrust level. This was due to problems with rough combustion during the buildup to full thrust.[62]

The pressure-fed tests exhibited surges in combustion-chamber pressure (known as "hard starts") that were powerful enough to blow up an engine. Walther Riedel, one of the German veterans, played an important role in introducing design modifications that brought this problem under control. The problem of rough combustion was new, however, and went beyond the German experience. It stemmed from combustion instability in the engine's single large thrust chamber. Ironically, the V-2's 18-pot motor had avoided this difficulty. Acting as preliminary burners, its numerous injector cups were too small to permit such instabilities.[63]

Following the successful full-thrust test of November 1950, it was not until March 1951 that problems of unstable combustion came under control.[64] However, this marked another milestone. For the first time, the Americans had encountered and solved an important problem that the Germans had not experienced. While combustion instabilities would recur repeatedly during

---

62. *Ibid.*, pp. 75-81.
63. *Threshold*, Summer 1991, pp. 52-63.
64. *Ibid.*, p. 53; Report AL-1347 (North American), p. 81.

subsequent engine programs, the work of 1950 and 1951 introduced NAA to methods for solving this problem.

By then, the design and mission of the Navaho had changed dramatically. The August 1949 detonation of a Soviet atomic bomb, the fall of China to communism, and the outbreak of the Korean War in mid-1950 combined to signal to the nation that the rivalry with the Soviet Union was serious and that Soviet technical capability was significant. The designers at North American, working with their Air Force counterparts, accordingly sought to increase the range of the Navaho to as much as 5,500 nautical miles, and thereby give it intercontinental capability.

At the Pentagon in August 1950, conferences among Air Force officials brought a redefinition of the program that set this intercontinental range of 5,500 miles as a long-term goal. A letter from Major General Donald L. Putt, director of research and development within the Air Materiel Command, became the directive instructing NAA to pursue this objective.[65] An interim version, Navaho II, with range of 2,500 nautical miles, seemed technically feasible. The full-range version, Navaho III, represented a long-term project that would go forward as a parallel effort.[66]

The 1,000-mile Navaho of 1948, with its Phase III engine, had amounted to a high-speed pilotless airplane fitted with both rocket and ramjet propulsion. This design, however, had taken approaches based on winged rockets to their limit. The new Navaho designs separated the rocket engines from the ramjets, assigned each to a separate vehicle, and turned Navaho into a two-stage missile. The first stage or booster, powered by liquid-fuel rockets, accelerated the missile to Mach 3 and 58,000 feet. The ramjet-powered second stage rode this booster during initial ascent—similar to the way in which the Space Shuttle rides its external tank today—and then cruised to its target at Mach 2.75 (about 1,800 mph.).[67]

Lacking the thrust to boost the Navaho, the 75,000-pound rocket motor stood briefly on the brink of abandonment. Its life, however, was only beginning. This engine was handed over to von Braun, who was at Redstone Arsenal

---

65. Letter, Maj. Gen. D. L. Putt to Commanding General, Air Materiel Command, 21 August 1950 (cited in Fahrney, *History*, p. 1297).
66. Report AL-1347 (North American), p. 88; Fahrney, *History*, pp. 1296-1297; Neal, *Navaho*, pp. 12-14.
67. "Standard Missile Characteristics: XSM-64 Navaho" U.S. Air Force, November 1, 1956, Air Force Museum, Wright-Patterson AFB, Ohio.

*Space Stations and Winged Rockets*

in Huntsville, Alabama, directing development of the Army's Redstone missile. With a range of 200 miles, this missile needed an engine. In March 1951, the Army awarded a contract to NAA for this rocket motor. Weighing less than

V-2 engine, left, and its successor developed for Navaho. (Rocketdyne)

half as much as the V-2's 18-pot engine (1,475 pounds versus 2,484), this motor delivered 34 percent more thrust than that of the V-2.[68]

For Navaho II, this basic engine would be replaced by a new one with 120,000 pounds of thrust. A twin-engine installation, totaling 240,000 pounds, provided the initial boost. For Navaho III, NAA upgraded the engine to 135,000 pounds of thrust and designed a three-engine cluster for that missile's booster.[69]

In 1954 and 1955, the Air Force and Army made a major push into long-range missiles—but these were not Navahos. Instead, they were the Air Force's Atlas, Titan, and Thor, along with the Army's Jupiter. When these new programs needed engines, however, it was again NAA that produced the rocket motors that would do the job. The Navaho's 135,000 pounds of thrust was upgraded to 139,000 and then again to 150,000 pounds. In addition to this, a parallel effort at Aerojet General developed very similar engines for the Titan.[70]

"We often talked about this basic rocket as a strong workhorse, a rugged engine," says Paul Castenholz, a test engineer who worked at Santa Susana. "I think a lot of these programs evolved because we had these engines. We anticipated how people would use them; we weren't surprised when it happened. We'd hear a name like Atlas with increasing frequency, but when it became real, the main result was that we had to build more engines and test them more stringently."[71]

The Navaho of 1948, designed as a winged rocket with ramjets, stood two steps removed from the missiles that later would go on to deployment and operational status. First, the versions of 1950 and after were designed and built as high-speed aircraft with a separate rocket booster. Subsequently, those versions were replaced by the Atlas and other missiles of that era.

Even though the Air Force cancelled the Navaho program in 1957, its legacy lived on. Bollay's research center, called the Aerophysics Laboratory, became the nucleus that allowed NAA to take the lead in piloted space flight. In 1955, this laboratory split into four new corporate divisions: Rocketdyne, Autonetics, the Missile Division, and Atomics International. Rocketdyne

---

68. *Threshold*, Summer 1991, p. 63.
69. Neal, *Navaho*, pp. 30-31; AAS History Series, vol. 20, pp. 133-144.
70. AAS History Series, vol. 13, pp. 19-35; vol. 20, pp. 133-144.
71. Author interview, Paul Castenholz, Colorado Springs, August 18, 1988.

## Space Stations and Winged Rockets

became the nation's premier builder of rocket engines. Autonetics emerged as a major center for guidance and control. The Missile Division, later renamed Space and Information Systems, built the Apollo spacecraft as well as the second stage of the Saturn V Moon rocket.[72]

The Navaho also left a legacy in its people. Sam Hoffman, who brought the 75,000-pound engine to success, presided over Rocketdyne as it built the main engines for the Saturn V. Paul Castenholz headed development of the J-2, the hydrogen-fueled engine that powered Saturn V's upper stages. John R. Moore, an expert in guidance, became president of Autonetics. Dale Myers, who served as Navaho project manager, went to NASA as Associate Administrator for Manned Space Flight.[73]

Navaho's engines, including those built in the parallel effort at Aerojet General, represented a third legacy. Using such engines, Atlas, Thor, and Titan were all successful as launch vehicles. Upper stages were added to Thor which evolved into the widely-used Delta. Additional upgrades raised the thrust of its engine to 205,000 pounds. A cluster of eight such engines, producing up to 1.6 million pounds of thrust, powered the Saturn I and Saturn I-B boosters, which flew repeatedly in both the Apollo and Skylab programs.[74] Between 1946 and 1950, the winged rockets of the Navaho program played a pioneering role, planting seeds that would flourish for decades in aerospace technology.

## The X-15: An Airplane for Hypersonic Research

During the 1940s and 50s, the nation's main centers for aeronautical research operated within a small federal agency, the National Advisory Committee for Aeronautics (NACA; it became the National Aeronautics and Space Administration, NASA, in 1958). After World War II, NACA and the Air Force became increasingly active in supersonic flight. Rocket-powered aircraft such as the Bell X-1 and the Douglas Skyrocket D-558 set the pace. The X-1 broke the sound barrier in 1947; the Skyrocket approached Mach 2 only

---

72. Murray, *Lee Atwood*, pp. 47, 56, 62-64, 71.
73. Author interviews: Eugene Bollay, Santa Barbara, California, January 24, 1989; Sam Hoffman, Monterey, California, July 28, 1988; Paul Castenholz, Colorado Springs, August 18, 1988; John R. Moore, Pasadena, California, May 28, 1996; Dale Myers, Leucadia, California, May 24, 1996.
74. *Threshold*, December 1987, pp. 16-23; AAS History Series, vol. 13, pp. 19-35.

four years later. Also, between 1949 and 1951, NAA designed a new fighter, the F-100, planning it to be the first jet plane to go supersonic in level flight.[75]

Supersonic aviation brought difficult problems in aerodynamics, propulsion, aircraft design, and stability and control in flight. Still, at least for flight speeds of Mach 2 and somewhat higher, it did not involve the important issue of aerodynamic overheating. Though fitted with rocket engines, the cited aircraft were built of aluminum, which cannot withstand high temperatures. At speeds beyond Mach 4 lay the realm of hypersonic flight, where problems of heating would dominate.

Nevertheless, by the early 1950s, interest in such flight speeds was increasing. This was due in part to the growing attention given to prospects for an intercontinental ballistic missile (ICBM), a rocket able to carry a nuclear weapon to Moscow. In December 1950, the Rand Corp., an influential Air Force think tank, reported that such missiles now stood within reach of technology. The Air Force responded by giving a study contract to the firm of Convair in San Diego, where, a few years earlier, the designer Karel Bossart had nurtured thoughts of such missiles. Bossart's new design, developed during 1951, called for the use of the Navaho's 120,000-pound-thrust rocket engine. The design was thoroughly unwieldy; it would stand 160 feet tall and weigh 670,000 pounds. Nevertheless, it represented a milestone. For the first time, the Air Force had an ICBM design concept that it could pursue using rocket engines that were already being developed.[76]

Among the extraordinarily difficult technical issues faced by the ICBM, the problem of reentry was paramount. Because an ICBM's warhead would reenter the atmosphere at Mach 20 or more, there was excellent reason to believe that it would burn up like a meteor. As early as 1951, however, the NACA aerodynamicist H. Julian Allen offered a solution. Conventional thinking held that hypersonic flight would require the ultimate in slender needle-nose shapes. Allen broke with this approach, showing mathematically that the best design would introduce a nose cone as blunt or flat-faced as possible. Such a shape would set up patterns of airflow that would carry most of the heat of reentry away from the nose cone, rather than delivering this heat to its outer surface.[77]

---

75. Ley, *Rockets*, pp. 423-425; Gunston, *Fighters*, pp. 170-171.
76. Neufeld, *Ballistic Missiles*, pp. 68-70.
77. Allen and Eggers, NACA Report 1381; Hansen, *Transition*, p. 3.

## Space Stations and Winged Rockets

There was further interest in hypersonics at Bell Aircraft Corp. in Buffalo. Here Walter Dornberger, who had directed Germany's wartime rocket development, was proposing a concept similar to Eugen Sänger's skip-gliding rocket plane. The design of the rocket (known as the Bomi—Bomber Missile) required a two-stage vehicle with each stage winged, piloted, and rocket-powered. Dornberger argued that Bomi would have the advantage of being able to fly multiple missions like any piloted aircraft, and it could be recalled once in flight. By contrast, an ICBM could fly only once and would be committed irrevocably to its mission once in flight.[78]

Bell Aircraft, very active in supersonic flight research, had built the X-1, which was the first through the sound barrier. Also, Bell Aircraft was building the X-1A that would approach Mach 2.5 and the X-2 that would top Mach 3.[79] Robert Woods, co-founder of the company and a member of NACA's influential Committee on Aerodynamics, had been a leader in the design of these aircraft. He also took a strong interest in Dornberger's ideas.

In October 1951, at a meeting of the Committee on Aerodynamics, Woods called for NACA to develop a new research airplane resembling the V-2, to "obtain data at extreme altitudes and speeds, and to explore the problems of reentry into the atmosphere." In January 1952, Woods wrote a letter to the committee, urging NACA to pursue a piloted research airplane capable of reaching beyond Mach 5. He accompanied this letter with Dornberger's description of Bomi. That June, at Woods's urging, the committee passed a resolution proposing that NACA increase its program in research aircraft to examine "problems of unmanned and manned flight in the upper stratosphere at altitudes between 12 and 50 miles."[80]

NACA already had a few people who were active in hypersonics, notably the experimentalists Alfred Eggers and John Becker, who had already built hypersonic wind tunnels.[81] At NACA's Langley Aeronautical Laboratory, Floyd Thompson, the lab's associate director, responded to the resolution by setting up a three-man study group chaired by Clinton Brown, a colleague of Becker. In Becker's words, "Very few others at Langley in 1952 had any

---

78. *Spaceflight*, vol. 22 (1980), pp. 270-272.
79. Miller, *X-Planes*, pp. 25-26, 37, 41-42.
80. AAS History Series, vol. 13, p. 296; Hansen, *Transition*, pp. 5-6.
81. Hallion, ed., *Hypersonic*, pp. xxxi-xxxv.

knowledge of hypersonics. Thus, the Brown group filled an important educational function badly needed at the time."[82]

According to Thompson, he was looking for fresh unbiased ideas and the three study-group members had shown originality in their work. Their report, in June 1953, went so far as to propose *commercial* hypersonic flight, suggesting that airliners of the future might evolve from boost-glide concepts such as those of Dornberger. At the more practical level, however, the group warmly endorsed building a hypersonic research aircraft. NACA-Langley already had a Pilotless Aircraft Research Division (PARD), which was using small solid-fuel rockets to conduct supersonic experiments. Brown's group now recommended that PARD reach for higher speeds, perhaps by launching rockets that could cross the Atlantic and be recovered in the Sahara Desert.[83]

PARD, a NACA in-house effort, went forward rapidly. In November 1953, it launched a research rocket that carried a test nose cone to Mach 5.0. The following October, a four-stage rocket reached Mach 10.4.[84] To proceed with a piloted research airplane, NACA's limited budget needed support from the Air Force. Here too there was cross-fertilization. Robert Gilruth, head of PARD and an assistant director of NACA-Langley, was also a member of the Aircraft Panel of the Air Force's Scientific Advisory Board. At a meeting in October 1953, this panel stated that "the time was ripe" for such a research airplane, and recommended that its feasibility "should be looked into."[85]

The next step came at a two-day meeting in Washington of NACA's Research Airplane Projects Panel. Its chairman, Hartley Soulé, had directed NACA's participation in research aircraft programs since the earliest days of the X-1 project in 1946. The panel considered specifically a proposal from Langley, endorsed by Brown's group, to modify the X-2 for flight to Mach 4.5. They rejected this concept, asserting that the X-2 was too small for hypersonic work. The panel members concluded instead that "provision of an entirely new research airplane is desirable."[86]

NACA's studies of such an airplane would have to start anew. In March 1954, John Becker set up a new group that took on the task of defining a

---

82. *Ibid.*, p. 381.
83. *Ibid.*, pp. 381-382; Hansen, *Transition*, pp. 6-9.
84. Hallion, ed., *Hypersonic*, p. lxiv.
85. *Astronautics & Aeronautics*, February 1964, p. 54.
86. *Ibid.*; Hansen, *Transition*, p. 9.

design. Time was of the essence; everyone was aware that the X-2 project, underway since 1945, had yet to make its first powered flight.[87] Becker stipulated that "a period of only about three years be allowed for design and construction." Hence NACA would move into the unknown frontiers of hypersonics using technology that was already largely in hand.[88]

Two technical problems stood out: overheating and instability. Because the plane would fly in the atmosphere at extreme speeds, it was essential that it be kept from tumbling out of control. As on any other airplane, tail surfaces were to provide this stability. Investigations had shown that these would have to be excessively large. A Langley aerodynamicist, Charles McLellan, came to the rescue. While conventional practice called for thin tail surfaces that resembled miniature wings, McLellan now argued that they should take the form of a wedge. His calculations showed that at hypersonic speeds, wedge-shaped vertical fins and horizontal stabilizers should be much more effective than conventional thin shapes. Tests in Becker's hypersonic wind tunnel verified this approach.[89]

The problem of overheating was more difficult. At the outset, Becker's designers considered that, during reentry, the airplane should point its nose in the direction of flight. This proved unacceptable because the plane's streamlined shape would cause it to enter the dense lower atmosphere at excessive speed. This would subject the aircraft to disastrous overheating and to aerodynamic forces that would cause it to break up. These problems, however, appeared far more manageable if the plane were to enter with its nose high, presenting its flat undersurface to the air. It then would lose speed in the upper atmosphere, easing both the overheating and the aerodynamic loads. In Becker's words, "It became obvious to us that what we were seeing here was a new manifestation of H. J. Allen's 'blunt body' principle. As we increased the angle of attack, our configuration in effect became more 'blunt.'"[90] While Allen had developed his principle for missile nose cones, it now proved equally useful when applied to hypersonic airplanes.

Even so, the plane would encounter far more heat and higher temperatures than any aircraft to date had received in flight. New approaches in the

---

87. *Astronautics & Aeronautics*, February 1964, p. 53.
88. Hallion, ed., *Hypersonic*, p. 1.
89. *Astronautics & Aeronautics*, February 1964, pp. 54, 56.
90. Hallion, ed., *Hypersonic*, p. 386.

structural design of these aircraft were imperative. Fortunately, Dornberger's group at Bell Aircraft had already taken the lead in the study of "hot structures." These used temperature-resistant materials such as stainless steel. Wings might be covered with numerous small and very hot metal panels resembling shingles that would radiate the heat away from the aircraft. While overheating would be particularly severe along the leading edges of the wings, these could be water-cooled. Insulation could protect an internal structure that would stand up to the stresses and forces of flight; active cooling could protect a pilot's cockpit and instrument compartment. Becker described these approaches as "the first hypersonic aircraft hot structures concepts to be developed in realistic meaningful detail."[91]

His designers proceeded to study a hot structure built of Inconel X, a chrome-nickel alloy from International Nickel. This alloy had already demonstrated its potential, when, during the previous November, it was used for the nose cone in PARD's rocket flight to Mach 5.[92] The hot structure would be of the "heat sink" type, relying on the high thermal conductivity of this metal to absorb heat from the hottest areas and spread it through much of the aircraft.

As an initial exercise, they considered a basic design in which the Inconel X structure would have to withstand only conventional aerodynamic forces and loads, neglecting any extra requirements imposed by absorption of heat. A separate analysis then considered the heat-sink requirements, with the understanding that these might greatly increase the thickness and hence the weight of major portions of the hot structure. When they carried out the exercise, the designers received a welcome surprise. They discovered that the weights and thicknesses of a heat-absorbing structure were nearly the same as for a simple aerodynamic structure.[93] Hence, a hypersonic research airplane, designed largely from aerodynamic considerations, could provide heat-sink thermal protection as a bonus. The conclusion was clear: piloted hypersonic flight was achievable.

The feasibility study of Becker's group was intended to show that this airplane indeed could be built in the near future. In July 1954, Becker presented the report at a meeting in Washington of representatives from NACA, the Air Force's Scientific Advisory Board, and the Navy. (The Navy, actively involved

---

91. *Ibid.*, p. 384.
92. *Ibid.*, p. lxiv.
93. *Astronautics & Aeronautics*, February 1964, p. 58.

with research aircraft, had built the Douglas Skyrocket.) Participants at the meeting endorsed the idea of a joint development program that would build and fly the new aircraft by drawing on the powerful support of the Pentagon.[94]

Important decisions came during October 1954, as NACA and Air Force panels weighed in with their support. At the request of General Nathan Twining, the Air Force Chief of Staff, the Aircraft Panel of the Scientific Advisory Board presented its views on the next 10 years of aviation. The panel's report paid close attention to hypersonic flight:

*In the aerodynamic field, it seems to us pretty clear that over the next ten years the most important and vital subject for research and development is the field of hypersonic flows.... This is one of the fields in which an ingenious and clever application of the existing laws of mechanics is probably not adequate. It is one in which much of the necessary physical knowledge still remains unknown at present and must be developed before we arrive at a true understanding and competence....*

*[A] research vehicle which we now feel is ready for a program is one involving manned aircraft to reach something of the order of Mach 5 and altitudes of the order of 200,000 to 500,000 feet. This is very analogous to the research aircraft program which was initiated ten years ago as a joint venture of the Air Force, the Navy, and NACA. It is our belief that a similar cooperative arrangement would be desirable and appropriate now.*[95]

In addition to this, NACA's Committee on Aerodynamics met in executive session to make a formal recommendation concerning the new airplane. The committee included representatives from the Air Force and Navy, from industry, and from universities.[96] Its member from Lockheed, Clarence "Kelly" Johnson, vigorously opposed building this plane, arguing that experience with earlier experimental aircraft had been "generally unsatisfactory." New fighter designs were advancing so rapidly as to actually outpace the performance of research aircraft. To Johnson, their high-performance flights had served mainly to prove the bravery of the test pilots. While Johnson pressed his views strongly, he was in a minority of one. The other committee mem-

---

94. AAS History Series, vol. 13, p. 299.
95. Hallion, ed., *Hypersonic*, pp. xxiii-xxix.
96. Hansen, *Transition*, pp. 11, 30 (footnote 22).

bers passed a resolution endorsing "immediate initiation of a project to design and construct a research airplane capable of achieving speeds of the order of Mach number 7 and altitudes of several hundred thousand feet."[97]

With this resolution, Hugh Dryden, the head of NACA, could approach his Air Force and Navy counterparts to discuss the initiation of procurement. Detailed technical specifications were necessary and would come, by the end of 1954, from a new three-member committee, with Hartley Soulé as the NACA representative. The three members used Becker's study as a guide in deriving the specifications, which called for an aircraft capable of attaining 250,000 feet and a speed of 6600 feet per second while withstanding reentry temperatures of 1200 degrees Fahrenheit.[98]

In addition to this, as NACA and the military services reached an agreement on procurement procedures, a formal Memorandum of Understanding came from the office of Trevor Gardner, Special Assistant for Research and Development to the Secretary of the Air Force. This document stated that NACA would provide technical direction, that the Air Force would administer design and construction, and that the Air Force and Navy would provide the funding. It concluded, "Accomplishment of this project is a matter of national urgency."[99]

Now the project was ready to proceed. Under standard Air Force practices, officials at Wright-Patterson Air Force Base would seek proposals from potential contractors. Early in 1955, the aircraft also received a name: the X-15. Competition between proposals brought the award of a contract for the airframe to NAA. The rocket engine was contracted to Reaction Motors, Inc.[100] The NAA design went into such detail that it even specified the heat-resistant seals and lubricants that would be used. Nevertheless, in many important respects it was consistent with the major features of the original feasibility study by Becker's group. The design included wedge-shaped tail surfaces and a heat-sink structure of Inconel X.[101]

The X-15 was to become the fastest and highest flying airplane until the space shuttle flew into orbit in 1981. In August 1963, the X-15 set an altitude record of 354,200 feet (67 miles), with NASA's Joseph Walker in the cockpit.

---

97. *Ibid.*, pp. 12-14.
98. *Ibid.*, p. 14; AAS History Series, vol. 8, p. 299.
99. Hallion, ed., *Hypersonic*, p. I-6.
100. *Ibid.*, pp. I-iv, 11-15.
101. *Astronautics & Aeronautics*, February 1964, p. 54.

*Space Stations and Winged Rockets*

X-15. (NASA)(E-5251)

Four years later, the Air Force's Captain William Knight flew it to a record speed of 4,520 miles per hour, or Mach 6.72.[102] In addition to setting new records, the X-15 accomplished a host of other achievements.

A true instrument of hypersonic research, in 199 flights it spent nearly nine hours above Mach 3, nearly six hours above Mach 4, and 82 minutes above Mach 5. Although the NACA and the Air Force had hypersonic wind tunnels, the X-15 represented the first application of aerodynamic theory and wind tunnel data to an actual hypersonic aircraft. The X-15 thus enhanced the usefulness of these wind tunnels, by providing a base of data with which to validate (and in some instances to correct) their results. This made it possible to rely more closely on results from those tunnels during subsequent programs, including that of the Space Shuttle.

The X-15 used movable control surfaces that substituted for ailerons. It also introduced reaction controls: small rocket thrusters, mounted to the air-

---

102. Hallion, ed., *Hypersonic*, pp. I-v, I-viii.

## THE SPACE SHUTTLE DECISION

craft, that controlled its attitude when beyond the atmosphere. As it flew to the fringes of space and returned, the X-15 repeatedly transitioned from aerodynamic controls to reaction controls and back again. Twenty years later, the Space Shuttle would do the same.

In another important prelude to the shuttle, the X-15 repeatedly flew a trajectory that significantly resembled flight to orbit and return. The X-15 ascended into space under rocket power, flew in weightlessness, then reentered the atmosphere at hypersonic speeds. With its nose high to reduce overheating and aerodynamic stress, the X-15 used thermal protection to guard the craft against the heat of reentry. After reentry, the X-15 then maintained a stable attitude throughout its deceleration, transitioned to gliding flight, and landed at a preselected location. The shuttle would do all these things, albeit at higher speeds.

The X-15 used a rocket engine of 57,000 pounds of thrust that was throttleable, reusable, and "man-rated"—safe enough for use in a piloted aircraft. The same description would apply to the more powerful Space Shuttle Main Engine.

The demands of the project pushed the development of practical hypersonic technology in a number of areas. Hot structures required industrial shops in which Inconel X could be welded, machined, and heat-treated. The pilot required a pressure suit for use in a vacuum. The X-15 required new instruments and data systems including the "Q-ball," which determined the true direction of airflow at the nose. Cooled by nitrogen, the "Q-ball" operated at temperatures of up to 3,500 degrees Fahrenheit and advised the pilot of the angle of attack suitable for a safe reentry.[103]

Like the Navaho, the X-15 also spurred the rise of people and institutions that were to make their mark in subsequent years. At NACA-Langley, the X-15 combined with the rocket flights of PARD to put an important focus on hypersonics and hypervelocity flight. Leaders in this work included such veterans as Robert Gilruth, Maxime Faget, and Charles Donlan.[104] A few years later, these researchers parlayed their expertise into leadership in the new field of piloted space missions. In addition to this, part of NACA-Langley split off to establish the new Manned Spacecraft Center in Houston as NASA's princi-

---

103. *Ibid.*, pp. 157-159; AAS History Series, vol. 8, p. 306; Miller, *X-Planes*, p. 110.
104. NASA SP-4308; see index references.

pal base for piloted space flight. Gilruth headed that center during the Apollo years, while Faget, who had participated in Becker's 1954 X-15 feasibility study, became a leading designer of piloted spacecraft.[105]

The X-15 program brought others to the forefront as well. At NAA the vice president of the program, Harrison "Stormy" Storms, became president of that company's Space Division in 1960. While Gilruth was running the Manned Spacecraft Center, Storms had full responsibility for his division's elements of Apollo: the piloted spacecraft and the second stage of the Saturn V Moon rocket.[106] In addition to this, Neil Armstrong, the first man to set foot on the Moon, was among the test pilots of the X-15.[107]

## *Lifting Bodies: Wingless Winged Rockets*

Although the X-15 emerged as a winged rocket par excellence, an alternate viewpoint held that future rocket craft of this type could have many of the advantages of wings without actually having any of these structures. Such craft would take shape as "lifting bodies," wingless and bathtub-shaped craft that were able to generate lift with the fuselage. This would allow them to glide to a landing. At the same time, such craft would dispense with the weight of wings, and with their need for thermal protection.

How can a bathtub generate lift, and fly? Lift is force that is generated when the aerodynamic pressure is greater below an aircraft than above it. Wings achieve this through careful attention to their shape; a properly-shaped aircraft body can do this as well. The difference is that wings produce little drag, whereas lifting bodies produce a great deal of drag. Hence the lifting body approach is unsuitable for such uses as commercial aviation, where designers of airliners seek the lowest possible drag. Space flight, however, is another matter.

The lifting body concept can be traced back to the work of H. Julian Allen and Alfred Eggers, at NACA's Ames Aeronautical Laboratory near San Francisco. Allen developed the blunt-body concept for a missile's nose cone, shaping it with help from Eggers. They then considered that a reentering body, while remaining blunt to reduce the heat load, might have a form that

---

105. NASA SP-4307; see index references.
106. Resumé of Harrison A. Storms.
107. Miller, *X-Planes*, p. 108.

would give lift, thus allowing it to maneuver at hypersonic speeds. The 1957 M-1 featured a blunt-nose cone with a flattened top. While it had some capacity for hypersonic maneuverability, it could not glide subsonically or land horizontally. It was hoped that a new shape, the M-2, would do these things as well. Fitted with two large vertical fins for stability, it was a basic configuration suitable for further research.[108]

Beginning in 1959, a separate line of development took shape within the Flight Dynamics Laboratory of Wright-Patterson Air Force Base. The program that developed sought to advance beyond the X-15 by building small hypersonic gliders, which would study the performance of advanced hot structures at speeds of up to 13,000 miles per hour, three-fourths of orbital velocity. This program was called ASSET—Aerothermodynamic/elastic Structural Systems Environmental Tests.[109]

The program went forward rapidly by remaining small. The project's manager, Charles Cosenza, directed it with a staff of four engineers plus a secretary, with 17 other engineers at Wright-Patterson providing support.[110] In April 1961, the Air Force awarded a contract to McDonnell Aircraft Corp. for development of the ASSET vehicle. McDonnell was already building the small piloted capsules of Project Mercury; the ASSET vehicle was also small, with a length of less than six feet. Not a true lifting body, it sported two tiny and highly-swept delta wings. Its bottom, which would receive the most heat, was a flat triangle. For thermal protection, this triangle was covered with panels of columbium and molybdenum. These would radiate away the heat, while withstanding temperatures up to 3,000 degrees Fahrenheit. The nose was made of zirconium oxide that would deal with temperatures of up to 4,000 degrees.[111]

Beginning in September 1963 and continuing for a year and a half, five of the six ASSET launches were successful. They used Thor and Thor-Delta launch vehicles, the latter being a two-stage rocket that could reach higher velocities. The boosters lofted their ASSETs to altitudes of about 200,000 feet. The spacecraft then would commence long hypersonic glides with ranges as great as 2,300 nautical miles. Onboard instruments transmitted data

---

108. Hallion, ed., *Hypersonic*, pp. 529, 535, 864-866.
109. *Ibid.*, pp. 449-450, 505.
110. *Ibid.*, p. 459.
111. *Ibid.*, pp. 451, 452, 464-469.

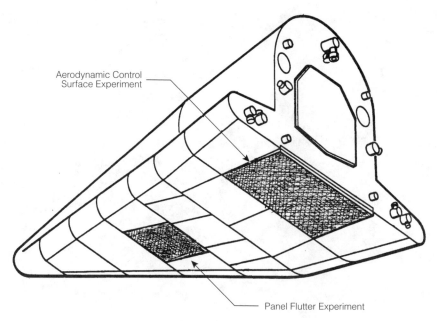

ASSET's use of metallic shingle-like panels as thermal protection permitted use of indivudual panels for specific experiments. (U.S. Air Force)

on temperature and heat flow. The craft were equipped to float following splashdown; one of them actually did this, permitting direct study of an advanced hot structure that had survived baptism by fire.[112]

The success of ASSET led to the development of Project PRIME—Precision Recovery Including Maneuvering Entry. Beginning in late 1964, the contract for this Air Force project went to the Martin Co., where interest in lifting bodies had flourished for several years. Unlike ASSET, PRIME featured true lifting bodies, teardrop-shaped and fitted with fins. PRIME was slated to ride the Atlas, which was more powerful than the Thor-Delta and could reach near-orbital speeds.[113]

Whereas ASSET had executed simple hypersonic glides, PRIME carried out the more complex maneuver of achieving crossrange, namely, flying far to the left or right of its flight path. Indeed, to demonstrate such reentry

---

112. *Ibid.*, pp. 504-519.
113. Hallion, *Path*, pp. 30-31.

# THE SPACE SHUTTLE DECISION

*The homebuilt M2-F1 lifting body, left, and the Northrop M2-F2. (NASA)(E-14339)*

maneuvering was its reason for being. PRIME did not attempt to produce data on heating, for ASSET had covered this point nicely, nor did it break new ground in its construction. Slightly larger than ASSET, it used a conventional approach for missile nose cones that featured an aluminum structure covered with a thermally-protective "ablative" layer that would carry away heat by vaporizing in a controlled fashion during reentry. The ablative material also served as insulation to protect the underlying aluminum.

With its peak speed topping 17,000 mph, PRIME could bridge the Pacific, flying from Vandenberg Air Force Base in California to Kwajalein, not far from New Guinea. In April 1967, during its best performance, PRIME achieved a crossrange of 710 miles, puting it within five miles of its target. A waiting recovery plane snatched PRIME in mid-air as it descended by parachute.[114]

ASSET and PRIME demonstrated the value of lifting bodies at the hypersonic end of the flight path: gliding, maneuvering, surviving reentry using

---

114. Hallion, ed., *Hypersonic*, pp. V-ii, V-iv, 702-703.

# Space Stations and Winged Rockets

Martin Marietta's X-24A, built for subsonic flight, duplicated the shape of PRIME, which flew at near-orbital velocity. (NASA)(E-18769)

advanced hot structures. Both types of craft, however, used parachutes for final descent, making no attempt to land like conventional aircraft. If lifting bodies were to truly have merit, they would have to glide successfully not only at hypersonic speeds but at the slow speed of an aircraft on a final approach to a runway. Under the control of a pilot, lifting bodies would have to maintain stable flight all the way to a horizontal touchdown.

These requirements led to a second round of lifting-body projects focusing on approach and landing. These projects went forward with ASSET and PRIME at the same time. R. Dale Reed, the initiator of this second round of projects, was a sailplane enthusiast, a builder of radio-controlled model airplanes, and a NASA engineer at Edwards Air Force Base. He had followed with interest the work at NASA-Ames on the M-2 lifting-body shape, and he resolved to build it as a piloted glider. He drew support from the local community of aircraft homebuilders. Designated as the M2-F1, the aircraft was built of plywood over a tubular steel frame. Completed in early 1963, the aircraft was 20 feet long and 13 feet across.

## THE SPACE SHUTTLE DECISION

The M2-F1 needed a vehicle that could tow it along the ground to help get it into the air for initial tests. The M2-F1, however, produced a lot of drag and needed a tow car with more power than NASA's usual vans and trucks. Reed and his friends bought a stripped-down Pontiac with a big engine and a four-barrel carburetor that could reach speeds of 110 mph. The car was turned over to a funny car shop in Long Beach for modification. Like any other flight-line vehicle it was sprayed yellow and "National Aeronautics and Space Administration" was added on its side. Initial piloted tow tests showed reasonable success, allowing the project to use a C-47, called the Gooney Bird, for true aerial tests. During these tests, the Gooney Bird towed the M2-F1 above 10,000 feet, then set it loose to glide to an Edwards AFB lake bed. Beginning in August 1963, the test pilot Milt Thompson did this repeatedly. Through these tests, Reed, working on a shoestring budget, showed that the M-2 shape, optimized for hypersonic reentry, could glide down to a safe landing.

During much of this effort, Reed had support from the NASA director at Edwards, Paul Bikle. As early as April 1963, he alerted NASA Headquarters that "the lifting-body concept looks even better to us as we get more into it." The success of the M2-F1 spurred interest in the Air Force as well, as some of its officials, along with their NASA counterparts, set out to pursue piloted lifting-body programs that would call for more than plywood and funny cars.[115]

NASA contracted with the firm of Northrop to build two such aircraft, the M2-F2 and HL-10. The M2-F2 amounted to an M2-F1 built to NASA standards; the HL-10 drew on an alternate lifting-body design by Eugene Love of NASA-Langley. This meant that both NASA-Langley and NASA-Ames would each have a project. In addition to this, Northrop had a penchant for oddly-shaped aircraft. During the 1940s, the company had built flying wings that essentially were aircraft without a fuselage or tail. With these lifting bodies, Northrop would build craft now that were entirely fuselage and lacked wings. The Air Force project, the X-24A, went to Martin Co., which built it as a piloted counterpart of PRIME, maintaining the same shape.[116]

All three flew initially as gliders, with a B-52 rather than a C-47 as the mother ship. The B-52 could reach 45,000 feet and 500 mph, four times the

---

115. NASA SP-4303, pp. 148-152.
116. Hallion, *Path*, pp. 29, 31-32.

altitude and speed of the old Gooney Bird.[117] It had routinely carried the X-15 aloft, acting as a booster for that rocket plane; now it would do the same for the lifting bodies. Their shapes differed, and as with the M2-F1, a major goal was to show that they could maintain stable flight while gliding, land safely, and exhibit acceptable pilot handling qualities.[118]

These goals were not always met. Under the best of circumstances, a lifting body flew like a brick at low speed. Lowering the landing gear made the problem worse by adding drag. In May 1967, the test pilot Bruce Peterson, flying the M2-F2, failed to get his gear down in time. The aircraft hit the lake bed at more than 250 mph, rolled over six times, and then came to rest on its back, minus its cockpit canopy, main landing gear, and right vertical fin. Peterson, who might have died in the crash, got away with a skull fracture, a mangled face, and the loss of an eye. While surgeons reconstructed his face and returned him to active duty, the M2-F2 needed surgery of its own. In addition to an extensive reconstruction back at the factory, Northrop engineers added a third vertical fin that improved its handling qualities and made it safer to fly. Similarly, while the rival HL-10 had its own problems of stability, it flew and landed well after receiving modifications.[119]

These aircraft were mounted with small rocket engines that allowed acceleration to supersonic speeds. This made it possible to test stability and handling qualities when flying close to the speed of sound. The HL-10 set records for lifting bodies by making safe approaches and landings from speeds up to Mach 1.86 and altitudes of 90,000 feet.[120] The Air Force continued this work through 1975, having the Martin Co. rebuild the X-24A with a long pointed nose, a design well-suited for supersonic flight. The resulting craft, the X-24B, looked like a wingless fighter-plane fuselage. It also flew well.[121]

In contrast to the Navaho and X-15 efforts, work with lifting bodies did not create major new institutions or lead existing ones in important new directions. This work, however, did extend that of the X-15 with the hot-structure flights of ASSET and the maneuvering reentries of PRIME. The piloted lifting bodies then demonstrated that, with the appropriate arrangements of fins,

---

117. Miller, *X-Planes*, p. 153.
118. *Ibid.*, p. 151; NASA SP-4303, p. 153.
119. NASA SP-4303, pp. 159, 161-162; *Spaceflight*, vol. 21, (1979), pp. 487-489.
120. NASA SP-4303, p. 162.
121. Miller, *X-Planes*, pp. 156-160.

*The X-24B, a lifting body capable of supersonic flight. (NASA)(E-25283)*

they could remain stable and well-controlled when decelerating through the sound barrier and gliding to a landing. They thus broadened the range of acceptable hypersonic shapes.

## Solid-Propellant Rockets: Inexpensive Boosters

The X-15 and lifting-body programs demonstrated many elements of a reusable launch vehicle in such critical areas as propulsion, flight dynamics, structures, thermal protection, configurations, instruments, and aircraft stability and control. However, the reason for reusability would be to save money, and an airplane-like orbiter would need a low-cost booster as a first stage. During the 1950s and 1960s, the Navy, Air Force, and NASA laid groundwork for such boosters by sponsoring pathbreaking work with solid propellants.

The path to such propellants can be traced back to a struggling firm called Thiokol Chemical Corp. Its initial stock-in-trade was a liquid polysulfide polymer that took its name (Thiokol) from the Greek for "sulfur glue" and could be cured into a solvent-resistant synthetic rubber. During World War II, it found limited use in sealing aircraft fuel tanks—a market that disappeared after 1945. Indeed, business was so slow that even small orders would draw the attention of the company president, Joseph Crosby.

When Crosby learned that California Institute of Technology (CIT) was buying five- and ten-gallon lots in a steady stream, he flew to California to investigate the reason behind the purchases. He found a group of rocket researchers, loosely affiliated with CIT, working at a place they called the Jet Propulsion Laboratory. They were mixing Crosby's polymer with an oxidizer and adding powdered aluminum for extra energy. They were using this new propellant in ways that would make it possible to build solid-fuel rockets of particularly large size.[122]

Crosby soon realized that he too could get into the rocket business, with help from the Army. While Army officials could spare only $250,000 per year to help him get started, to Crosby this was big money. In 1950, Army Ordnance gave him a contract to build a rocket with 5,000 pounds of propellant. A year and a half later it was ready, with a sign on the side, "The Thing." Fourteen feet long, it burned for over forty seconds and delivered a thrust of 17,000 pounds.[123]

The best solid propellants of the day were of the "double base" type, derived from the explosives nitroglycerine and nitrocellulose. Some versions could be cast in large sizes. These propellants, however, burned in a sudden rush, and could not deliver the strong, steady push needed for a rocket booster. The new Thiokol-based fuel emerged as the first of a type that performed well and burned at a reasonable rate. These fuels drew on polymer chemistry to form as thick mixtures resembling ketchup. Poured into a casing, they then polymerized into resilient rubbery solids.[124]

The Navy also took an interest in solid propellants, initially for use in anti-aircraft missiles. In 1954, a contractor in suburban Virginia, Atlantic Research, set out to achieve further performance improvements. Two company scientists, Keith Rumbel and Charles Henderson, focused their attention on the use of powdered aluminum. Other researchers had shown that propellants gave the

---

122. *Fortune*, June 1958, p. 109.
123. *Ibid.*, p. 190; *Thiokol's Aerospace Facts*, July-September 1973, p. 10; *Saturday Evening Post*, October 1, 1960, p. 87.
124. Huggett et al., *Solid*, pp. 125-128; Ley, *Rockets*, pp. 171-173, 193, 436-438; Cornelisse et al., *Propulsion*, pp. 170-174.

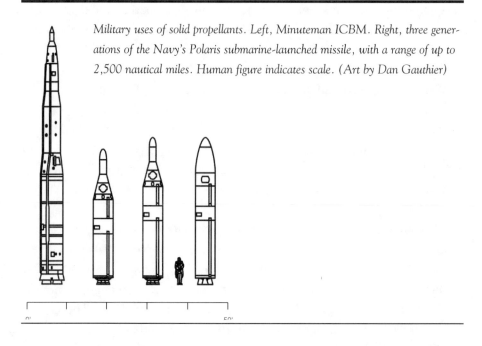

*Military uses of solid propellants. Left, Minuteman ICBM. Right, three generations of the Navy's Polaris submarine-launched missile, with a range of up to 2,500 nautical miles. Human figure indicates scale. (Art by Dan Gauthier)*

best performance with an aluminum mix of five percent; higher levels caused a falloff. Undiscouraged, Rumbel and Henderson decided to try mixing in really large amounts. The exhaust velocity, which determines the performance of a rocket, took a sharp leap upward. By early 1956, they confirmed this discovery with test firings. Their exhaust velocities, 7,400 feet per second and greater, compared well with those of liquid fuels such as kerosene and liquid oxygen.[125]

By then the Navy was preparing to proceed with Polaris, a program that sought to send strategic missiles to sea aboard submarines. Initial design concepts were unpleasantly large; a submarine would be able to carry only four such missiles, and the submarine itself would be excessive in size. The breakthrough in propellants coincided with an important advance that markedly reduced the weight of thermonuclear weapons. Lighter warheads meant smaller missiles. These developments combined to yield a solid-fueled Polaris missile that was very compact. Sixteen of them would fit into a conventional-sized submarine.[126]

---

125. Baar and Howard, *Polaris!*, pp. 31-32.
126. *Journal of Spacecraft and Rockets*, vol. 15 (1978), pp. 265-278.

The new propellants, and the lightweight warheads, also drew interest within the Air Force, though its needs contrasted sharply with those of the Navy. Skippers could take time in firing undersea missiles, for a submarine could hide in the depths until it was ready for launch. Admirals, however, preferred solid fuels over liquids because they presented less of a fire hazard. While the Air Force was prepared to use liquid propellants in its ICBMs, these would take time to fuel and prepare for launch—and during that time they would lie open to enemy attack. With solid propellants, a missile could be fueled in advance and ready for instant launch. Moreover, such a missile would be robust enough to fire from an underground chamber. Prior to launch, that chamber would protect the missile against anything short of a direct nuclear hit.

Lieutenant Colonel Edward Hall, who had midwifed the birth of the Navaho during the 1940s, now played a leading role in this newest project. He was the propulsion officer on the staff of Major General Bernard Schriever who was responsible for the development of the Atlas, Titan, and Thor. Hall developed a passionate conviction that an Air Force counterpart of Polaris would offer considerable advantage in facing the Soviet ICBM capability. At the outset of the new project, he addressed the problem of constructing very large solid-fuel charges, called grains. He could not draw on the grains of the Polaris for that missile had grains of limited size.

Hall gave contracts to all of the several solid-fuel companies that were in business at that time. Thiokol's Crosby, who had lost the Polaris contract to Aerojet General, now saw a chance to recoup. He bought a large tract of land near Brigham City, Utah, a remote area where the shattering roar of rockets would have plenty of room to die away. In November 1957, his researchers successfully fired a solid-fuel unit with 25,000 pounds of propellant, the largest to date.

Meanwhile, Hall had taken charge of a working group that developed a preliminary design for a three-stage solid-fuel ICBM. Low cost was to be its strong suit, for Hall hoped to deploy it in very large numbers. Early in 1958, with the test results from Thiokol in hand, Hall and Schriever went to the Pentagon and pitched the concept to senior officials, including the Secretary of Defense. But while that missile, named the Minuteman, might be launched on a minute's notice, it would take most of 1958 to win high-level approval for a fast pace of development.

## THE SPACE SHUTTLE DECISION

Barely two years later, in early 1961, the Minuteman was ready for its first flight from Cape Canaveral. It scored a brilliant success as all three stages fired and the missile flew to full range. The Air Force proceeded to raise the Minuteman to the status of a crash program. The first missiles were operational in October 1962, in time for the Cuban Missile Crisis. Because its low cost made it the first strategic weapon capable of true mass production, the Air Force went on to deploy 1,000 of the Minuteman rockets.[127]

The Air Force and NASA also prepared to build solid-fuel boosters of truly enormous size for use with launch vehicles. In contrast to liquid rockets that were sensitive and delicate, the big solids featured casings that a shipyard—specifically, the Sun Shipbuilding and Dry Dock Company, near Philadelphia—would manufacture successfully.

The Minuteman's first stage had a 60-inch diameter. In August 1961, United Technology Corp. fired a 96-inch solid rocket that developed 250,000 pounds of thrust. The following year saw the first 120-inch tests—twice the diameter of the Minuteman—that reached 700,000 pounds of thrust. The next milestone was reached when the diameter was increased to 156 inches, the largest size compatible with rail transport. During 1964, both Thiokol and Lockheed Propulsion Co. fired test units that topped the million-pound-thrust mark.

Large rocket stages can be moved by barges over water as well as by land. Aerojet was building versions with 260-inch diameters. It took some doing just to ignite such a behemoth. The answer called for a solid rocket that itself developed a quarter-million pounds of thrust, producing an eighty-foot flame that would ignite the inner surface of the big one all at once. This igniter rocket needed its own igniter, a solid motor that weighed a hundred pounds and generated 4,500 pounds of thrust. The 260-inch motor was kept in a test pit with its nozzle pointing upward. In February 1966, a night firing near Miami shot flame and smoke a mile and a half into the air that was seen nearly a 100 miles away. In June 1967, another firing set a new record with 5.7 million pounds of thrust.[128]

At NASA's Marshall Space Flight Center, a 1965 study projected that production costs for a 260-inch motor would run to $1.50 per pound of

---

127. Emme, ed., *History*, pp. 155-159; Neufeld, *Ballistic Missiles*, pp. 227-230, 237, 239; *Fortune*, June 1958, pp. 190-192.
128. *Quest*, Spring 1993, p. 26; *Astronautics*, December 1961, p. 125; November 1962, p. 81; *Astronautics and Aerospace Engineering*, November 1963, p. 52; *Astronautics & Aeronautics*, February 1965, pp. 42-43.

## Space Stations and Winged Rockets

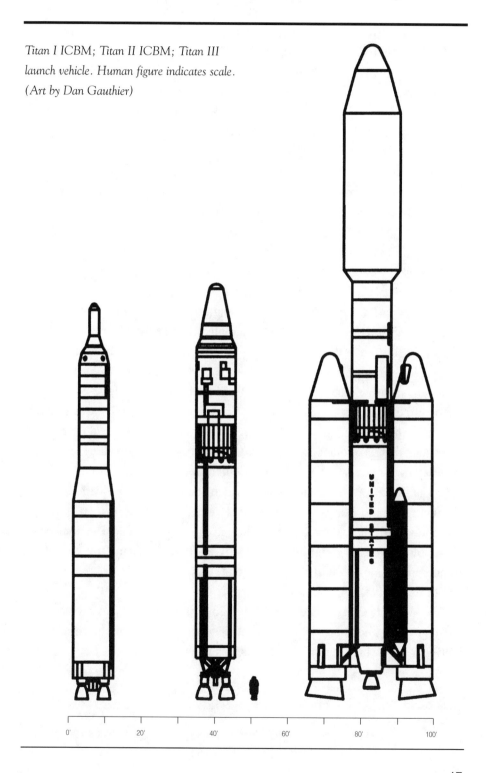

*Titan I ICBM; Titan II ICBM; Titan III launch vehicle. Human figure indicates scale. (Art by Dan Gauthier)*

weight, or roughly a dollar per pound of thrust. This contrasted sharply with the liquid-fueled Saturn V, which, with 7.5 million pounds of thrust versus 6 million for the big solid, was in the same class. Even without its Apollo moonship, however, the Saturn V cost $185 million to purchase, over thirty times more than the 260-inch motor. By 1966, NASA officials were looking ahead already to sizes as large as 600 inches, noting that "there is no fundamental reason to expect that motors 50 feet in diameter could not be made."[129]

Meanwhile, the Air Force not only was testing big solids but it was preparing to use them operationally as part of the Titan program which, in a decade, had evolved from building ICBMs to assembling a launch vehicle of great power. At the outset, Titan I was a two-stage ICBM project that ran in parallel with Atlas and used similar engines in the first stage. While it was deployed as a weapon, it was never used to launch a spacecraft or satellite.[130]

The subsequent Titan II represented a major upgrade as the engine contractor, Aerojet General, developed new engines that markedly increased the thrust in both stages. It too reached deployment, carrying a heavy thermonuclear warhead with a yield of nine megatons. By lightening this load somewhat, the Titan II was able to thrust a payload into orbit repeatedly. In particular, during 1965 and 1966, the Titan II carried 10 piloted Gemini spacecraft, each with two astronauts. Their weight ran above 8,300 pounds.[131]

The Air Force's Titan III-A added, to the Titan II, a third stage (the "transtage") which enhanced its ability to carry large payloads. It never served as an ICBM, but worked as a launch vehicle from the start. In particular, it served as the core for the Titan III-C, which flanked that core with a pair of 120-inch solid boosters. The rocket that resulted had more than a casual resemblance to the eventual Space Shuttle, which would use two somewhat larger solid boosters in similar fashion. After lifting the Titan III-C with 2.36 million pounds of thrust, its boosters then fell away after burnout, leaving the core to ignite its first stage, high in the air.

The Titan III-C had a rated payload of 23,000 pounds. NASA replaced the transtage with the more capable Centaur upper stage, which used liquid hydrogen as a high-energy fuel. This version, the Titan III-E Centaur,

---

129. *Astronautics & Aeronautics*, January 1966, p. 33; NASA budget data, February 1970.
130. Emme, ed., *History*, pp. 145, 147.
131. NASA SP-4012, vol. II, pp. 83-85; *Quest*, Winter 1994, p. 42; Thompson, ed., *Space Log*, vol. 27 (1991), p. 87.

## Space Stations and Winged Rockets

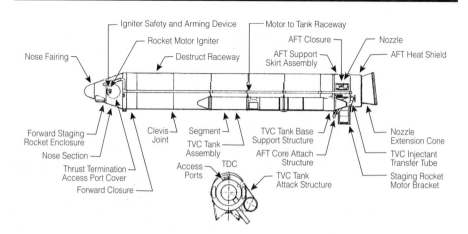

Solid rocket motor for the Titan III-C. (AIAA)

increased the payload to 33,000 pounds. Martin Marietta, the Titan III contractor, also proposed to delete the third stage while increasing the thrust of both the solid boosters and the core. This version, the Titan III-M, was never built, but it would have lifted a payload of 38,000 pounds.[132]

Hence during the 1960s, the X-15, ASSET, PRIME, lifting body and solid-booster efforts all combined to provide a strong basis for the Space Shuttle program. Such a program might build an orbiter in the shape of a lifting body with a hot structure for thermal protection. Piloted and crewed, it could maneuver during atmosphere entry, ride through the heat of reentry with its nose up, then transition to gliding flight and fly to a landing, perhaps at Edwards Air Force Base. Moreover, long before those early projects had reached completion (and even before some of them were underway), the Air Force set out to build a mini-shuttle that would ride a Titan III-C to orbit and then return. This project was called Dyna-Soar and, later, the X-20.

## *Dyna-Soar: A Failure in Evolution*

During the mid-1950s, with the Bomi studies of Bell Aircraft in the background and the X-15 as an ongoing program, a number of people eagerly carried out further studies that sought to define the next project beyond the

---

132. NASA SP-4012, vol. III, pp. 38-42; *Quest*, Fall 1995, p. 18; AAS History Series, vol. 13, pp. 19-35.

X-15. The ideas studied included Hywards (a piloted hypersonic boost-glide research aircraft), the Robo (Robot Bomber), and two reconnaissance vehicles, the System 118-P and the Brass Bell. With so many cooks in the kitchen, the Air Force needed a coordinated program in order to produce something as specific as the X-15. Its officials were in the process of defining this program when, in October and November 1957, the Soviet Union launched the world's first satellites. Very quickly, hypersonic flight became one of the means by which the U.S. might turn back the challenge from Moscow.

Having read the work of Sanger, hypersonic specialists knew of his ideas for skipping entry as a way to extend the range of a suborbital aircraft. The Air Force described this maneuver as "dynamic soaring." The craft that would do this acquired the name Dyna-Soar. By early 1958, this idea was being studied seriously by a number of aeronautical contractors with the clear understanding that the Air Force intended to request proposals and build a flying prototype. In June 1958, the Air Force narrowed the competition to two contenders: Boeing and a joint Bell Aircraft and Martin Co. team.[133]

By then, Dyna-Soar was caught up in the first round of a controversy as to whether this craft should be the prototype of a bomber. While the powerful Air Research and Development Command (ARDC) firmly believed that Dyna-Soar should be the prototype of a piloted military spaceplane, it found it difficult to point to specific military missions that such a craft could carry out. For nuclear weapons delivery, the Air Force was already building the Atlas, Titan, and Thor. For strategic reconnaissance, the Central Intelligence Agency had launched, in 1958, a program that aimed to build automated camera-carrying satellites and put the first ones into orbit in as little as one year.[134]

Air Force Headquarters, however, with support from the Office of the Secretary of Defense, refused to consider weapon-system objectives unless ARDC could define suitable military missions. Early in 1959, Deputy Secretary of Defense Donald Quarles wrote that his approval was only "for a research and development project and did not constitute recognition of Dyna-Soar as a weapon system."

In April, the Defense Director of Research and Engineering, Herbert York, made a clear statement of the program's objectives. Its primary goal

---

133. AAS History Series, vol. 17, pp. 255-259.
134. *Ibid.*, p. 260; Ruffner, ed., *Corona*, pp. 3-14.

would involve hypersonic flight up to a speed of 15,000 miles per hour, which would fall short of orbital velocity. The vehicle would be piloted, maneuverable, and capable of landing at a preselected base. York also threw a bone to ARDC, stating that it could pursue its own goal of testing military systems—provided that such tests did not detract from the primary goal. ARDC officials hastened to affirm that there would be no conflict. They promptly issued System Requirement 201, stating that Dyna-Soar would "determine the military potential of a boost-glide weapon system."[135]

In November 1959, the contract award went to Boeing. Two weeks later, the Air Force's Assistant Secretary for Research and Development, Joseph Charyk, said "not so fast." He was well aware that the project already faced strong criticism because of its cost, as well as from Eisenhower Administration officials who opposed space-based weapon systems. In addition to this, a number of technical specialists doubted that the concept could be made to work. Charyk therefore ordered a searching reexamination of the project that virtually re-opened the earlier competition. In April 1960, the Aerospace Vehicles Panel of the Air Force Scientific Advisory Board gave Dyna-Soar a go-ahead by approving Boeing's design concept, with minor changes.

During the next three and a half years, the program went forward as its managers reached for higher performance. The 1960 plan called for the use of a Titan I as the launch vehicle. Because the Titan I lacked the power to put it in orbit, the Dyna-Soar would fly suborbital missions only. Over the next year and a half, however, the choice of booster changed to the Titan II and then the powerful Titan III-C. A new plan, approved in December 1961, dropped suborbital flights and called for "the early attainment of orbital flight, with the Titan III booster."

This plan called initially for single-orbit missions that would not require the craft to carry an onboard retro-rocket for descent from orbit. Instead the booster, launched from Cape Canaveral, would place the craft on a trajectory that would re-enter the atmosphere over Australia. It then would cross the Pacific in a hypersonic glide, to land at Edwards Air Force Base. In May 1962, the plan broadened anew to include multi-orbit flights. Dyna-Soar now would ride atop the Titan III transtage that would inject it into orbit and remain attached to serve as a retro-rocket at mission's end.[136]

---

135. AAS History Series, vol. 17, p. 260.
136. *Ibid.*, pp. 261-269.

# THE SPACE SHUTTLE DECISION

Mockup of Dyna-Soar displayed in 1962. (Boeing)(P-30793)

The piloted Dyna-Soar spacecraft also emerged with highly-swept delta wings and two upturned fins at the wingtips. With a length of 35 feet, it lacked an onboard rocket engine and provided room for a single pilot only. Like ASSET, it relied on advanced hot structures, with a heat shield of columbium, well insulated, atop a main structure built from a nickel alloy that had been developed for use in jet engines.[137] In September 1962, a full-scale mockup was the hit of the show at an Air Force Association convention in Las Vegas. In addition to this, the Air Force named six test pilots who would fly Dyna-Soar as its astronauts.[138]

---

137. *Ibid.*, pp. 277-279.
138. *Ibid.*, p. 269.

## Space Stations and Winged Rockets

The question of military missions raised its head again when in mid-1961 the new Defense Secretary, Robert McNamara, directed the Air Force to justify Dyna-Soar on military grounds. Air Force officials discussed orbital reconnaissance, rescue, inspection of Soviet spacecraft, orbital bombardment, and use of the craft as a ferry vehicle. While McNamara found these reasons unconvincing, he nevertheless remained willing to let the program proceed as a research effort, dropping all consideration of a possible use of the craft as a weapon system. In an October 1961 memo to President Kennedy, McNamara proposed to "re-orient the program to solve the difficult technical problem involved in boosting a body of high lift into orbit, sustaining man in it and recovering the vehicle at a designated place."[139]

This reorientation gave the project another two years of life. With its new role as an experimental craft, it was designated by Air Force Headquarters as the X-20. In this new role, however, the program could not rely on a military justification; it would have to stand on its value as research. By 1963, this value was increasingly in question. ASSET, with its unpiloted craft, was promising to demonstrate hypersonic gliding entry and hot-structure technology at far lower cost. In the realm of piloted flight, NASA now was charging ahead with its Gemini program. Air Force officials were expecting to participate in this program as well.

These officials still believed that their service in time would build piloted spacecraft for military purposes. In March 1963, McNamara ordered a study that would seek to determine whether Gemini or the X-20 could better serve the role of a testbed for military missions. The results of the study gave no clear reason to prefer the latter.

In October, Air Force officials, briefing the President's Scientific Advisory Committee, encountered skepticism in this quarter as well. Two weeks later, McNamara and other senior officials received their own briefing. McNamara asked what the Air Force intended to do with the X-20 after using it to demonstrate maneuvering reentry. He insisted he could not justify continuing the project if it was a dead-end program with no ultimate purpose.

He canceled the program in December, stating that the purpose of the program had been to demonstrate maneuvering reentry and precision landing. The X-20 was not to serve as a cargo rocket, could not carry substantial pay-

---

139. *Spaceflight*, vol. 21 (1979), pp. 436-438.

## THE SPACE SHUTTLE DECISION

loads, and could not stay in orbit for long-duration missions. He could not justify continuing with the program because it was costly and would serve "a very narrow objective."[140]

At that moment, the program, well past the stage of paper studies, called for the production of 10 X-20 vehicles. Boeing had completed nearly 42 percent of the necessary tasks. While McNamara's decision drew hot criticism, he had support where it counted; the X-20 did not. Eugene Zuckert, the Air Force Secretary, continued to endorse the program to the end, but the project had little additional support among the Pentagon's civilian secretaries. In the Air Force, the Space Systems Division (SSD) was to conduct pilot training and carry out the flights. Support for the X-20, however, was lukewarm both at the SSD and at Aerospace Corp., its source of technical advice. General Bernard Schriever, commander of the ARDC,[141] was also lukewarm. So was his deputy commander for aerospace systems, Lieutenant General Howell Estes.[142]

This was the life and death of the Dyna-Soar. From its demise one can draw several conclusions. By 1963, the program's technical feasibility was no longer in question; it was just a matter of putting the pieces together. Although aerospace vehicles were continuing to evolve at a rapid pace, no technical imperative existed that could call the X-20 into existence. The program needed a mission, a justification sufficiently compelling to win political support from high-level officials. Dyna-Soar demonstrated that even though the means were in hand to pursue the development of a vehicle resembling the Space Shuttle, such a project would stand or fall on its merits. To be built, it would require a reason capable of attracting and winning endorsement from presidential appointees and other leaders at the highest levels.

---

140. AAS History Series, vol. 17, pp. 271-275.
141. Redesignated Air Force Systems Command in 1961.
142. *Ibid.*, p. 275; Hallion, ed., *Hypersonic*, p. II-xvii.

# CHAPTER TWO

# NASA's Uncertain Future

## *Technology Bypasses the Space Station*

During the 1950s, as Walt Disney and *Collier's* presented the space station and shuttle concept to the American public, the rapid pace of technical development was making it obsolete before it could ever be built. The concept had taken form in an era when radio was the only well-developed electronic technology. It was easy, therefore, to imagine that space flight would demand large orbiting crews to conduct satellite communications, weather observation, and military reconnaissance. Like a base in Antarctica, the space station would support these crews with comfortable accommodations inside a centralized facility.

This point of view appeared not only in the writings of Wernher von Braun, but in the work of his fellow visionary Arthur C. Clarke. In 1945, Clarke proposed building communications satellites in geosynchronous orbit, at an altitude of 22,300 miles. They would circle the Earth every 24 hours, to remain fixed in position in the sky:

> *Using material ferried up by rockets, it would be possible to construct a "space-station" in such an orbit. The station could be provided with living quarters, laboratories and everything needed for the comfort of its crew, who would be relieved and provisioned by a regular rocket service.... Since the gravitational stresses involved in the structure are negligible, only the very lightest materials will be necessary and the station could be as large as required.*

## THE SPACE SHUTTLE DECISION

> *Let us now suppose that such a station were built in this orbit. It could be provided with receiving and transmitting equipment...and could act as a repeater to relay transmissions between any two points on the hemisphere beneath, using any frequency which will penetrate the ionosphere.*[1]

Even then, in 1945, rocket researchers were broadening the use of radio by introducing telemetry: the automated transmission of instrument readings. Telemetry developed in the technology of weather balloons, which could carry meteorological instruments to high altitudes. By transmitting the instrument readings, telemetry eliminated the need to physically recover the instruments following a long flight. In addition to this, weather balloons (and rockets) required equipment of minimal weight. During World War II, telemetry was used actively during test flights of the V-2. After that war, when von Braun brought his V-2s to the U.S. and carried out a program of instrumented flights in New Mexico, telemetry again played an important role.[2]

In space flight, telemetry made it possible to envision automated spacecraft. As part of the *Collier's* series, von Braun offered a proposal for such a craft in 1953. It was to carry rhesus monkeys, along with a TV camera for observation of clouds and weather patterns. *Collier's* called it a "baby space station," describing it as the "first step in the conquest of space." Chesley Bonestell, in his lyric style, portrayed it in a closeup view, soaring high over New York City.

This spacecraft, however, would serve as a prelude to the full-size space station; in no way would it represent a substitute. In von Braun's words, "We scientists can have the baby rocket within five to seven years if we begin work now. Five years later, we could have the manned space station." Though the automated spacecraft could carry a TV camera, "most of the weather research must await construction of a man-carrying space station."[3]

Two other technical developments allowed automated satellites to come into their own. The first was the development of electronic circuits that had long life. This drew on work at Bell Telephone Laboratories, where the first transistors took form. Bell Labs also introduced the solar cell, a thin wafer of silicon that could transform sunlight directly into electric current. In addition

---

1. Pierce, *Beginnings*, pp. 38-39.
2. Ley, *Rockets*, pp. 263-265.
3. *Collier's*, June 27, 1953, pp. 33-40.

The Colliers series introduced a concept for an automated Earth satellite, described as a "baby space station." (Don Davis collection)

to this, while Arthur Clarke wrote of communications satellites, it was another of Bell Labs' specialists, John Pierce, who developed the invention that allowed these spacecraft to emerge as working technology. This was the traveling-wave tube, an electronic amplifier that could work with a broad range of frequencies.[4]

In his 1945 paper, Clarke was more able to envision frequent space supply flights in high orbits than to foresee electronic circuitry that would operate routinely and reliably for years, without maintenance. Crews in their orbiting stations would spend a great deal of time replacing vacuum tubes. The situation was not much different in 1953, when von Braun proposed his "baby space station." He envisioned a time in orbit of only 60 days, which was about as much as he could expect given the limits of circuitry at that time. As early

---

4. Bernstein, *Three Degrees*, pp. 75-75, 91-95, 102-105; Pierce, *Beginnings*, pp. 1-9; *New Yorker*, September 21, 1963, pp. 60-66.

as 1958, however, the Vanguard 1 satellite demonstrated the prospect of long life. Though it lacked instruments and carried only a radio transmitter, it was powered by solar cells and was able to transmit for over six years.[5]

Another important development brought the advent of spacecraft that could operate autonomously and return from orbit. This project, known as the Corona program, was run by the Central Intelligence Agency, with Lockheed as the contractor. Their spacecraft, called Discoverer, was able to stabilize while in orbit and point a lens at the Earth below. It also operated an automated camera, winding the exposed film into a protected cassette. At an appropriate moment, the spacecraft then released a reentry capsule that fired a retro-rocket. The capsule deployed a parachute to land within a specified target area. Air Force cargo planes were then often able to snatch the capsule in mid-air.[6]

It took over a dozen satellite launches before the CIA got this complex system to work successfully. While the first launch, Discoverer 1, flew in February 1959, it was not until Discoverer 13 and 14, in August 1960 that the program achieved success.[7] Its significance then was undeniable. The analyst Jeffrey Richelson described space reconnaissance as "one of the most significant military technological developments of this century and perhaps in all history. Indeed, its impact on postwar international affairs is probably second only to that of the atomic bomb. The photo-reconnaissance satellite, by dampening fears of what weapons the other superpower had available and whether military action was imminent, has played an enormous role in stabilizing the superpower relationship."[8]

These developments—telemetry, long-life electronics, onboard autonomy—completely changed the prospects for space flight. No longer would it be necessary to build von Braun's 7000-ton cargo rockets or to support large crews in orbiting stations. Instead, the nation would proceed by developing launch vehicles from the ICBMs and similar missiles that the military was building for defense purposes. Satellites would take shape as instrumented craft of modest size and weight. In turn, the space station ceased to hold the attention of visionaries such as von Braun, who went on to influence policy.

---

5. Emme, ed., *History*, p. 138; Thompson, ed., *Space Log*, vol. 27 (1991), p. 50.
6. Ruffner, ed., *Corona*, pp. 3-39.
7. *Ibid.*, pp. 16-24.
8. Richelson, *Secret Eyes*, p. 265.

*NASA's Uncertain Future*

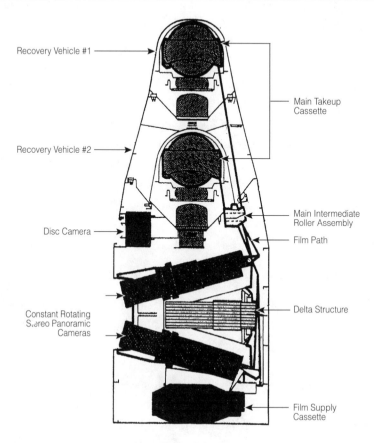

Cutaway view of the Corona satellite reconnaissance system. (Central Intelligence Agency)

Rather than emerging as a matter of urgency for the near future, the space station became something that might be built in the distant future.

In May 1961, President Kennedy committed NASA and the nation to a major effort in piloted space flight that had nothing to do with a space station. The goal, instead, was to land astronauts on the Moon. In doing this, NASA completely bypassed the classic approach of first building a space station and then using it as a base or staging area for the lunar mission. Instead, as a single Saturn V rocket carried a complete moonship with a crew of three, NASA went for the Moon in one fell swoop,.

The concept of an orbiting station, however, did not go away. If it now offered no obvious path for use in space applications, the space station still promised considerable value as a science center, supporting astronomy and studies of the Earth. Kennedy's effort aimed at a Moon landing; it was easy to imagine a permanent base on the Moon. A space station, in Earth orbit, could demonstrate and test many critical technologies. As an essential prelude to an eventual mission to Mars, it also could test the ability of astronauts to remain healthy when living for long periods in zero gravity.

The architecture of such stations also changed. The concept of a big rotating wheel fell by the wayside, in favor of designs that could fit atop a rocket as a single payload. The Saturn V could carry close to 300,000 pounds to orbit,[9] a capacity that spurred far-reaching thoughts. After 1965, attempts by NASA officials to use this capacity led to the development of a space station called Skylab.

## *Apollo Applications: Prelude to a Space Station*

The ubiquitous von Braun played a key role in initiating this new effort, not because he succeeded in convincing senior NASA officials of the merits of a space station, but rather because he knew that his staff would soon need new work. During the 1960s, he was director of NASA's Marshall Space Flight Center, where large launch vehicles were a specialty. As he stated in 1962, "we can still carry an idea for a space vehicle… from the concept through the entire development cycle of design, development, fabrication, and testing." His domain included the Michoud Assembly Facility near New Orleans, where complete Saturn V first stages were assembled. It also included the nearby Mississippi Test Facility, where these five-engine stages could operate as complete units on a test stand.[10]

The development of the Saturn V set the pace for the entire Apollo program. This Moon rocket, however, would have to reach an advanced state of reliability before it could be used to carry astronauts. The Marshall staff also was responsible for development of the smaller Saturn I-B that could put a piloted Apollo spacecraft through its paces in Earth orbit. Because both rockets would have to largely complete their development before Apollo could hit its stride, von Braun knew that his center would pass its peak of activity and

---

9. NASA SP-4012, vol. III, p. 27.
10. NASA SP-4208, p. 4; NASA SP-4206; see index references.

*George E. Mueller, NASA Associate Administrator for Manned Space Flight in 1968. (NASA)*

would shrink in size at a relatively early date. He would face large layoffs even while other NASA centers would still be actively preparing for the first mission to the Moon.[11]

At NASA Headquarters in Washington, D.C., George E. Mueller (pronounced "Miller"), Associate Administrator for Manned Space Flight, understood von Braun's situation for he had helped to create it. Mueller had been vice president of the firm of Space Technology Laboratories in Los Angeles, a division of TRW and a prime source of technical support for the Air Force's principal missile programs. Mueller had been deeply involved in the Minuteman ICBM effort, and had pushed successfully for "all-up testing," during which that missile fired all three stages and flew to its full range on its first flight.

---

11. NASA SP-4208, p. 5.

## THE SPACE SHUTTLE DECISION

Coming to NASA in 1963, he quickly became convinced that he could do the same with the Saturn V. von Braun had used a cautious step-by-step approach in developing the earlier Saturn I, flight-testing only the first stage before committing to flights of the complete two-stage launch vehicle. Mueller decided that similar caution in flight testing of Saturn V's three stages would push the first lunar landing into the next decade. He won von Braun's consent to allow Saturn V to fly "all-up" on its first flight by firing all three of its stages.[12]

This quickened the pace of development on the Apollo program, making it likely that the Saturn V would become available at a relatively early date. It also hastened the day when von Braun's center would largely complete its work and face layoffs. Mueller's decision, however, also made it likely that surplus Saturn-class rockets would become available for purposes other than direct support of Moon landings.

In August 1965, Mueller set up a new Saturn-Apollo Applications Program Office. The Saturn I-B emerged as an early focus for attention. This powerful rocket conducted only a limited program of developmental flights for Apollo before giving way to the much larger Saturn V. The Saturn I-B's second stage, the S-IVB, had a liquid-hydrogen propellant tank with a volume of nearly 10,000 cubic feet. There was interest in turning the S-IVB into an orbiting workshop. Mueller later stated that this would match the volume of "a small ranch house. The kind I can afford to buy."

By early 1967, the program called for an initial mission featuring two launches. The first would carry an Apollo spacecraft with its crew of three; the second would launch the workshop, mounted to an airlock and docking adapter. The S-IVB, modified for use in orbit, was to sprout large solar panels along with two floors within the 21-foot wide hydrogen tank. These floors would provide living quarters and work areas. The flight crew would rendezvous with the workshop and dock with the adapter. Inside the spent fuel tank, these astronauts would find an empty, bare-walled space that would require four days of fitting-out to turn into habitable living quarters. The crew would then stay in space for 28 days conducting biomedical tests as their principal activity. A subsequent mission to the workshop would bring a fresh crew to live in space for 56 days.[13]

---

12. *Ibid.*, pp. 6-7; NASA SP-4012, vol. II, pp. 54-58.
13. NASA SP-4208, pp. 20-21, 26-27, 53-55.

## NASA's Uncertain Future

In addition to Mueller's powerful Office of Manned Space Flight, a separate NASA program center, the Office of Space Science and Applications (OSSA), made its own contribution to the new post-Apollo effort. Within the field of space science, OSSA supported solar astronomy, using spacecraft to observe the Sun at ultraviolet and x-ray wavelengths that do not penetrate the atmosphere. In 1962 and 1965, two Orbiting Solar Observatory spacecraft returned a great deal of useful data and sparked interest in an advanced automated solar observatory. Such plans fit the cyclic activity of the Sun itself, which, every 11 years, rises to a peak in the number of sunspots, radiation levels, and magnetic activity. The next such peak was to occur in 1969, leaving ample time for development of the new spacecraft.

OSSA's plans fit the solar cycle much better than the budget cycle. OSSA had little clout, and the demands of Apollo were all-consuming; pressed by its budgetary needs, scientific satellites tended to fall by the wayside. The head of OSSA, Homer Newell, was undismayed. Though his advanced automated observatory failed to win support and had to be canceled, Newell saw that he could seek an even more ambitious solar observatory by hitching his wagon to the star of piloted space flight. Working with Mueller, Newell developed a concept for an Apollo Telescope Mount (ATM), as a second important component of Apollo Applications.

This ATM took shape as a substantial spacecraft in its own right. Requiring its own Saturn I-B to carry it aloft, it also called for its own set of solar panels that would unfold to form a large cross. The program plan called for it to rendezvous with the orbiting workshop early in the 56-day second mission. The astronauts would move it into position and install it as part of the complete space laboratory. With a dozen instruments, the ATM would test the ability of astronauts to conduct useful scientific research by operating sophisticated equipment in orbit.[14]

These missions were to herald a major program. Released in March 1966, NASA's initial schedule envisioned 26 launches of the Saturn I-B and 19 of the Saturn V. Flight hardware would include three S-IVB stages intended for on-orbit habitation, four ATMs, and three more capable space stations that would ride atop the Saturn V. The Bureau of the Budget (BoB), an arm of the White House, was not encouraging. Bureau officials were concerned that

---

14. *Ibid.*, pp. 36-37, 69-71.

# THE SPACE SHUTTLE DECISION

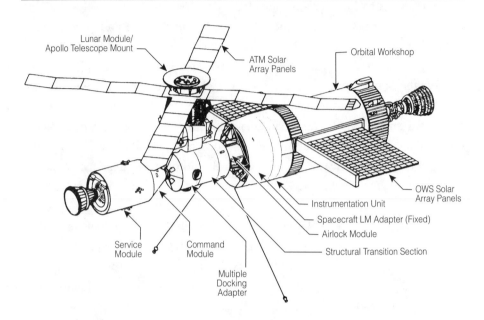

*Apollo Applications wet workshop, derived from an S-IVB upper stage. Note rocket engine at right. (NASA)*

Apollo Applications might wastefully duplicate an Air Force program, the Manned Orbiting Laboratory. In addition to this, with Apollo reaching the peak of its funding, those officials were in no mood to allow NASA to launch another costly program.

Initial discussions focused on the budget request for FY 1967 that President Lyndon B. Johnson would present to Congress early in 1966. Mueller hoped at first for $450 million, with over $1 billion in FY 1968. Bureau of the Budget officials preferred to start by offering $100 million, though they were willing to listen to arguments for $250 million. This part of NASA's budget included Apollo. To keep it on schedule, Mueller had to put Apollo Applications under a particularly severe squeeze with only $42 million (less than a tenth of his initial budget mark) for FY 1967.[15]

The FY 1968 budget brought more of the same. Initial discussions between NASA and the BoB chopped the request from $626 million to $454

---

15. *Ibid.*, pp. 42-43; NASA SP-4011, p. 71.

million, a sum that at least would get the program off to a good start. In his budget message to Congress, Johnson endorsed this figure with an argument that would be heard again in subsequent years: "We have no alternative unless we wish to abandon the manned space capability we have created." Though Johnson and the BoB were now on board, Congress, which cut the authorization to $347 million, was not. Not even the appropriation—more bad news at $300 million—was safe, as the NASA Administrator, James E. Webb, transferred part of it to other activities. Apollo Applications was left with only $253 million, the lowest level Mueller could accept.[16]

It nevertheless was enough, barely, to get the program under way and turn it into something more than a design exercise. As serious engineering activity got under way, however, designers came to realize that they were pursuing an approach marked with pitfalls. The approach continued to call for a "wet workshop," a propulsive stage that would then serve as living and working quarters while in orbit. After reaching orbit, however, astronauts would have to convert the empty fuel tank into these quarters and install a good deal of equipment. As studies proceeded, it became increasingly doubtful that all this could be done.

The alternative would be to build the space station as a "dry workshop" with no provision for use as a rocket stage. Unable to propel itself into orbit, the dry workshop would need the heavy lifting power of a Saturn V. That rocket's payload capacity would make it possible to incorporate the ATM from the outset, rather than having to bring it up on a separate flight. The complete, well-integrated space station could undergo tests and verification on the ground.

While studies of a dry workshop were being conducted at the same time as those of the wet version, they were never endorsed by NASA Administrator James Webb. The sticking point was the need for a Saturn V. The historians Charles Benson and David Compton note that "it had taken all of Webb's power of persuasion to convince Congress and the BoB that Apollo required at least 15 Saturn V launch vehicles, and he would tolerate no suggestion that any could be used for something else."[17] When Webb resigned from NASA in October 1968, he took his objections with him. In addition to this, in

---

16. NASA SP-4208, pp. 53, 86-87.
17. *Ibid.*, pp. 105-109.

December 1968, Apollo 8 carried three astronauts on a successful flight that orbited the Moon and returned safely. This was only the third flight of a Saturn V, making it highly plausible that it would indeed be possible to spare one of those behemoths for Apollo Applications.[18]

With the mounting technical problems of the wet workshop approach, Mueller became convinced that it simply was not practical. Hence, only a dry workshop could save the program. The new NASA Administrator Thomas Paine became convinced in 1969 that it was necessary to make the switch. His decision was subject only to the success of Apollo 11, the planned first lunar-landing mission. He signed the project-change document on July 18, while Apollo 11 was en route to the Moon. Four days later, with the landing accomplished and the astronauts homeward bound, the Apollo Applications program manager, William Schneider, sent telexes to the NASA centers that directed them to proceed with the dry workshop.

Program cutbacks, however, had taken their toll. Apollo Applications, initially conceived as a long-running extension of Apollo, was down to a single workshop supported by three astronaut crews flying the Saturn I-B. There was hope for a second workshop that would carry different equipment. The program needed a new name; a committee considered close to a hundred possibilities, including "Socrates" and "LSD." The winning name, "Skylab," came from Lieutenant Colonel Donald Steelman, an Air Force officer on duty with NASA. The new name, which replaced Apollo Applications, was formally adopted in February 1970.[19]

## *Space Station Concepts of the 1960s*

There was only a single Skylab orbiting workshop in existence. Though NASA had built a second model, there were no funds to launch this spacecraft, and it wound up on display at the National Air and Space Museum.[20] To this day, Skylab remains the closest thing to a true space station that NASA has ever built and operated. Nevertheless, it represented no more than a small step toward that goal.

---

18. NASA SP-4012, vol. II, p. 61.
19. NASA SP-4208, pp. 107-110, 112, 114-115.
20. *Ibid.*, p. 353.

Skylab grew out of Apollo Applications, which merely sought to make good use of Apollo launch vehicles and equipment. Though the Skylab spacecraft strongly modified the standard S-IVB rocket stage, its design was heavily constrained. The 22-foot diameter of Skylab followed from the diameter of the S-IVB, even though the Saturn V could accommodate payloads of up to 33 feet across. Similarly, although Skylab included the ATM as part of its package, its total weight, 165,000 pounds, fell well short of the lifting power of the Saturn V. These restrictions arose because the dry workshop, which used the Saturn V, developed out of the wet workshop, which was to have used the much smaller Saturn I-B.[21]

In addition, Skylab was not permanently inhabited. It supported three crews in orbit, during 1973 and 1974, who stayed respectively for 28, 59 and 84 days. Though the last such mission continues to hold the record for duration in U.S.-built spacecraft, Soviet and Russian cosmonauts have stayed in orbit for up to 437 days in the Mir station. Following the return of the third Skylab crew, in February 1974, NASA made no further attempt to use this valuable facility. Skylab's orbit, left to decay, caused it to burn up in the atmosphere in July 1979.[22]

In spite of its limitations and its shrinking budgets, Apollo Applications was important. Not just a paper study, it was a true and funded program, with a project office at NASA Headquarters that stood alongside similar offices for Gemini and Apollo.[23] It thus gave considerable hope to those in both NASA and the industry who were carrying out studies for the *next* space station. During the 1960s, a number of studies sought to define such a station.

NASA's Langley Research Center took an early interest in such studies, setting up a space station office within its Applied Mechanics and Physics Division. Early work, from 1959 to 1962, focused anew on the rotating-wheel configuration. At the outset, the Langley designers considered a range of shapes that could rotate to provide artificial gravity. Like Potočnik and von Braun before them, they decided the wheel was best. With a radius of 75 feet, it would rotate at four revolutions per minute, producing two-fifths of normal gravity.

Langley then contracted with North American Aviation (NAA) to carry out further studies. A prime question was how to fit so large a structure into

---

21. *Ibid.*, pp. 107-108; Thompson, ed., *Space Log*, vol. 27 (1991), p. 137.
22. Thompson, *Space Log*, vol. 27 (1991), pp. 137, 138, 141; vol. 31 (1995), p. 68.
23. NASA SP-4208, pp. 20-21.

# THE SPACE SHUTTLE DECISION

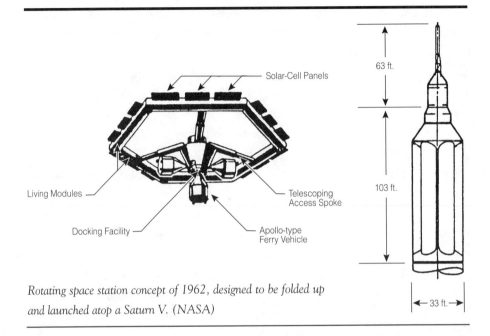

*Rotating space station concept of 1962, designed to be folded up and launched atop a Saturn V. (NASA)*

the cargo volume of a Saturn V. NAA changed the wheel to a hexagon composed of six long cylinders joined at their ends. These would fold into a package 103 feet long by 33 feet in diameter. Once in orbit, mechanical screw jacks would unfold the hinged parts. The complete space station would include a hub with a docking facility for Apollo spacecraft. With telescoping spokes joining the hub to the hexagon, the station's volume of 45,000 cubic feet would accommodate up to 36 crew members.[24]

In size between Potocnik's concept of 1928 and von Braun's of 1952, NASA's concept represented a brilliant attempt to bring the rotating wheel into an era in which major tasks, including piloted flight to the Moon, would be carried out in space. Even so, it was behind the times. The project's emphasis on artificial gravity was better suited to an earlier age when large crews were expected to live in comfort. At the same time, by 1960, tasks that were to be conducted by astronauts were ready for automated electronics. In addition to this, by 1963 it was clear that studies of human physiology during extended durations in weightlessness would represent an important rationale for a space station. Subsequent concepts reflected these changes.

---

24. *Ibid.*, pp. 9-10; AAS History Series, vol. 14, pp. 80-83.

## NASA's Uncertain Future

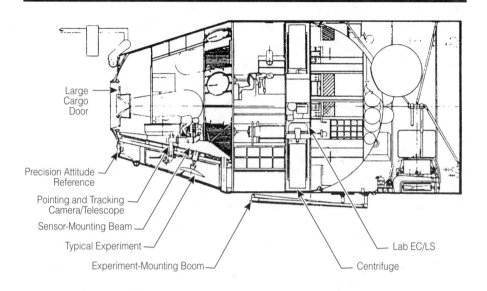

*Cutaway view of MORL. (Douglas Aircraft)*

Langley's next round of studies, called the Manned Orbiting Research Laboratory (MORL), rejected the rotating wheel once and for all. Late in 1963, Douglas Aircraft won this study contract. In many ways, MORL illustrated what Apollo Applications might have accomplished if it had been given high priority and ample funding.

Rather than seeking to support large crews in the comfort of artificial gravity, MORL emphasized small crews that would live in weightlessness in versatile, compact stations. The basic station was to fly atop a Saturn I-B and hence had that rocket's diameter of 22 feet. Weighing 30,000 pounds at launch, MORL would enclose 9,000 cubic feet of internal volume, with a crew of six. Each astronaut would serve a six-month tour of duty. A modified Apollo spacecraft, riding its own Saturn I-B, would carry supplies along with new three crew members to the space station.

Specialized equipment would enhance the usefulness of MORL. It would carry astronomical telescopes. A crew-tended radar would support large-scale topographical mapping. Douglas Aircraft also proposed to install a nine-lens camera system for observation of the Earth's surface and weather at a variety of wavelengths. With astronauts tending a lab full of plants, animals and bacteria, additional modules would research new fields such as life sciences. The addition of other such modules would allow the basic station to expand to house nine astronauts rather than the original six. Selected crew members would remain in orbit for as long as a year.

Use of the Saturn V would enable the MORL to fly in orbits as high as 23,000 miles while continuing to receive resupply. The MORL would be able to fly to lunar orbit to map the Moon's surface. It would be able to land on the Moon and to serve as a base. Serving as a test bed for systems intended for use in a piloted mission to Mars, MORL also might evolve into an important element of a spacecraft built to carry out such a mission.[25]

At the Manned Spacecraft Center (renamed the Johnson Space Center in 1973) in Houston, other investigators agreed that a space station could represent an intermediate step toward a mission to Mars. That center had its own space station group that had contracted with the Space Division of the Boeing Co. to conduct the pertinent study. Completed in 1967, that study envisioned a Mars spaceship that also could serve as an Earth-orbiting station.

The Mars ship would take the form of a two-deck module, 22 feet in diameter, with room for both crew members and equipment. For use as a space station, the vehicle would add a second module, together with a central section, midway along the station, that could accommodate the docking of two Apollo spacecraft. With a weight of 248,000 pounds, this complete station would ride a Saturn V to orbit. It would support a crew of eight, with these astronauts flying on the Saturn I-B, in Apollo craft modified to carry four rather than the usual three people. Two such launches would provide the initial staff. Subsequent flights every 90 days would bring fresh crew members as well as new supplies. The station would remain continuously occupied for two years.

Without resupply or revisit en route, the Mars mission would also last two years. Mission designers would chop the space station in two, retrieving the basic two-deck module and staffing it with a crew of four. After being placed in orbit by a single Saturn V launch, additional Saturn V flights would carry fully-fueled S-IVB stages to boost the Mars ship toward its destination. While it would fly past and not land on that planet or even orbit it, the mission would drop off planet probes, landers, and an orbiter during this flyby. During the

---

25. *Astronautics & Aeronautics*, March 1967, pp. 34-46; NASA SP-4308, pp. 293-300.

# NASA's Uncertain Future

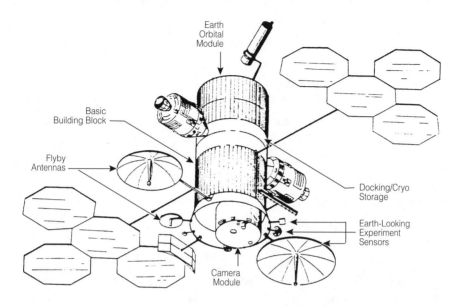

*Boeing space station concept of 1967. (Boeing)*

close approach to that planet, the flight to Mars would culminate in an 11-day period of intense crew activity followed by the long voyage home.[26]

Not everyone agreed that a space station should serve as a way station for flight to Mars. An alternate viewpoint stressed the usefulness of such stations for science alone. This view found support at NASA's Marshall Space Flight Center. A 1966 study there noted that a proper science station could not be all things to all people. It was argued that different sciences would impose characteristic demands that would be mutually incompatible.

Astronomy in space, for example, would require gamma-ray, x-ray, optical, and radio telescopes. These would have to point in fixed directions during their observations, maintaining stability to within 0.001 degrees. A due-east launch from Cape Canaveral could put them in orbit, with an inclination to the equator of 28 degrees. By contrast, observation of the Earth's surface and weather would ideally require a polar orbit that demands more energy at launch. An Earth-observing station would have to turn slowly to point continually downward, rather than stare at a fixed position in space. It could work

---

26. Report D2-114012-1 (Boeing).

with a stability of 0.05 degrees. Biomedical experiments, including long-duration studies of the human response to weightlessness, would be even less demanding. Able to work in any orbit, they would dispense with the costly control systems necessary for pointing and stabilization.

The Marshall study thus called for two stations, each with a crew of nine and a lifetime of five years for the station. They would fly to orbit atop the Saturn V. One station, supporting astronomy, would fly due east from the Cape. The second station, supporting meteorology and Earth observations, would not use the hard-to-reach polar orbit, but would achieve an intermediate inclination of 55 degrees. This inclination would still permit coverage of the world's major land masses. Biologists and life-science specialists, not requiring a specific orbit, could build a specialized module that could fly as part of either station.[27]

It is important to note that these studies lacked the support of a NASA Headquarters program office similar to that of Apollo Applications after 1965. These studies, however, did have the attention of center directors. In 1963, the original MORL Studies Office reported directly to Floyd Thompson, the director of NASA-Langley.[28] Similarly, it was no secret that Wernher von Braun, director of NASA-Marshall, had a strong and ongoing interest in space stations. With no one at Headquarters who was ready to take those studies and push for their fulfillment, the space station represented only a possible new direction for NASA. In no way was there a commitment to pursue that direction.

In addition, these studies reflected the characteristic point of view that space stations could offer intrinsic advantages. In 1968, Robert Gilruth, director of the Manned Spacecraft Center, defined such a station as "a site in space developed to support men, experimental equipment, and operations permanently and to take advantage of the favorable economies of size, centralization, and permanency—in terms of power, volume, instruments, communications, data reduction, and logistics."[29] This amounted to an assertion that those "favorable economies" actually existed, a point from which both Congress and the Budget Bureau soon would differ.

Likewise, it was not easy to assume that space stations would win support on their merits for use in science. The concepts of the day anticipated the rou-

---

27. AAS History Series, vol. 14, pp. 83-86.
28. NASA SP-4308, p. 294.
29. *Astronautics & Aeronautics*, November 1968, p. 54.

tine use of the Saturn I-B with the Apollo spacecraft for resupply and crew rotation. The Apollo 7 mission, which had flown atop the Saturn I-B in 1968, cost $145 million. Two years later, a single flight of a Saturn V with its moonship would cost up to $375 million. By contrast, in FY 1970, the National Science Foundation, which sponsors a broad range of basic research in a large number of fields, received a budget of $440 million.[30] Indeed, it would take a true believer to assert that a Saturn V, even with an Apollo mission, could offer the scientific return of a year's worth of grants from the NSF to the nation's universities and research centers.

This point was not lost on the advanced-planning designers who were nurturing their space stations. They saw that the expensive Saturn V might not remain the only way to launch a large station; a reusable launch vehicle might cut costs while offering even greater lifting power. In addition to this, it might prove feasible to dispense with the Saturn I-B, replacing it with a low-cost launcher of intermediate size. A number of specialists pursued these hopes during the 1960s, as they allowed their imaginations to run free. In pursuing their designs, they laid a considerable amount of groundwork for the serious studies for a practical space shuttle that followed.

## *Early Studies of Low-Cost Reusable Space Flight*

No one could deny that space flight was expensive. Launch vehicles flew only once. There was no way to reuse them; they launched their payloads and then splashed into the ocean. A Saturn I-B came to $45 million, excluding its Apollo spacecraft and flight operations; a Saturn V cost $185 million. For these rockets to carry three astronauts cost as much as $60 million per person.[31]

Advocates of reusable launch vehicles said that using throwaway Saturns was tantamount to flying a planeload of passengers across the Atlantic and having that airliner fly only once. It is a measure of the truly enormous cost of space flight that this comparison was off by three orders of magnitude. The Boeing 727, a popular jet of the 1960s, had a sticker price of $4.2 million. It carried 131 passengers. Had each such plane made only a single flight, the cost of a ticket would have been some $30,000.[32] The corresponding price for

---

30. NASA budget data, February 1970; *Science*, 5 February 1971, p. 460.
31. NASA budget data, February 1970.
32. Serling, *Legend*, p. 186; *Pedigree*, p. 58.

a ticket on a Saturn V was 2,000 times greater. A more appropriate if less exact simile came from *Newsweek* in 1961.[33] It compared the space race to the potlatch ceremony of the Kwakiutl tribe of the Pacific Northwest, whose members vie to throw the most valuable objects into a fire. Clearly, the nation was unlikely to persist in this celestial potlatch unless it had the most compelling of reasons.

An initial step toward reusability came at NASA-Marshall during 1961 and 1962, where engineers sought to learn whether a high-performance rocket engine could survive a dunking in seawater. They worked with the H-1, a standard engine from Rocketdyne that went on to power the Saturn I-B. Following immersion, investigators dismantled the engine, checked its parts for corrosion, reassembled it, and ran it successfully on a test stand. Thus, it was proven that this powerful engine, rated at 187,000 pounds of thrust, could withstand a bath in seawater and return to service.[34]

The next question was whether a Saturn-class first stage could be recovered for reuse. There was considerable interest in using a flexible and deployable wing invented by Francis Rogallo of NASA-Langley. The "Rogallo wing" later found its niche as a type of hang glider, allowing enthusiasts to fly from clifftops and soar on uprising air like birds. Advocates hoped to use it as a directional parachute, permitting a large booster to descend by gliding to a designated recovery point.

Studies showed that this approach would not work with existing first stages such as the Saturn I-B. Because they had not been designed for recovery, they lacked the storage room for the furled Rogallo wing.[35] Thus, it would not be possible to introduce reuse by the simple approach of mounting a deployable wing to a Saturn booster. Studies funded by NASA-Marshall, under the name "50- to 100-Ton Payload Reusable Orbital Carrier," showed, however, that NASA might achieve better results by installing fixed wings on the Saturn V's first stage.

The new first stage would use that booster's standard engines, adding landing gear, a pilot compartment, insulation to protect against the heat of atmosphere reentry, and large wings, sharply swept, with big vertical fins at the tips. These modifications would add 300,000 pounds of weight. The

---

33. *Newsweek*, January 2, 1961, p. 42.
34. Akridge, *Space Shuttle*, pp. 8-9; NASA SP-4012, vol. II, p. 56.
35. Akridge, *Space Shuttle*, p. 9; *Astronautics & Aeronautics*, August 1968, pp. 50-54.

second stage, however, would retain its full lifting power. Thus, the payload would be decreased by only 20 percent.

Smaller winged rockets also drew interest, as analyses showed that even with parachutes, recovery of any craft at sea would be both costly and clumsy. Leonard Tinnan, a manager at North American Aviation (NAA), wrote that "in comparing parachute or other so-called 'simple' means of booster recovery with the 'sophisticated' fixed-wing approach, for example, it becomes rather easy to demonstrate that the former is economically superior—if the time and costs associated with the mid-ocean retrieval and refurbishment of booster stages, and the impact of corresponding extension of turnaround time, are omitted or minimized. In the final analysis, however, all such factors must be fully considered."[36]

A review of design concepts of the early 1960s shows that engineers were of two minds on approaches to reuse. The prospect of aircraft-type operation tantalized a number of these people, with the X-15 offering inspiration by flying routinely in flight test. Designers expected that their reusable launch vehicles would fly often. For this they would need wings and runways because recovery at sea would hamper frequent flight schedules. Other investigators wanted reusable launchers that would carry far more payload than a Saturn V. Far too large for wings, such leviathans would have to come down in the ocean.

Perhaps the largest of these reusable launchers was the Nexus. The work of a group at General Dynamics led by Krafft Ehricke, the Nexus was to represent the next leap beyond the Saturn V, carrying up to eight times more payload. Fully fueled, it would weigh 24,000 tons, as much as an ocean-going freighter. It would carry a 1,000 tons to orbit, allowing it to launch a spaceship bound for Mars. This behemoth would have a diameter of 202 feet with its height approaching that of the Washington Monument. It would fly as a single-stage launch vehicle. Fully recoverable, it would touch down in the ocean following a return from orbit. Parachutes would slow its descent. Retrorockets, firing during the last seconds, would assure a gentle landing.[37]

Others hoped to develop new types of engines. The years since World War II had brought enormous advances in turbojets, rockets, and ramjets. By 1960, all three offered tested paths to high-speed flight. With such further develop-

---

36. *Astronautics*, January 1963, pp. 50-56.
37. *Astronautics & Aeronautics*, January 1964, pp. 18-26.

## THE SPACE SHUTTLE DECISION

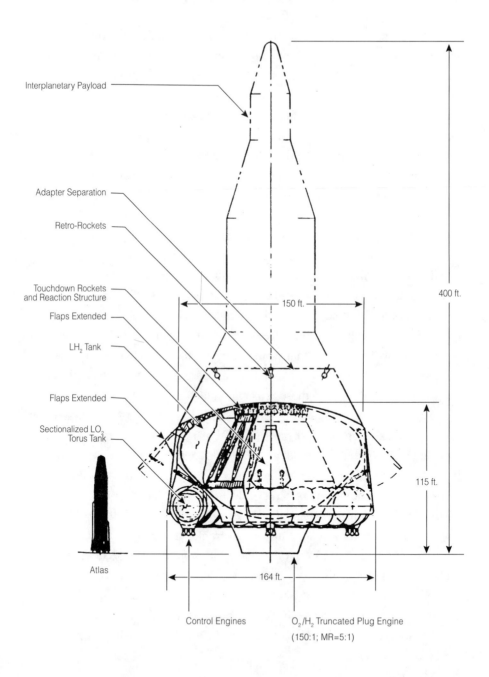

NEXUS heavy-lift booster concept. Atlas ICBM at lower left indicates scale. (Krafft Ehricke)

ments in the offing, advocates of advanced propulsion saw their prospects in two novel concepts: LACE (Liquid Air Cycle Engine), an airbreathing rocket; and the scramjet, a hypersonic jet engine.

LACE sought to overcome the requirement that a rocket must carry its oxygen as a heavy quantity of liquid in an onboard tank. Instead, this concept sought to allow a rocket to get its oxygen from air in the atmosphere. Because rocket engines operate at very high pressure, no air compressor could compress the ambient air so as to allow it to flow into a thrust chamber. If the air could be liquefied, however, it would form liquid air, which could be pumped easily to high pressure. LACE sought to do this by passing the incoming air through a heat exchanger that used supercold liquid hydrogen, chilling the air into liquid form. The engine then would use the hydrogen and liquefied air as propellants.[38]

This approach drew strong interest at Marquardt Co., a Los Angeles propulsion-research firm. In tests at Saugus, California, in 1960 and 1961, Marquardt engineers successfully demonstrated a LACE design that used heat exchangers built by Garrett AiResearch. A film of those tests, shown at a conference of the Institute of the Aeronautical Sciences in March 1961, shows liquid air coming down in a torrent, as seen through a porthole. Marquardt went on to operate test engines with thrusts of up to 275 pounds. During these tests, LACE performed twice as well as conventional hydrogen-fueled rockets.

There were further innovations as well. Four-fifths of air is nitrogen, which does not burn. The presence of this nitrogen reduced the performance of LACE by cooling the exhaust and demanding extra liquid hydrogen to accomplish liquefaction. Oxygen, however, liquefies at 90 degrees Kelvin while nitrogen liquefies at the lower temperature of 77 degrees Kelvin. Thus, by carefully controlling the heat-exchange process, oxygen in the air could be liquefied preferentially. This represented a topic for further research. In 1967, at General Dynamics, a test of this concept demonstrated 90 percent effectiveness in excluding the nitrogen.[39]

While LACE represented a new direction in rocket research, the scramjet represented advances in the design of the ramjet. Ramjet engines showed their power during the 1950s when the Lockheed X-7, an unpiloted missile,

---

38. Heppenheimer, *Hypersonic*, pp. 15-16.
39. *Ibid.*, p. 16; *Aviation Week*, May 8, 1961, p. 119. Film courtesy of William Escher, Kaiser Marquardt, Van Nuys, California.

reached Mach 4.31 or 2,881 miles per hour setting a record for the flight of airbreathing engines.[40] This was close to the speed limit of a ramjet. Air in such a ramjet, flowing initially at supersonic speeds, had to slow to subsonic velocity in order to burn the fuel. When it slowed, an engine became hot and lost power.

For a ramjet to reach speeds well beyond Mach 4, this internal airflow would have to remain supersonic. This would keep the engine cool and prevent it from overheating. This also imposed the difficult problem of injecting, mixing, and burning fuel in such a supersonic airflow. Nevertheless, a number of people hoped to build such an engine, which they called a scramjet.[41]

Scramjet advocates included Alexander Kartveli, the vice president for research and development at Republic Aviation, and Antonio Ferri, a professor at Brooklyn Polytechnic Institute. During World War II, Ferri had been one of Europe's leading aerodynamicists and had directed Italy's premier research facility, a supersonic wind tunnel. Kartveli was one of America's leading airplane designers, crafting such fighter aircraft as the F-84 and the F-105. During the 1950s, his focus was on another proposed fighter, the XF-103. It was to use a ramjet to reach speeds of Mach 3.7 (2,450 mph) and altitudes of 75,000 feet.[42]

Ferri, who worked as a consultant on this project, formed a close friendship with Kartveli. They complemented each other professionally, Kartveli studying issues of aircraft design, Ferri emphasizing the details of difficult problems in aerodynamics and propulsion. As they worked together on the XF-103 they each stimulated the other to think bolder thoughts. Among the boldest put forth first by Ferri, and then supported by Kartveli with more detailed studies, was the idea that scramjet-powered aircraft would have no natural limits to speed or performance. They could fly to orbit, reaching speeds of Mach 25.[43]

In the Air Force, concepts such as LACE and scramjets drew support from Weldon Worth, technical director at the Aero Propulsion Lab of Wright-Patterson Air Force Base. Beginning in about 1960, Worth built up a program of basic research called Aerospaceplane. Not aiming at actually building an

---

40. Miller, *X-Planes*, p. 72.
41. Heppenheimer, *Hypersonic*, pp. 12-14.
42. *Ibid.*, pp. 10-12; Gunston, *Fighters*, pp. 184, 193-195.
43. *Republic Aviation News*, September 9, 1960, pp. 1, 5.

airplane that would fly to orbit, the program pursued design studies and propulsion research that might lead to such aircraft in the distant future. The propulsion efforts were often very basic. When, in November 1964, Ferri succeeded in getting a scramjet to deliver thrust, it was impressive enough to merit an Air Force news release. Ferri went on to set a goal of 644 pounds of thrust for his test engine; he managed 517 pounds, 80 percent of his goal.[44]

Aerospaceplane was too hot to keep under wraps. A steady stream of leaks brought continuing coverage in the trade magazine *Aviation Week*.[45] At the *Los Angeles Times*, the aerospace editor Marvin Miles developed his own connections, which led to banner headlines: "Lockheed Working on Plane Able to Go Into Orbit Alone"; "Huge Booster Not Needed by Air Force Space Plane."[46] The Air Force's Scientific Advisory Board (SAB) was not amused. As early as December 1960, it warned that "too much emphasis may be placed on the more glamorous aspects of the Aerospaceplane resulting in neglect of what appear to be more conventional problems."

By 1963, with hype outrunning achievement, the SAB had had enough. In October, it declared that "today's state-of-the-art is inadequate to support any real hardware development, and the cost of any such undertaking will be extremely large.... [T]he so-called Aerospaceplane program has had such an erratic history, has involved so many clearly infeasible factors, and has been subjected to so much ridicule that from now on this name should be dropped. It is also recommended that the Air Force increase the vigilance that no new program achieves such a difficult position."[47] Soon after, the Aerospaceplane died as a formal program. The scramjet, however, continued to live as NASA-Langley pursued an experimental program, the Hypersonic Research Engine, that continued well into the 1970s.[48]

Amid the gigantism of the Nexus and the far-out futurism of Aerospaceplane, there were those who were content to envision winged craft powered by conventional rocket engines. Here, too, the exuberance of the day

---

44. Heppenheimer, *Hypersonic*, pp. 14-17; Hallion, ed., *Hypersonic*, pp. 948-952; news release, USAF Aeronautical Systems Division, November 12, 1964. Scramjet test data from Louis Nucci, General Applied Science Laboratories, Inc., Ronkonkoma, New York.
45. *Aviation Week*: October 31, 1960, p. 26; December 26, 1960, pp. 22-23; June 19, 1961, pp. 54-62; November 6, 1961, pp. 59-61; April 23, 1962, pp. 26-27. See also *Missiles and Rockets*, May 22, 1961, p. 14.
46. *Los Angeles Times*: November 3, 1960, p. 3A; January 15, 1961, front page.
47. Hallion, ed., *Hypersonic*, p. 951.
48. *Ibid.*, pp. 747-842; Heppenheimer, *Hypersonic*, pp. 17-20.

sometimes found expression in concepts of heroic size, such as the Astroplane of Aerojet-General. This concept included wings that would carry liquid hydrogen, much as the wings of airliners carry jet fuel. The Astroplane would have a wingspan of 423 feet and a length of 260 feet, excluding its payload. Carrying up to 220 tons of cargo, it would weigh 5,000 tons at liftoff, and would rise into the air with twice the thrust of a Saturn V.[49]

There were several design exercises, however, that projected modest size and near-term technology. One such concept, the Astro from Douglas Aircraft, was a two-stage fully-reusable launch vehicle with a payload of 37,150 pounds. Both stages of the Astro were designed as lifting bodies and would burn hydrogen and oxygen, using rocket engines that were already under development. The project engineers saw no problem with reuse of such rockets, noting that one of their engines, the Pratt & Whitney RL-10, had already "been operated more than 9,000 seconds with more than 50 restarts."

Nevertheless, these engineers also shared the enthusiasm of the times. Written in 1963, their paper on the Astro anticipated that this vehicle could be operational "in the 1968-70 period." Each flight would cost $1.5 million. In readying the second stage for a reflight, turnaround time "would range between 2.5 and 5 days, based on a two-shift operation." The Astro would fly 240 times per year.[50]

The era's exuberance was understandable; it had taken less than 35 years to advance from Lindbergh in Paris to astronauts in orbit. It was expected that this pace would continue. Amid the plethora of new possibilities, however, promising ideas sometimes were lost in the shuffle. This happened to Martin Marietta's Astrorocket concept of 1964. In the light of subsequent events, the concept seems to have offered a glimpse of the future, not only because the design was highly futuristic but because it clearly foreshadowed a class of design concepts that later stood in the forefront between 1969 and 1971.

With a planned liftoff weight of 1,250 tons, Astrorocket was to be intermediate in size between the Saturn I-B and the Saturn V. It was a two-stage fully-reusable design, with both stages having delta wings and flat undersides. These undersides fitted together at liftoff, belly to belly. The designers of Astrorocket were no clairvoyants; rather, they drew on the background of

---

49. *Astronautics & Aeronautics*, January 1964, pp. 35-41.
50. *Ibid.*, pp. 42-51.

# NASA's Uncertain Future

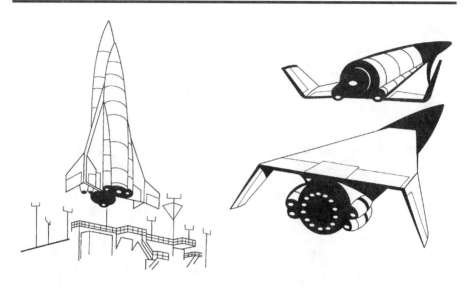

*Martin Marietta's Astrorocket concept. (Art by Dennis Jenkins)*

Dyna-Soar and studies at NASA-Ames of winged reentry vehicles.[51] The design studies of 1969-1971 followed the same approach, featuring two-stage fully-reusable configurations and a strong preference for delta wings.

Unfortunately, Astrorocket was at least five years ahead of its time. It failed to win support from NASA, the Air Force, and even its own management of Martin Marietta. That firm would continue to pursue studies of reusable launch vehicles, but these would not be Astrorockets.

"Let a hundred flowers bloom, let a hundred schools of thought content," said China's Chairman Mao in 1956.[52] Studies of future space transportation were certainly blossoming. The field, however, needed vigorous pruning to define the most promising approaches. Wielding their garden shears, a number of investigators began to address some key questions.

Was it worth waiting for the scramjet? While its performance far surpassed that of even the best rockets, its development would take time and its prospects were not certain. Even accepting that the next generation of launch vehicles would continue to use rockets, there was the question of whether

---

51. Hallion, ed., *Hypersonic*, pp. 952-954.
52. *Oxford*, p. 328.

such craft should take off horizontally, like an airplane. A booster, heavy with propellant, would need large, massive wings to do this. The vehicle, however, might ride a rocket-powered sled that would accelerate to several hundred miles per hour, at no cost to the booster in onboard fuel.

In 1962, NASA-Marshall set out to address such issues through design studies. The first step was to set standards for the design of launch-vehicle concepts. Each concept had to carry ten passengers or ten tons of cargo. Aircraft-type approaches were paramount, with Marshall stating that contractor designs "should be compatible with a philosophy used in the development of supersonic commercial jet aircraft and should offer a potential commercial application in the late 1970s, such as operating the vehicle over global distances for surface-to-surface transport of cargo and personnel."

This study, called "Reusable Ten Ton Orbital Carrier Vehicle," awarded contracts of $428,000 to Lockheed and of $342,000 to NAA. From June 1962 to December 1963, designers looked at two-stage fully-reusable configurations that put fixed wings on both stages, and carried through separate designs for both vertical and horizontal launch. They also considered concepts that drew on the Air Force's Aerospaceplane, with advanced airbreathing engines to provide propulsion in the first stage.

Subsequent studies investigated additional alternatives and pursued design issues in greater depth. In 1965, General Dynamics defined a concept for a reusable second stage that had the shape of a lifting body; both that firm and Lockheed conducted studies of first stages that could carry such a second stage. First-stage concepts continued to cover both vertical and horizontal launch. When using airbreathing engines, design choices ranged from conventional turbojet engines to scramjets. At General Dynamics the possibilities included LACE, for which that company had an active experimental program.

These studies concluded that, without exception, rocket engines were preferable to airbreathers for first-stage propulsion. A leader in these efforts, Max Akridge of NASA-Marshall wrote that "the economic advantage for the rocket engine was always about the same as the developmental cost of the airbreathing engine." Similarly, vertical takeoff proved to offer an advantage over horizontal launch because the cost of developing a rocket sled was not offset by lower weight and cost in the flight vehicle.

These studies defined the preferred approach of NASA-Marshall's Future Projects Office which called for a two-stage fully-reusable launch vehicle,

# NASA's Uncertain Future

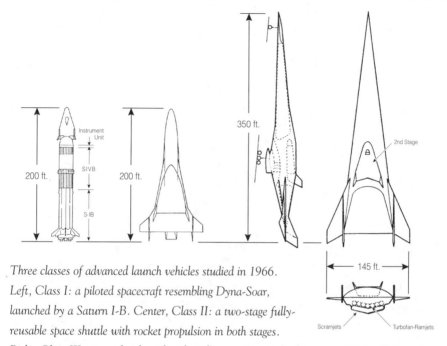

*Three classes of advanced launch vehicles studied in 1966. Left, Class I: a piloted spacecraft resembling Dyna-Soar, launched by a Saturn I-B. Center, Class II: a two-stage fully-reusable space shuttle with rocket propulsion in both stages. Right, Class III: space shuttle with airbreathing engines in the first stage. (U.S. Air Force)*

with both stages having fixed wings and rocket propulsion. The work also established the technical feasibility of such vehicles. NASA's Manned Spacecraft Center also adopted this approach, and NASA as a whole proceeded to hold to such designs until 1971.[53]

A dissenting word came from the Air Force, where people were in no hurry to define a single class of concepts. At Wright-Patterson Air Force Base, the Flight Dynamics Laboratory emerged as a center for such studies. The FDL, conducting two design exercises during 1965, drew the interest of the Aeronautics and Astronautics Coordinating Board, a joint NASA-Air Force committee. In August 1965, this board set up a subpanel that spent the next year reviewing technology and design concepts for reusable launch vehicles. The subpanel issued its report in September 1966.

Rather than focus on a single type of craft, the subpanel took the view that advancing technology would permit increasingly capable designs to emerge

---

53. Akridge, *Space Shuttle*, pp. 5, 16-19; *Aviation Week*, March 26, 1962, pp. 20-21; Report LR 18790 (Lockheed); Report GD/C-DCB-65-018 (General Dynamics); Nau, *Comparison*.

83

in the relatively near future. By 1974, the nation might have a vehicle, called Class I, in which a small reusable spacecraft would ride atop an expendable booster. The Saturn I-B could serve as this booster; Martin Marietta's proposed Titan III-M was another possibility, as was a new booster derived from the 260-inch solid rocket motor that was then being tested. Essentially, the spacecraft would be tantamount to an updated version of the Dyna-Soar. In turn, two-stage fully-reusable configurations (counted as Class II), such as those of NASA-Marshall, could be available by 1978. By 1981, the prospects could broaden to include Class III, featuring horizontal takeoff and a first stage powered by scramjets.

Like others in the field, the authors of this report were optimistic. NASA's eventual Space Shuttle would fall into Class I, with two solid boosters, an expendable propellant tank, and a reusable orbiter. However, it would not fly until 1981, the year in which this subpanel expected to see an operational scramjet. Nevertheless, the work of this subpanel was significant for three reasons.

It brought reusability into the realm of ongoing collaborations between NASA and the Air Force. It was a reminder that development of a new Dyna-Soar was a quick route to reusability. In addition to this, in the words of the report's summary, "It is important to note that no single, most desirable vehicle concept could be identified by the Subpanel for satisfying future DoD and NASA objectives." The Air Force would not follow the lead of NASA-Marshall by focusing attention on a single design approach; the hundred flowers would continue to bloom.[54]

## Two Leaders Emerge: Max Hunter and George Mueller

While many were talking about airline-type space operations, few had the professional background that would allow them to do much about it. Most managers and senior designers had entered the realm of space flight by way of the Pentagon's missile program of the 1950s. Few of them had working knowledge of the standard methodology for determining the operating costs of commercial airliners, as published initially in 1940 and subsequently adopted by the Air Transport Association.

---

54. Hallion, ed., *Hypersonic*, pp. 964-978; Ames, chairman, *Report*.

## NASA's Uncertain Future

At Lockheed Missiles and Space Company (LMSC), Max Hunter was one of the few people in the industry with an intimate knowledge of both airline economics and of launch-vehicle design. Earlier in his career, working at Douglas Aircraft, he had spent two and a half years dealing with the performance of transport aircraft. In those days, Douglas ruled the skies with its DC-6 and DC-7 airliners. For some time, Hunter was in charge of all calculations on their performance and economics. He then joined the Thor missile project and served as chief design engineer. Rebuilt with upper stages, the Thor became the Delta launch vehicle and emerged as NASA's most widely used booster.

This background allowed Hunter to approach the problem of low-cost space transportation from a fresh perspective. Existing studies left him dissatisfied; he writes that "by the end of 1963 the state of recoverable rockets was terrible." He disliked two-stage fully-reusable concepts which to him meant building two vehicles to do the work of one, with the smaller of the two—the second stage—being the one that counted. He also felt that the technology of scramjets or single-stage-to-orbit concepts lay far in the future. By March 1964, however, he had the germ of a new idea: the stage-and-a-half configuration.

This new idea was to consist of a reusable core fitted with large expendable tanks that would hold most of the propellant. The core would carry everything that was costly and important: payload, crew, engines, electronics, onboard systems. With a heat shield on its underside, it would achieve complete reuse. The tankage would consist of simple and inexpensive aluminum shells that would carry liquid hydrogen and liquid oxygen. They would fall away during the ascent to orbit, leaving the core to continue with the mission.

Hunter went to work at Lockheed in the fall of 1965. On his first day, he was asked if there was anything he thought should be done that was not being done already. He responded with an internal company memo on orbital transportation, which drew the attention of a number of senior managers. These included Eugene Root, the president of LMSC, who provided the internal company support that allowed Hunter to begin to pursue his ideas. He proceeded to take his gospel to meetings of professional societies, and won funding from the Air Force. He particularly emphasized that the economic model of the Air Transport Association, though developed for airliners, could apply as well to rocket transports.

## THE SPACE SHUTTLE DECISION

Paradoxically, two-stage fully-reusable vehicles promised launch costs as low as one-third of Hunter's approach—but only when flying up to a hundred times per year. Because it had a far lower development cost for 10 or fewer flights per year, the stage-and-a-half had a decided advantage. In Hunter's words, "its development can consequently be justified at an earlier point in time with a smaller number of missions."[55]

While Hunter gave an airline industry view of airplane-type space operations, NASA's George Mueller, head of the Office of Manned Space Flight, was promoting such concepts as well. His domain included all of Apollo; he also was a strong proponent of space stations, and he was pushing vigorously for a strong Apollo Applications program. Looking to the future, he understood that low-cost space flight would be essential for viable space stations.

As a first step, in December 1967, he invited a number of NASA and industry specialists to a one-day symposium, held in January at NASA Headquarters. Because much of the data from industry was proprietary, Mueller limited attendance to representatives of government agencies. Even so, some 80 people, most of them from NASA and the Air Force, attended the conference. The symposium proceedings give a clear view of the topic at the end of 1967, when the field was alive with ideas but when no single design approach had come to the forefront. In addition to this, those proceedings presented design solutions that, four years later, would show up in the final Space Shuttle configuration.

Martin Marietta was the most conservative, pitching its Titan III-M along with a small reusable spacecraft, similar to the Dyna-Soar, that would carry six people. This was the quintessential Class I design (featuring an expendable booster) that NASA and the Air Force had identified in their 1966 joint study. The Titan III-M was to rely on twin 120-inch solid boosters, slightly smaller than the solid rockets that, 13 years later, would boost the operational Space Shuttle.

Those rockets were not built as single units, but rather as a stack of segments, like short lengths of pipeline that are bolted together at their flanges. Manufacturers such as Thiokol filled each segment with the solid propellant, then sent them off by highway or railroad. Such segmented rockets were much easier to transport than the unsegmented type; the segments could be stacked and joined at the launch site, using putty to fill the gaps.

---

55. Hunter, *Origins*. Reprinted in part in *Earth/Space News*, November 1976, pp. 5-7.

## NASA's Uncertain Future

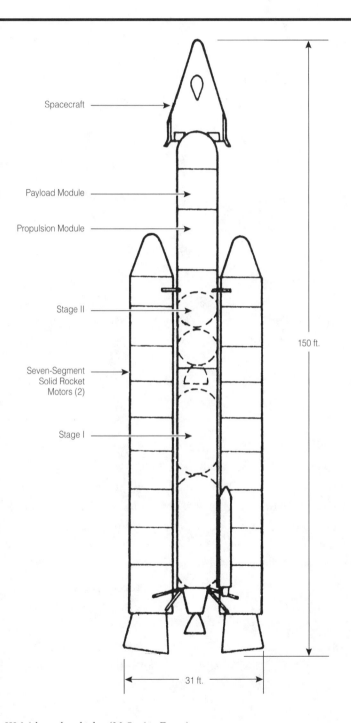

Titan III-M launch vehicle. (U.S. Air Force)

# THE SPACE SHUTTLE DECISION

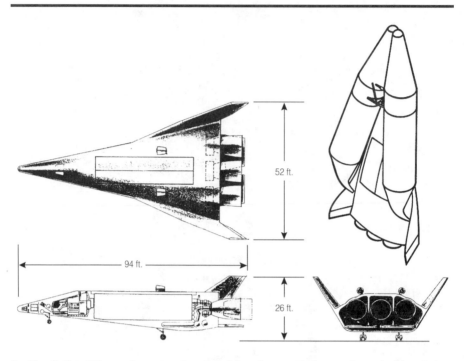

*Lockheed's Star Clipper: three-view drawing of the orbiter, with the complete vehicle, including propellant tanks, in upper right. (Lockheed; Dan Gauthier)*

The standard Titan III-C used five-segment solid rockets, each 85 feet long with a thrust of 1,180,000 pounds. For the Titan III-M, these rockets were to grow to seven segments, each 112 feet in length with a thrust of 1,508,000 pounds. The first stage was also to grow in length, to hold more propellant, while receiving liquid-fueled engines with 11 percent more thrust. The combination would carry 38,000 pounds to orbit from Cape Canaveral, or 32,000 pounds from Vandenberg Air Force Base.[56]

Lockheed presented Max Hunter's configuration. Called Star Clipper, it featured a core vehicle in the form of a lifting body, triangular in shape. The expendable propellant tanks would be 156 inches in diameter (the limit for highway or rail transport) and would join at the front, running along the sides of the core. The vehicle's avionics would include an automated on-

---

56. Akridge, *Space Shuttle*, p. 35; Schnyer and Voss, *Review*, pp. 15-16, 40-47; *Quest*, Fall 1995, pp. 18-19; *Astronautics*, August 1961, pp. 22-25, 50-56.

board checkout system, similar to those on airliners. Lockheed managers claimed that the Star Clipper could lift off within one hour after arrival at the launch pad.[57]

McDonnell Aircraft, recently merged to form McDonnell Douglas, had built the piloted Mercury and Gemini spacecraft, and had been studying new launch-vehicle concepts for six years. Like Lockheed, it had adopted the stage-and-a-half approach, again with a reusable core flanked by expendable propellant tanks. Known as Tip Tank, this concept would carry 12 astronauts, sitting side by side like passengers in first class. The core again had the shape of a lifting body, but McDonnell went one better than Lockheed by proposing to add small wings that would fold within the fuselage and snap out for use in landing. These wings then would help the craft to handle better during the landing approach, when conventional lifting bodies tended to dive toward a runway at speeds of several hundred miles per hour.[58]

The Lockheed and McDonnell Douglas concepts counted as only partially reusable, because their external tanks would not be recovered. During 1971, this became the configuration NASA would adopt; the shuttle orbiter would take shape as a core vehicle of the type Hunter had recommended. Its propellants would go into a big expendable tank, with two large solids flanking this tank in the fashion of the Titan III-M. Hence as early as 1967, the basic elements of the eventual shuttle not only were well known but had influential advocates among NASA's contractors.

At that early date, however, there was no reason to pick this approach over others that also had their advocates. The two-stage fully-reusable concept continued to shine, and General Dynamics, with Air Force support, had been studying a version called the Triamese. It would feature a standard vehicle fitted with rocket engines and a pilot compartment. Like the core of McDonnell Douglas' Tip Tank, it was tantamount to a lifting body with deployable wings. Three such vehicles, identical in shape, would fit together to make a complete launch system. The middle vehicle would carry the payload and would serve as the core; the other two would serve as tankage, carrying most of the propellant. This standardization represented an attempt to save money during development, for then it would not be necessary to

---

57. Schnyer and Voss, *Review*, pp. 17-22.
58. *Ibid.*, pp. 35-39.

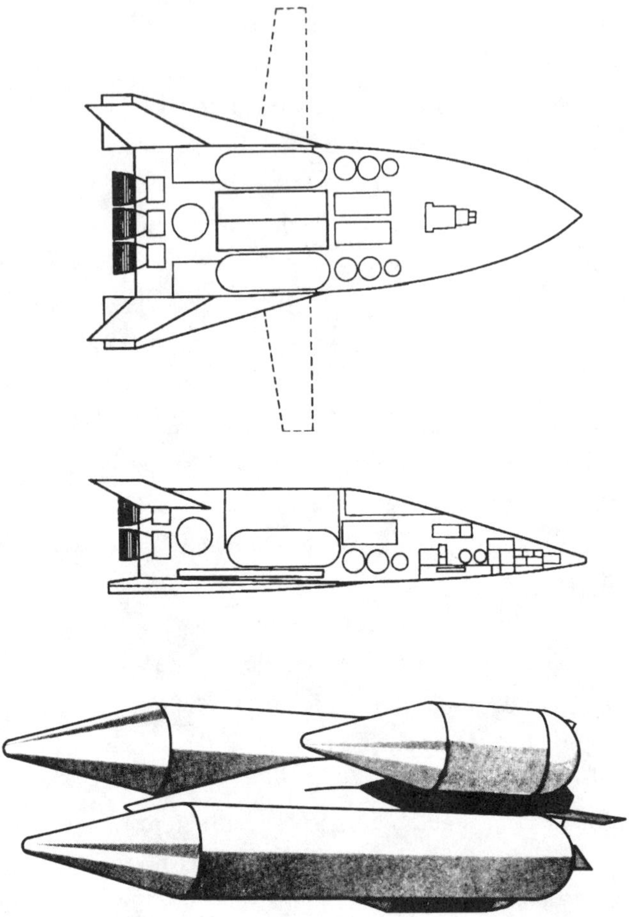

# THE SPACE SHUTTLE DECISION

*McDonnell Douglas'
Tip Tank: top and side
views of the orbiter,
showing foldout wings,
and complete vehicle
with propellant tanks.
(Dennis Jenkins;
NASA)*

develop a reusable first stage with a design of its own. In the Triamese approach, all three vehicles would reenter and return to a runway.[59]

General Dynamics did not present this concept at Mueller's symposium, but instead discussed five alternatives, ranging from the Titan III-M to a two-stage fully-reusable configuration. The company showed, again, that the former had a low development cost but a high cost per flight; the latter had the highest development cost but the lowest per-flight cost. Though these conclusions were not new, they too pointed a path to the future.

---

59. Reports GDC-DCB-67-031, GDC-DCB-68-017 (both from General Dynamics).

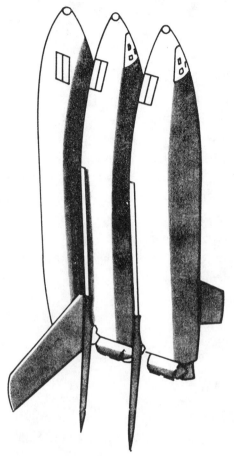

*Triamese concept of General Dynamics. (NASA)*

These conclusions addressed the issue of designing a reusable launch vehicle to meet economic criteria. If the criterion was to achieve the lowest possible cost per flight, thus attaining true airline-like operation, then one would go with the two-stage fully-reusable, even though this approach carried high development cost. If the most important goal was to achieve minimum development cost, then one would choose the Titan III-M. Stage-and-a-half configurations appeared intermediate, both in development and in launch costs. In sum, one could choose a level of reusability so as to balance between these two types of cost. As its space shuttle concepts matured, NASA would spend much of 1971 seeking this balance.

## THE SPACE SHUTTLE DECISION

The General Dynamics presentation offered more. Within the industry, it was widely appreciated that piloted aircraft cost much less to develop than missiles or expendable launch vehicles. The reason was that missiles demanded extensive and costly ground tests to assure that they would fly properly, with no pilot at the controls. By contrast, the development of aircraft took full advantage of their reusability. Test pilots could start with simple exercises in taxiing and takeoff, then reach toward higher speeds and greater levels of performance, in step-by-step programs. At each step, the aircraft would come back, where engineers could study it carefully and correct deficiencies. Such flight testing was far less costly than ground tests.

General Dynamics then drew on recent experience with the X-15 and the Atlas ICBM, arguing that piloted craft could maintain this advantage even as rocket-powered vehicles of extreme performance. The X-15 and Atlas had both gone through development in the late 1950s; their empty weights were similar, and both mounted rocket engines that came to their respective contractors as government-furnished equipment. Although the X-15 was more complex than Atlas, it had less than half the development cost because it too followed the step-by-step approach to flight test, with its test pilots often taking action to save the vehicle from disaster. Indeed, the X-15 would likely have been destroyed on as many as a third of its flights had there been no pilot aboard.[60] Test pilots thus served as inexpensive substitutes for the automated systems that might have been required to take their place.

The reusable concepts of the day, and those that followed during 1968 and 1969, were often referred to as Integral Launch and Reentry Vehicles. The Air Force, in particular, used that designation in its own work.[61] Mueller adopted a different term, calling such vehicles space shuttles. The term had appeared now and then through the years. For example, Philip Bono of Douglas Aircraft had offered a concept called the ROMBUS (Reusable Orbital Module, Booster, and Utility Shuttle). Dating to 1963, it resembled the immense Nexus, and its mission was similar. Walter Dornberger, who had proposed to build Bomi during the 1950s, lately had been writing of a "recoverable and reusable space transporter, or shuttle." He described it as "an economical space plane capable of putting a fresh egg, every morning, on the

---

60. Schnyer and Voss, *Review*, pp. 28-34; *Astronautics*, January 1963, p. 53.
61. Jenkins, *Space Shuttle*, p. 56; Hallion, ed., *Hypersonic*, p. 995.

table of every crew member of a space station circling the globe."[62] Mueller now made the term his own, fully aware that the space shuttle was to shuttle to and from such a station.

In August 1968, in London, he received an award from the British Interplanetary Society and gave a prepared address in which he pledged his troth to the shuttle as NASA's next goal:

*I believe that the exploitation of space is limited in concept and extent by the very high cost of putting payload into orbit, and the inaccessibility of objects after they have been launched. Therefore, I would forecast that the next major thrust in space will be the development of an economical launch vehicle for shuttling between Earth and the installations, such as the orbiting space stations which will soon be operating in space....*

*These space stations will be used as laboratories in orbit and will provide the facilities to study and understand the nature of space. They will provide observatories to view the sun, the planets and the stars beyond the atmospheric veil of earth. Stations in orbit will provide bases for continuous observation of the earth and its atmosphere on an operational basis—for meteorological and oceanographic uses, for earth resource data gathering and evaluation, for communications and broadcasting and ground traffic control....*

*One of the applications of these stations that has intrigued planners for many years has been their use as fuel and supply bases, and as transfer points enroute to high or distant orbits, to lunar distance, or toward the planets....*

*Essential to the continuous operation of the space station will be the capability to resupply expendables as well as to change and/or augment crews and laboratory equipment.... Our studies show that using today's hardware, the resupply cost for a year equals the original cost of the space station....*

*Therefore, there is a real requirement for an efficient earth-to-orbit transportation system—an economical space shuttle.... The shuttle ideally would be able to operate in a mode similar to that of large commercial air transports and be compatible with the environment of major airports.... The cockpit of the space shuttle would be similar to that of the large intercontinental jet aircraft, containing all instrumentation essential to complete on-board checkout.... Interestingly enough, the basic design described above*

---

62. *Astronautics & Aeronautics*, January 1964, pp. 28-34; November 1965, pp. 88-94.

*for an economical space shuttle from earth to orbit could also be applied to terrestrial point-to-point transport....*

*Barron Hilton, whose hotels ring the earth, has suggested that a Hilton resort hotel in low earth orbit would offer unique attractions. Looking at the earth from space, seeing sunrise and sunset every 90 minutes, floating in the zero g of weightlessness, are all unearthly experiences. More seriously, lack of gravity lightens the load on the heart and certain other organs, so that the Orbiting Resort might also be a health spa....*

*The Space Shuttle is another step toward our destiny, another hand-hold on our future. We will go where we choose—on our earth—throughout our solar system and through our galaxy—eventually to live on other worlds of our universe. Man will never be satisfied with less than that.*[63]

This was not your usual speech by a government official. Napoleon may have spoken often of "destiny," but even within NASA, an agency not known as a home for shrinking violets, such talk was slightly out of the ordinary at least. It helped that Mueller was talking to his fellow enthusiasts and was speaking in London, where his presentation was not likely to receive hostile fire from the *Washington Post*. Mueller's hopes, however, contrasted sharply with recent experience, wherein NASA had tried and failed to define an ambitious Apollo Applications effort as a major post-Apollo program. The agency's budget was on a sharp downhill slide, and NASA was nowhere near the bottom. Indeed, it had not begun to see the bottom.

## NASA and the Post-Apollo Future

Before federal bureaucrats such as Mueller could grapple with human destiny, they first had to face the more prosaic question of what NASA would do after landing astronauts on the Moon. The first significant interest in this issue came in January 1964, when President Johnson, in office for barely two months, sent a letter to NASA Administrator James Webb.

The background to this letter involved a program of the Atomic Energy Commission called NERVA (Nuclear Engine for Rocket Vehicle Application) that was developing a nuclear-powered rocket engine. While NASA did not

---

63. Mueller, *Address*, August 10, 1968.

*NASA's Uncertain Future*

*James E. Webb, NASA Administrator between 1961 and 1968. (NASA)*

need it for Apollo, such an engine might prove useful indeed in any follow-on program of piloted flight to Mars. The program had strong support from Senator Clinton Anderson (D-New Mexico), chairman of the Senate space committee; it also had the support of Webb. Its opponents, however, included President Kennedy's science advisor, Jerome Weisner. Weisner convinced Johnson to ask NASA to identify the future missions that would require NERVA's power.

Johnson took up this and other issues in his letter to Webb. Could NASA list possible space objectives beyond those already approved? What supporting research and development would these new goals require? How much of NASA's current work, particularly in the development of launch vehicles such as the Saturn V, could support such future programs?

An old hand at Washington politics, Webb smelled a rat. He later described this as "part of a power play rather than a desire for proposals. It was an effort to put us on the defensive and to make us commit ourselves to certain missions which they could then attack." Accordingly, Webb did not reply immediately, but set up a committee that proceeded to take its sweet time in preparing a response. Meanwhile he mollified Johnson with interim replies, listing possible future missions but declining to choose among them.[64]

Events that summer showed that Webb was wise to be cautious. As far back as 1962, the Future Projects Office at NASA-Marshall had contracted with several major aerospace firms for initial studies of piloted planetary missions, including landing on Mars. These studies continued during subsequent years. Then, in mid-1964, the new presidential science advisor, Donald Hornig, asked Webb to present an estimate of the cost of a piloted Mars landing that might follow Apollo.

The initial estimate, internal to NASA, was $32 billion. An internal review added $5 billion for program contingencies and forwarded the total of $37 billion to Webb. He accepted some further additions that hiked the cost to $50 billion, and gave this figure to Hornig. Hornig doubled it to $100 billion, on his own initiative, and gave this new estimate to a Congressional committee. The next day, newspapers quoted one congressman as stating that the piloted Mars mission would cost $200 billion, amounting to 40 years of NASA's budgets at the 1965 rate of $5 billion per year. In the words of an observer, "In only one week, a well developed estimate of $37 billion was multiplied into a $200 billion program."[65]

A year after receiving his initial request, Webb finally gave a full reply to Johnson's letter in a report written in February 1965. It amounted to a verbose exercise in saying little that was new or significant and saying it at considerable length, while offering no targets for skeptics. The report reviewed recent and current NASA activities in detail, and included three single-page lists of future possibilities. These lists resembled pages from a book index, lacking any trace of description, estimated cost, schedule, or priority. In an outstanding display of political adroitness, the report called for "a continued balanced program" that would "not impose unreasonably large demands upon the Nation's resources." No one could oppose such recommendations; they were on a par with supporting motherhood and apple pie.

---

64. NASA SP-4102, p. 243; Logsdon, *Apollo*, Chapter 1, pp. 27-28.
65. AAS History Series, vol. 17, pp. 421-429.

## NASA's Uncertain Future

Webb's report drew questions within the Senate space committee, which complained that "alternatives are presented, but no criteria are given as to how a selection would be made." That was just as Webb intended; he was not about to take the initiative in offering a plan that critics could attack. He would have been quite willing to have the President take the lead, as Kennedy had done in supporting Apollo in 1961. Johnson, however, also preferred to keep his options open. In March 1965, he told his advisor Jack Valenti that he did not intend to make a new Kennedy-style commitment in space: "I think I would have more leeway and running room by saying nothing, which I would prefer."[66]

The historian Arthur Levine notes that two years later, Webb explained to him just why he had finessed Johnson's initial request:

> *First, the announcement by NASA in the mid-1960s of a long-term goal would make the agency vulnerable. It would provide ammunition to critics, who would be able to shoot down the proposed program as being too expensive or impractical, thereby raising the possibility that long-range technology developments tied to the announced goal would be cut out. This in turn would cripple the agency's ability to support the Apollo and other advanced missions that depended on a strong base of advancing technology.*
>
> *Second, should NASA announce a long-term post-Apollo goal, critics would claim that the lunar landing was simply an interim goal, subordinate to the new effort. For example, if NASA announced that the post-Apollo goal should be a manned Mars landing, the Apollo program for a moon landing would be relegated to a secondary position. This would raise the possibility of cutting support for Apollo, thus jeopardizing the program or stretching it out. In the event of subsequent change in national opinion on the worth of the long-range goal, both the lunar landing and the more distant goal might never be realized.*
>
> *Third, the major effort required for planning, proposing, and defending a new long-range goal would tie up the energies of top NASA leadership and key scientists and engineers, diverting them from concentrating on making Apollo a success.*[67]

---

66. NASA SP-4102, p. 243; Jack Valenti to Lyndon Johnson, March 30, 1965 (Lyndon Johnson Presidential Library, Austin, Texas); Smith, chairman, *Summary*. Reprinted in NASA SP-4407, vol. I, pp. 473-490.
67. Levine, *Future*, pp. 118-119.

## THE SPACE SHUTTLE DECISION

The last point addressed the fact that there was no consensus, even within NASA itself, as to NASA's next goal. George Mueller, head of the Office of Manned Space Flight, had his eye on a piloted mission to Mars. The two most powerful center directors, von Braun at NASA-Marshall and Robert Gilruth of the Manned Spacecraft Center, preferred a different objective: a space station. Mueller also liked space stations and was well aware of their usefulness as preparations for Mars. von Braun and Gilruth, however, saw space stations as major elements of a program that, diverging sharply from one that would aim at Mars, would focus on activities in Earth orbit.

Nevertheless, during 1965 and 1966, the beginnings of a post-Apollo future began to take shape. Not surprisingly, its major features were in line with the initiatives that Webb had suggested in his report to Johnson. Apollo Applications emerged, strongly backed by Mueller. For Mars, attention focused on an ambitious automated mission called Voyager that would orbit that planet and then send craft to land on its surface, looking with instruments for signs of life. Plans for Voyager flourished for a time. While initial designs called for use of the Saturn I-B, in October 1965 its officials decided instead to try for the much larger Saturn V.[68]

In addition to this, even though Webb was unwilling to carry through a serious plan for NASA's future, the President's Science Advisory Committee (PSAC) proved willing to do it for him. This blue-ribbon panel was potentially a source of clout; it operated within the Executive Office of the President, and received support from another White House group, the Office of Science and Technology. In February 1967, the PSAC issued a major report, *The Space Program in the Post-Apollo Period*. John Newbauer, editor of the trade journal *Astronautics & Aeronautics*, wrote that it "should prove the pivot for policy discussions for some time to come." He described it as "the most cohesive and solid appraisal of space-program goals since the Space Act itself," which led to the founding of NASA in 1958.[69]

The PSAC report did not endorse anything so specific as piloted flight to Mars. Nevertheless, it proposed an organizing theme: "a program directed ultimately at the exploration of the planets by man." The report defined this as "a

---

68. NASA SP-4102, p. 147; Logsdon, *Apollo*, Chapter 1, pp. 17-18.
69. *Astronautics & Aeronautics*, March 1967, p. 20.

balanced program based on the expectation of eventual manned planetary exploration." The program would pursue several intermediate goals including continued lunar missions by astronauts; long-duration piloted flights, at first through Apollo Applications and later in a true space station; and "a strongly upgraded program of early unmanned exploration of the nearby planets."

The PSAC was certainly not in NASA's pocket; its report pulled no punches. It criticized the Apollo Applications wet workshop: "some doubts arise about man's ability to carry out extensive construction efforts in space. The requirement that man actually construct his laboratories in space in these initial applications may constitute a serious impediment to their development." A true space station might represent "a more effective use of funds." The panel endorsed building a single wet workshop, if only as an initial step: "The launch vehicle and spacecraft for this experiment are already on order, and the opportunity for 28- and 56-day flights in 1968 should be taken."

In other areas, the report was more favorable: "In the period after the initial two Apollo lunar landings we recommend that a sustained program of lunar exploration...continue manned expeditions at the rate of between one and two per year." The PSAC recommended "that the Saturn V vehicle continue to be produced," and that "the post-Apollo Saturn V production rate be fixed at 4 systems per year."

On Voyager: "We recommend an expanded commitment to the Voyager planetary lander program, pointing toward a soft landing of a Surveyor-type module on Mars in 1973." As a prelude to Apollo, a program called Surveyor was seeking to conduct soft landings of automated spacecraft on the Moon, and had scored its first success the previous June.

On a space station:

*We recommend that programs of studies and advanced developments be initiated promptly with the objective of a launch in the mid 1970's of the first module of a space station for very prolonged biological studies of man, animals, and other organisms in earth orbit. Such a station should be designed with consideration of its possible role in support of earth orbital astronomy.*

On future launch vehicles:

*The payload capabilities of the [Saturn I-B] are not significantly superior to those of the Titan III-M, while the launch costs of the [Saturn I-B] are about*

> *double those of the Titan III-M.... Because of the continuing requirements for manned and man-attended systems we visualize that an important problem will be posed for a long time by the cost associated with taking men to and from orbit.... For the longer range, studies should be made of more economical ferrying systems, presumably involving partial or total recovery and reuse.*[70]

The report did not give NASA everything it might have wanted, even in dealing with projects that were achievable in the short-term. It endorsed only a modest Apollo Applications effort, as noted. It ignored NERVA, though that program was proceeding smartly with its nuclear engine and offered a promising source of propulsion for a piloted mission to Mars. The PSAC also recommended delaying a commitment to a true space station until 1971 or 1972, although its advocates hoped for such a decision as early as 1968.[71] Yet by endorsing construction of this station "in the mid-1970's," and by openly embracing Mars as a long-term goal, the PSAC endorsed a program that went well beyond what NASA in fact would be able to pursue.

While Mars was in the ascendancy at the PSAC, NASA's hopes were about to prove star-crossed. The agency had been charging ahead with Apollo; in January 1967 it had a Saturn I-B on a pad at Cape Canaveral that was being readied to launch a mission into orbit. Late that month, the astronauts Gus Grissom, Ed White, and Roger Chaffee were conducting a pre-launch exercise atop that rocket, within their spacecraft. A fire broke out; the men could not escape, and they perished before help could reach them.[72]

In the aftermath of this fire, plans for the future went on hold while NASA struggled to win success with Apollo. There also was bad news elsewhere in Washington and in the nation. In January, the President had presented the federal budget for Fiscal Year 1968, anticipating a deficit of $8 billion. The Vietnam War, however, was escalating rapidly. By August, when the estimate was close to $30 billion, Johnson asked Congress to approve a 10 percent income-tax surcharge to keep it from rising further.

The summer of 1967 also brought major riots. Looters in Newark plundered stores on a massive scale; snipers fired from rooftops, and fires blazed high. The city's 1,400 police officers could not control the situation. Speaking

---

70. *Ibid.*, pp. 20-22; Long, chairman, *Space Program*.
71. Logsdon, *Apollo*, p. I-32.
72. *Ibid.*, Chapter 1, pp. 37-38; Chaikin, *Man*, pp. 11-26.

of "a city in open rebellion," New Jersey's governor called in the National Guard. At the peak, almost half of the city was in the hands of the rioters. The upheavals raged for five days; 27 people lost their lives.

Detroit blew a week later; the next 11 days saw 1,600 fire alarms. Three miles of Grand River Avenue, a major thoroughfare, burned to the ground. Some sections of downtown resembled the burned-out German cities of World War II. Forty-three people died; over 7,000 were arrested; 5,000 were left homeless.[73]

"Conditions have greatly changed since I submitted my January budget," the President admitted. "Because the times have placed more urgent demands upon our resources, we must now moderate our efforts in certain space projects." In the House, an appropriations subcommittee reopened hearings on the NASA budget, and proceeded to make deep cuts in virtually every program except Apollo.

With cities burning, taxes rising, and the Vietnam War escalating, NASA proceeded to shoot itself in the foot. In a stunning display of tactlessness, the Manned Spacecraft Center invited 28 companies to bid on a study of piloted flyby missions to Mars and Venus, beginning in 1975. When this announcement created an uproar, MSC withdrew its request. It was too late. In Congress, the view took hold that the automated Voyager project should be canceled because it was the first step toward a needless extravagance: a piloted mission to Mars.

The final cut in NASA's budget came to $511 million, a reduction of 10 percent. Voyager was canceled, being eliminated in conference with the Senate. Apollo Applications, budgeted at $454 million in the January presidential request, ended with $253 million. The conferees spared Apollo, voting funds to allow this program to recover in the wake of the fire at Cape Canaveral. The cuts, however, hit hard at future programs.[74]

Voyager did not remain dead for long. Within days of its formal cancellation, NASA officials began discussing a follow-on concept that was approved by the president in the budget for FY 1969. The new project had the name Viking, and its mission remained the same: to orbit Mars with automated spacecraft, place landers gently on the surface, and look for signs of

---

73. Logsdon, *Apollo*, p. I-46; Manchester, *Glory*, pp. 1079-1081.
74. Logsdon, *Apollo*, Chapter 1, pp. 46-47; NASA SP-4102, p. 148.

life. Viking, however, would not ride a Saturn V; it would use the Titan III-Centaur. While this was certainly a splendid launch vehicle, it had less than one-eighth the lifting power of its much larger cousin.[75]

That summer's near-debacle confirmed Webb's belief that even a modest post-Apollo planning effort could backfire badly. With Apollo continuing to reign supreme in a time of cutbacks, Webb took to raiding the Apollo Applications budget by reprogramming some of its funds. In June 1968, he told his center directors that this program was nothing more than "a surge tank for Apollo." In this fashion, he took from the future to meet the needs of the present. Above all else, Apollo *had* to succeed.[76]

That program's peak funding had come in FY 1965. That year also saw NASA's appropriation peak at $5.25 billion. After this, the budget slid downward; the appropriation for FY 1969, which began the previous July, was $3.953 billion, a drop of 25 percent. NASA's in-house employment stayed close to the FY 1965 level of 33,000 positions. The contractors, however, were having a hard time of it; their personnel had fallen by half, from 377,000 to 186,000.[77] Unless NASA could take hold of something new and major, it was likely to shrink to insignificance.

Mueller had hoped that Apollo Applications could come to the forefront as this new program. Already in 1968, it was clear that this would not happen. The agency had spent several years trying to pursue such a route to the future, without success. More was involved here than budget cuts per se. Congress and the Administration had imposed those cuts because NASA had failed to make a persuasive case for its plans. Nor was NASA able to propose anything as compelling as Apollo.

Apollo, above all, had the beauty of simplicity. Everyone knew of science-fiction visions of astronauts on the Moon. The program's goal was succinct: to carry out the lunar landing during the decade of the 1960s, and to bring its explorers back safely. As von Braun stated in 1964, "Everybody knows what the Moon is, everybody knows what this decade is, and everybody can tell a live astronaut who returned from the Moon from one who didn't."[78]

---

75. NASA SP-4012, vol. III, pp. 27, 40-41, 213-219.
76. NASA SP-4208, pp. 86-87, 104; NASA SP-4102, p. 254.
77. NASA budget data, February 1970.
78. *U.S. News & World Report*, June 1, 1964, p. 54.

Apollo Applications lacked this compelling character. In the end, it was a program with no compelling central focus. It offered only modest initiatives: solar astronomy, flights with durations of weeks, medical studies, and opportunities to use Saturn-class rockets that otherwise might go to waste. The historian John Logsdon writes that, according to program critics, these initiatives "were designed to fit the specific features of the Apollo and Saturn hardware. The missions suggested were not necessarily those deserving highest priority, and modified Apollo/Saturn equipment was not necessarily the most effective way of carrying out those missions."[79] Here was enough to support a single orbital workshop, but not enough to compete with something as historic as putting the first man on the Moon.

An opportunity, however, did exist to plan once again with boldness. The PSAC report had danced around this, proposing nothing more than "the expectation of eventual manned planetary exploration." That was not NASA's style; the agency had established itself by literally reaching for the Moon, not by resting content with an expectation that astronauts would get there someday. The new goal was there for anyone who would dare to pursue it, to seize it. One could see it in the night sky, glowing redly; one could name this goal with a single word: Mars. During 1969, NASA would seek seriously to establish a piloted expedition to this planet as the basis for the agency's future.

---

79. Logsdon, *Apollo*, p. I-26.

CHAPTER THREE

# Mars and Other Dream Worlds

A key component of early Space Shuttle plans was its linkage to a possible mission to Mars as the next major NASA undertaking. During 1967 and 1968, the Atomic Energy Commission (AEC) reached key milestones in propulsion on the road to Mars. In tests in Nevada, the AEC conducted successful demonstrations of nuclear reactors built for use in rocket propulsion and showed that its contractors were ready to develop a flight-rated engine suitable for piloted missions to that planet.[1]

## *Nuclear Rocket Engines*

The AEC's nuclear-rocket program could trace its beginnings to December 1953, when the nuclear scientist Robert Bussard published an article on this topic in the classified *Journal of Reactor Science and Technology*. His paper stirred interest, and led to the initiation of an experimental effort called Project Rover at Los Alamos, New Mexico. Initial work aimed at building a succession of rocket reactors named Kiwi after the flightless bird of New Zealand.

The basic approach followed Bussard's proposal, calling for a compact reactor built of graphite, which withstands high temperatures and actually gains strength when heated. Hydrogen, flowing through channels in this reactor core, would receive heat from the reactor and reach temperatures of

---

1. NASA SP-4012, vol. II, pp. 487-488.

several thousand degrees. This gas would then expand and flow through a nozzle, to produce thrust.

Such a rocket appeared highly promising because it offered the greatest possible exhaust velocity and hence the best performance. It would do this by taking advantage of hydrogen's low molecular weight: two, in appropriate units, compared with 18 for water vapor and 44 for carbon dioxide. Molecules of low weight fly faster, and hence yield a higher exhaust velocity; for this purpose, hydrogen is best.[2]

The first version of Kiwi was heavy and produced only 70 megawatts (MW) of power, a modest amount. When it ran for five minutes in mid-1959, however, it suggested strongly that nuclear propulsion indeed was worth pursuing. Some NASA officials had already been following this work; now they joined with their AEC counterparts to set up a joint program office. Los Alamos managers laid plans for advanced Kiwi reactors that would aim at 1,000 megawatts. In addition to this, the joint office set a follow-on goal of developing a flight-rated engine called NERVA (Nuclear Engine for Rocket Vehicle Application). In June 1961, NASA and the AEC chose Aerojet General as the prime contractor for development of the complete nuclear engine, with Westinghouse, an experienced builder of reactors, as the principal subcontractor.[3]

In November 1962, during a test of a new and promising Kiwi, disaster struck. The analyst James Dewar writes that "paralleling the rapid increase in power was a rapid increase in the frequency of flashes of light from the nozzle. On reaching 500 MW, the flashes were so spectacular and so frequent that the test was terminated and shut-down procedures begun. Quick disassembly confirmed that the flashes of light were reactor parts being ejected from the nozzle. Further disassembly and analysis revealed that over 90 percent of the reactor parts had been broken, mostly at the core's hot end."

Harold Finger, head of the joint office, decided that there would be no further hot tests until the cause of the failure was found and carefully fixed. The failure was found to have been caused by vibrations produced by gas flowing through the core, which cracked the uranium fuel elements. It took over a year and a half of new designs to restore confidence in the project. In the end,

---

2. *Astronautics*, December 1962, pp. 32-35; *Astronautics & Aeronautics*, May 1968, pp. 44, 45; Halliday and Resnick, *Physics*, pp. 516-519.
3. *Astronautics & Aeronautics*, June 1965, p. 42; NASA SP-4012, vol. II, pp. 478-480, 484-485.

*Mars and Other Dream Worlds*

vibration-free reactor cores proved to be attainable. In August 1964, another Kiwi ran for eight minutes at 900 MW, with complete success. The engineers then restarted it and successfully ran it again at full power. This series of tests

NERVA *nuclear rocket under test. (Smithsonian Institution Photo No. 75-13750)*

demonstrated an effective exhaust velocity of 24,450 ft/sec, far more than any chemically-fueled rocket could achieve.[4]

Other work improved the non-nuclear parts of the rocket. The AEC test facility included a liquid-hydrogen pump that served well during the ground tests, but was unsuitable for flight. During 1965, however, workers assembled a complete nuclear engine that included a hydrogen-cooled nozzle as well as a flight-type turbopump. Tests of this engine began that December and reached full power in March 1966. This was the first operation of a nuclear rocket with major components representative of a flight-rated engine.

Subsequent work returned anew to reactor development, emphasizing long-duration tests as well as high power. In December 1967, an experimental version of NERVA carried through a 60-minute endurance run at rated temperature and full power, 3,630 degrees Fahrenheit and 1,100 MW. In addition to this, Los Alamos was developing a new class of reactors called Phoebus, rated at 5,000 MW. A June 1968 test ran for over 30 minutes, with 12 minutes at or above 4,000 MW.

By then plans were in hand for a true flight engine, with 1,560 MW of power and temperatures of 4,040° F. It would produce a thrust of 75,000 pounds with an exhaust velocity of 26,500 ft/sec, nearly twice that of the best hydrogen-oxygen rocket then available. This version of NERVA would not take off from the ground, but would serve in upper stages. The plans called for developing this engine through a Preliminary Flight Rating Test, a pre-flight qualification. It would be ready for actual space missions soon after 1975.[5]

The rapid pace of advances in Nevada contrasted painfully with the lack of plans in Washington. With NASA having no approved post-Apollo future, it was quite possible to anticipate a time when Aerojet might build a well-tested NERVA, ready for flight, only to find that NASA had no reason to use it. NASA's prospects did not improve during 1968, as the agency launched a new attempt to plan its future.

This new planning effort began that February when NASA Administrator Webb named Homer Newell to direct it. Newell, who had headed the Office of Space Science and Applications, had been promoted to Associate

---

4. AAS History Series, vol. 12, pp. 109-124; *Astronautics & Aeronautics*, June 1965, pp. 34-35, 42-46.
5. *Astronautics & Aeronautics*, May 1968, pp. 42-53; NASA SP-4012, vol. II, pp. 487-488.

## Mars and Other Dream Worlds

Administrator in October 1967, giving him agency-wide responsibility. He hoped to prepare proposals that could influence the FY 1970 budget request that would go to the Bureau of the Budget (BoB) late in September. He proceeded to set up 12 working groups, drawing on a broad range of NASA specialists. A Planning Coordination Group (PCG) would direct the working groups' activities; a Planning Steering Group then would choose alternatives and pass recommendations to the Administrator.

This effort accomplished little. The head of the PCG noted "a definite failure to pull together among the key program offices, science and manned space flight."[6] Newell himself admitted that the results

> *were not up to the standards of boldness and imagination expected at the beginning of the cycle, or worthy of our first decade in space. It is probable that the agency had become so conditioned to retreat over the past two years that an intellectual conservatism pervaded the planning.... The total effort in terms of forward motion was pedestrian, even timid.*[7]

Willis Shapley, NASA's Associate Deputy Administrator, spoke of "Homer Newell's monumentally bureaucratic planning process. The number of new ideas that were injected—well, I think the Space Shuttle was really the only one that I can remember." For instance,

> *all the planetary missions sounded about the same. Somebody might have thought of some other instrument here and there, but in terms of forward NASA planning, everybody was just projecting exactly what the next step in his own little segment was.*[8]

Moreover, even as NASA was pouring old wine into new bottles, the Budget Bureau was turning up its nose at the proffered vintage. NASA's plans emphasized long-duration piloted flights, but a BoB staff paper responded with skepticism:

> *It is difficult to conceive of any use short of a manned planetary expedition that would require men to operate in orbit for more than 30 days. Most*

---

6. Logsdon, *Apollo*, Chapter 2, pp. 7-10; NASA SP-4102, pp. 256-257.
7. Memo, Homer Newell to George Low, February 9, 1970.
8. John Mauer interview, Willis Shapley, October 26, 1984, pp. 8-9.

# THE SPACE SHUTTLE DECISION

*scientific endeavors that require the collection of data by means of space flight can be accomplished by unmanned systems at considerably less expense than the manned flight systems.*

More broadly, this staff paper saw little reason to continue with post-Apollo piloted space flight, other than competition with the Soviet Union:

*Reasons for proceeding other than competition include enhancing the national prestige, advancing the general technology, or simply faith that manned space flight will ultimately return benefits to mankind in ways now unknown and unforeseen. None of these secondary arguments can be quantified and most are difficult to support.*

*The case for continuation of a manned space flight effort after Apollo is one of continuing to advance our capability to operate in space on a larger scale, for longer duration, for ultimate purposes that are unclear.*[9]

The NASA appropriation for FY 1970 came in at $3.697 billion. This was very close to the Administration's request to Congress. Nevertheless, it represented another step on that agency's downward road.[10]

"Do not go gentle into that good night," wrote the poet Dylan Thomas, in writing of elderly people facing death. "Rage, rage against the dying of the light." NASA's light was not yet dying, not with the piloted Moon landings immediately ahead. But under new leadership, this agency was ready to rage with vigor against the slow demise that seemed to be marked out as its fate.

## A New Administrator: Thomas Paine

During 1966 and 1967, Webb worked with Robert Seamans as his deputy. As Administrator, Webb dealt with NASA's external environment, including Congress, the White House, and the Budget Bureau. Seamans had held high-level NASA positions since 1960. As Deputy Administrator, he served as the agency's general manager; all line and staff offices reported to him. He left NASA early in 1968 to take a professorship at Massachusetts Institute of

---

9. Budget Bureau, "National Aeronautics and Space Administration—Highlights Summary," October 30, 1968. Reprinted in NASA SP-4407, Volume I, pp. 495-499.
10. NASA SP-4102, p. 188.

Thomas Paine. (NASA)

Technology. His replacement, Thomas Paine, took over his post and became the new Deputy Administrator.[11]

The son of a naval commodore, Paine served in World War II as a radar and engineering officer aboard the submarine *USS Pompon*. The experience stayed with him; decades later, trapped in a boring meeting, he would fill the time by drawing a sketch of his submarine under way on the surface and ready for action. After receiving a Ph.D. in physical metallurgy at Stanford University in 1949, he joined the General Electric (GE) Research Laboratory in Schenectady, New York. In this company, he rose to manager of engineering applications. In 1963, he returned to the west coast and became head of TEMPO, a GE think tank in Santa Barbara, California. From this position, he went on to become Webb's deputy at NASA in January 1968.[12]

Like a good Navy man, he hit the ground running. Though still recovering from the Apollo fire during that year, NASA was pushing forward in

---

11. *Ibid.*, pp. 309, 310; Logsdon, *Apollo*, p. I-14.
12. Biographical data, Thomas O. Paine papers, Library of Congress. These papers include an example of a submarine sketch.

expectation of meeting President Kennedy's goal of a piloted lunar landing by the end of 1969. The Apollo spacecraft, well along in development, would soon be ready to carry its crew of three. The Saturn V, in flight test, soon would be ready as well. However, a vital element of Apollo, the lunar module, was encountering delays. This spacecraft, with room for two astronauts, was to carry out the actual landing on the Moon.

The Saturn V was to carry a complete moonship, comprised of both an Apollo spacecraft and a lunar module, with the latter being flight-tested in both Earth and lunar orbits before it could qualify for the demanding task of a lunar landing. Its delay in development, however, raised the prospect that a Saturn V might be ready for launch, with only the Apollo spacecraft qualified for flight as its payload.

At the Manned Spacecraft Center in Houston, George Low, head of the Apollo Spacecraft Program Office, was ready to accept this. He recommended that the mission leave the lunar module on the ground, but send the Apollo spacecraft into lunar orbit, allowing its crew to circle the Moon repeatedly before breaking out of that orbit to return to Earth.

In August 1968, Webb and Mueller—both out of the country attending a United Nations conference in Vienna—left Paine in Washington to mind the store. Sam Phillips, the Apollo program director, told him of Low's proposal. Paine, who found the concept exciting, gave it his full support. He then tried to sell it to Webb, as both he and Phillips talked to their Administrator via overseas telephone.

Webb was shocked at the audacity of the idea, and yelled, "Are you out of your mind?" They had not even flown a piloted Apollo spacecraft in Earth orbit. In addition to this, Webb viewed the lunar module as a lifeboat that could save the crew of a lunar mission if their Apollo spacecraft were to become disabled.[13] Any piloted lunar mission would be dangerous; to fly without a lunar module would make it more so. In turn, the deaths of additional astronauts, in the wake of the Apollo fire, would shake NASA to its foundations.

Nevertheless, Webb did not say no; he left the door open. When Paine then strengthened his argument by sending Webb a long cable, Webb grudgingly agreed to consider this proposal, at least for purposes of planning. Events now played into Paine's hands. In September, the Soviet Union carried out an

---

13. This happened during the Apollo 13 mission, in April 1970.

important lunar mission, Zond 5. This spacecraft looped around the Moon, returned to Earth, reentered the atmosphere, came down in the Indian Ocean, and was recovered. Two turtles were aboard, and they came back safely. An impressed Webb described this flight as "the most important demonstration of total space capacity up to now by any nation."

Zond 5 raised the stakes. All along the goal of Apollo had been to beat Moscow to the Moon; yet by sending a cosmonaut in place of the turtles, the Soviets could still win the race with another Zond mission. While Zond would only loop around and not land on the Moon, if cosmonauts were to do this, they would become the first pilots to fly to the Moon. Subsequent Apollo landings then would appear merely as following in Soviet footsteps.

During that same September, Webb announced that he would step down as NASA Administrator. He had held that post since 1961; he now would turn it over to Paine, who would serve as Acting Administrator until the next president, due to be elected in November, could name a new head of NASA. Webb's resignation took effect early in October and left Paine free to make decisions as opposed to more recommendations.

In that same month, the Apollo 7 mission successfully flew with three astronauts in Earth orbit, as they tested their spacecraft during an 11-day mission. This exorcised the ghosts of the Apollo fire, and led within weeks to the commitment Paine had sought. Having flown successfully in Earth orbit, the Apollo spacecraft, sans lunar module, indeed would fly to the Moon on the next mission. The flight that resulted, Apollo 8, carried the astronauts Frank Borman, James Lovell, and William Anders. On Christmas Eve 1968, much of the world listened as a radio circuit carried their voices, live from lunar orbit:

> Anders: *In the beginning God created the heaven and the earth. And the earth was without form and void, and darkness was upon the face of the deep....*
>
> Lovell: *And God called the light Day, and the darkness he called Night. And the evening and the morning were the first day....*
>
> Borman: *And God said, "Let the waters under the heavens be gathered together in one place, and let the dry land appear." And it was so....*[14]

---

14. NASA SP-4205, pp. 256-260; Logsdon, *Apollo*, Chapter 2, pp. 10-14; Levine, *Future*, pp. 101-102; Chaikin, *Man*, pp. 56-59; Heppenheimer, *Countdown*, pp. 237-239, 243-244; NASA SP-4102, pp. 257-258, 311.

# THE SPACE SHUTTLE DECISION

The year 1968 had been one of war and upheaval, as public bitterness over Vietnam drove the powerful Lyndon Johnson from the presidency. Nevertheless, *Time* picked the crew of Apollo 8 as its Men of the Year, and assessed the significance of their mission:

> *What the rebels and dissenters ask will not be found on the moon: social justice, peace, an end to hypocrisy—in short, Utopia. But to the extent that the rebels really want a particular kind of tomorrow—rather than simply a curse on, and an escape from, today—the moon flight of Apollo 8 shows how that Utopian tomorrow could come about. For this is what Westernized man can do. He will not turn into a passive, contemplative being; he will not drop out and turn off; he will not seek stability and inner peace in the quest for nirvana. Western man is Faust, and if he knows anything at all, he knows how to challenge nature, how to dare against dangerous odds and even against reason. He knows how to reach for the moon.*
>
> *That is Western man, and with these qualities he will succeed or fail.*[15]

Apollo 8 reflected Paine's leadership and initiative, which he had displayed even while Webb still headed NASA. That mission also reflected a characteristic boldness, a willingness to reach for new horizons; this too was part of Paine's approach. He would display such boldness time and again during his tenure, as he pushed his colleagues to think more daring thoughts.

In several respects, however, Paine's position was weaker than that of Webb. Webb had served the Truman Administration as director of the Budget Bureau from 1946 to 1949, and as Under Secretary of State from 1949 to 1952. Though Republicans held the White House during the subsequent eight years, Webb's background made him a charter member of the Democrats' shadow government, ready to receive an important sub-cabinet post when they regained the presidency in 1960.

Paine had no such background. He held no record of government service, or even of involvement in the space program as a technical manager. In the words of the historian John Logsdon, "he was as new to the ways of Washington as James Webb had been a master of them." Paine had obtained

---

15. *Time*, January 3, 1969, p. 17.

his initial selection as Deputy Administrator because he had offered himself for appointment to a high-level Washington position, nature unspecified. The head of the Civil Service Commission found Paine's name in a list of executives who had expressed interest in receiving such posts, and passed it on.

With the new president, Richard Nixon, free to name his own NASA Administrator, Paine submitted a pro forma resignation upon Nixon's inauguration in January 1969. Nixon's staff offered the post to candidates that included General Bernard Schriever, who had built the Air Force's big missiles; Simon Ramo, a co-founder of TRW who had provided Schriever with vital technical support; and Patrick Haggerty, head of Texas Instruments. When all declined, Nixon decided to stick with Paine. Continuing as Acting Administrator, Paine received Senate confirmation as Administrator, without qualification, in March 1969.[16]

There was less to this than met the eye, for in no way did Nixon intend to endorse Paine's bold approach to space flight. "He was not committed to space," recalls Hans Mark, director of NASA's Ames Research Center and later an Air Force Secretary. "Nixon had no real interest in it. He didn't want to be the president that would kill our space program, but he had no personal interest in it at all."

Mark assesses Paine as "a rank failure" because he was

*a Democrat in a Republican administration. Just to give you an idea of why I think Nixon didn't give a damn about the space program, he didn't go out and look for a strong Administrator. What better way to have a pliant NASA than to have a Democrat sitting there exposed to his people? When Paine was confirmed as Administrator under Nixon, my reaction was, "Oh, my God—nobody is going to pay any attention to us."*

Willis Shapley recalls that Paine initially

*had expected to be fired because he was a liberal Democrat. And Nixon delayed in replacing the Administrator of NASA for a long time. Suddenly, they realized that with the Apollo program coming up and a reasonable chance that it might have failed spectacularly, they wanted to distance them-*

---

16. NASA SP-4102, p. 309; Logsdon, *Apollo*, pp. I-14, II-12, III-2.

# THE SPACE SHUTTLE DECISION

*selves from the Apollo program. That led to, by pretty straightforward political logic, "All right, this was Kennedy's program; it's going to be Kennedy's failure, and here's a liberal Democrat, Tom Paine. All right, he can be the fall guy. Also, we can't find anybody else for the job." So that's how Tom Paine became Administrator of NASA.*[17]

## *Space Shuttle Studies Continue*

Amid drama near the Moon and change in Washington, NASA plugged ahead, as space shuttle studies continued to receive their modest share of attention. As early as October 1966, a meeting in Houston brought together officials from NASA Headquarters, NASA-Marshall, and the Manned Spacecraft Center. They began to plan a joint study of Space Shuttles to provide logistics for a space station. A month earlier, a joint Air Force-NASA study had concluded that "no single, most desirable vehicle concept could be identified... for satisfying future DoD and NASA objectives."[18] The participants in the Houston meeting hoped to create a united front within NASA. Noting the substantial number of studies already available, Daniel Schnyer, representing the Office of Manned Space Flight, declared that "we have a vast store of knowledge to draw on, and should now be able to get together and decide on an agency concept for the entire logistic system."

This meeting was sufficiently noteworthy for Max Akridge, a representative from NASA-Marshall, to describe it as "the beginning of the space shuttle as such."[19] Little came of it; as George Mueller learned when he hosted his symposium in January 1968, various groups and designers were still pursuing their individual approaches. Still, NASA Marshall would now work in tandem with the Manned Spacecraft Center. This joint approach was in the forefront when, in October 1968, managers from these two centers launched a new round of space shuttle studies.

Why could NASA not order up a definitive treatment of some particular concept, such as Lockheed's Star-Clipper, and be done with it? NASA needed such studies because there was no way to get such a definitive treatment, at

---

17. John Mauer interview, Willis Shapley, October 26, 1984, p. 7; interview, Hans Mark, Austin, Texas, October 16, 1991, pp. 9-10.
18. Ames, chairman, *Report*, p. 1.
19. Akridge, *Space Shuttle*, pp. 25-26.

least not with the modest sums available to underwrite individual studies. Like professors reviewing a graduate student's dissertation, senior managers could always read the report of a particular study and raise new questions, new topics for further examination. Similarly, like grad students who look for jobs wherein they can continue their dissertation research, engineers and managers were eager to carry forward with new design exercises and further analyses. They would continue to do this as long as funding remained available.

George Mueller was the man behind the new activity. In the wake of the 1968 election, his hopes were high. In mid-December, just prior to the launch of Apollo 8, he talked with Wernher von Braun and chided him: "You'd been telling me that my space shuttle was in the future and you needed an interim system," such as the Titan III-M. Mueller predicted that the new president would want to go "all out," adding that "this may be the big program for Nixon." Von Braun replied, "If Nixon wants to spend $3 billion, who am I to say no?"[20]

Initial activity revolved around a formal Request for Proposal, issued October 30. It represented an invitation for interested companies to describe the studies they hoped to conduct. Responses were due by the end of November. In the parlance of the day, the spacecraft under consideration were described anew as Integrated Launch and Reentry Vehicles, with the new studies representing a continuation of work funded by the Air Force under that designation.

The study contracts, signed at the end of January 1969, called for all designs to follow a common set of ground rules. The Statement of Work defined the basic mission as the resupply of a space station, in an orbit with 55-degree inclination. Each vehicle was to carry 12 people, as passengers and crew, or a payload of 25,000 pounds. Contractors were also to present variations of their basic designs, to accommodate payloads ranging from 5,000 to 50,000 pounds. Payload bays were to provide cargo volume of at least 3,000 cubic feet; when returning from orbit, these vehicles were to carry at least 2,500 pounds. Designers were to seek to achieve aircraft-like checkout and ease of maintenance, including readiness for rapid launch in as little as 24 hours.

North American Aviation (NAA), which now had the name of North American Rockwell after having merged with the firm of Rockwell International, took a contract. It was to study new low-cost expendable

---

20. *Ibid.*, pp. 47-48.

boosters with reusable upper stages. Having ruled out existing expendables such as the Saturn I-B or Titan III-C, NASA welcomed new concepts.

Lockheed was told to concentrate its effort on new studies of stage-and-a-half configurations such as the Star-Clipper. The new topics would include use of solid- or liquid-fueled booster stages that would ignite at liftoff. (Here was another glimpse of the future; the final Shuttle configuration, three years later, would use such solid boosters.) In addition to this, Lockheed was to develop its own versions of the Triamese concept for comparison with those of General Dynamics (GD).

GD focused on two-stage fully-reusable approaches, excluding stage-and-a-half designs. GD's configurations therefore demanded a "flyback first stage," capable of returning to the launch site, and this firm's engineers were to consider unpiloted concepts. The Triamese approach was to remain an important though not predominant focus of effort. At the same time, NASA officials raised a new question: Is there a way to design a flyback first stage that could develop into a Triamese vehicle?

McDonnell Douglas, home of the Tip Tank stage-and-a-half concept, was to study it anew. Like Lockheed, this firm was to look at the use of booster stages; in addition to this, its investigators would study fully-reusable concepts. These became central to this company's efforts as the study progressed during 1969.

At first glance, these contracts promised more of the same. One could see a few new faces in the crowd. Martin Marietta failed to win a study contract while North American Rockwell, builder of the Apollo spacecraft and the second stage of the Saturn V, won instead. The promised concepts, however, amounted to new variations on old themes. No one was offering anything so original as Max Hunter's stage-and-a-half approach, and each of the four studies had only $300,000 in funding.[21]

In the light of subsequent events, one notes particularly the strong emphasis on space station logistics. The study requirements placed little weight on contractor concepts that would specialize in carrying automated spacecraft as payloads. Yet there was a reason for this focus on space station logistics: Tom Paine wanted such a station, and was pushing hard in Washington to win approval to build it.

---

21. *Ibid.*, pp. 46-53, 55-57.

## Space Shuttle Policy: Opening Gambits

George Mueller believed strongly in convening advisory boards, and in sponsoring studies of future programs. One of his ongoing panels, the Science and Technology Advisory Committee (STAC), sought to bring top-level scientists into consultation with his Office of Manned Space Flight (OMSF). Many scientists were strongly skeptical of the OMSF and its works; they regarded Apollo as a costly extravagance, and argued instead for less costly automated spacecraft. STAC scientists, who at least were not vocal in their opposition, gave OMSF access to high-quality scientific advice. It also offered a counter to criticism from the scientific community. The chairman of STAC, Charles Townes, was a physicist who had shared a Nobel Prize for work leading to the invention of the laser.[22]

In December 1968, a month after the election, Townes agreed to chair an advisory group that would make recommendations to the new administration on space policy. The group's members included Robert Seamans, who had been NASA's Deputy Administrator. Other members, including Townes, had served on the President's Science Advisory Committee panel that had prepared the 1967 report, *The Space Program in the Post-Apollo Period*.

Townes's group included at least one true believer: Francis Clauser, vice-chancellor of the University of California at Santa Cruz, who was about to take over as chairman of the college of engineering at Caltech.[23] In a letter appended to the final report, Clauser urged Nixon to

> *chart a bold program...I think our rate of development can be considerably more rapid than presented in the task force report. For example, I believe we can place men on Mars before 1980. At the same time we can develop economical space transportation which will permit extensive exploration of the moon.*

On the whole, however, the new report was considerably more cautious in tone than the PSAC review of 1967. On NASA's future, Townes' panelists agreed that "we do not recommend a commitment now to a large space sta-

---

22. Logsdon, *Apollo*, p. I-16.
23. *Ibid.*, p. I-23; Long, chairman, *Space Program*, p. v; *Astronautics & Aeronautics*, May 1969, p. 35; Townes, chairman, *Report*, letter attached. Reprinted in NASA SP-4407, vol. I, p. 512.

tion, extensive development of 'low-cost boosters,' or a manned planetary expedition." The panel called for "a new look...at the balance between the manned and unmanned segments of the NASA space program." Asserting in effect that the existing program was badly out of balance, the report proposed redress: "an active and successful manned program for several years while at the same time steadily decreasing the level of funding for manned space flight to perhaps $1.25 billion by fiscal 1972."

On a space station:

*We are against any present commitment to the construction of a large space station.... The "manned space station" concept, proposed as a program for the late 1970s, is on much more doubtful ground. It is much too ambitious to be consistent with the present clear needs for continued exploration of man's usefulness in space. On the other hand, it is not obviously an effective way of continuing to demonstrate for prestige purposes our manned space capability.... It therefore seems premature to make any firm program decisions regarding the proposed manned space station.*

On a space shuttle:

*The unit costs of boosting payloads into space can be substantially reduced, but this requires an increased number of flights, or such an increase coupled with an expensive development program. We do not recommend initiation of such a development.*

On piloted flight to Mars:

*The great majority of the task force is not in favor of a commitment at present to a manned planetary lander or orbiter.... It would be undesirable to define at this time a new goal that is both very ambitious in scope and highly restrictive in schedule, for example a manned landing on Mars before 1985, even though such a goal might be achievable. Such a commitment, adopted now, might inhibit our ability to establish a proper balance between the manned space program and the scientific and applications programs.*[24]

---

24. Townes, chairman, *Report*. In NASA SP-4407, vol. I, pp. 499-512.

## Mars and Other Dream Worlds

This report, released in January 1969 prior to the inauguration, was not to Paine's liking. In a written critique, he noted

> *its repeated opposition to the word "commitment." We must not commit, the report says, to a space station, to low-cost space transportation, to manned planetary exploration. I can understand this reluctance to make commitments, but I cannot sympathize with or accept it. I understand that the word "commitment" means to many scientists the type of commitment we made to Apollo, but I do not agree with those who regret or deplore that commitment. They see only its disadvantages....*
>
> *We have been frustrated too long by a negativism that says hold back, be cautious, take no risks, do less than you are capable of doing. I submit that no perceptive student of the history of social progress doubts that we will establish a large laboratory in earth orbit, that we will provide a practical system for the frequent transfer of men and supplies to and from such a laboratory, that we will continue to send men to the Moon, and that eventually we will send men to the planets. If this is true, now is the time to say so. Now is the time for the President of the United States to say, "This country will establish a scientific laboratory in earth orbit. This country will develop a practical space transportation system. This country will send men to the planets."*[25]

By then, Richard Nixon was in the White House and was dealing with the federal budget. Preparing a budget for a particular fiscal year (FY) took time; though elected in 1968 and inaugurated in 1969, the first budget that would be truly Nixon's own would cover FY 1971. At the moment he was dealing with the budget for FY 1970 that would begin in mid-1969. Like other new presidents, Nixon, seeking to contrast his own financial prudence with the spendthrift ways of his predecessor, ordered his department heads to look for ways to make cuts.

The new Director of the Bureau of the Budget, Robert Mayo, wrote a government-wide letter to those heads of agencies on January 23, asking them to review their portions of President Johnson's FY 1970 budget and to propose areas where spending might be reduced. Paine took this as an invitation to press instead for an increase. A month later, Paine replied to Mayo with a

---

25. Letter, Paine to DuBridge, May 6, 1969.

# THE SPACE SHUTTLE DECISION

Robert Mayo, right, with Nixon and Treasury Secretary, David M. Kennedy. (National Archives)

seven-page letter, single spaced. Its crux required little more than a single line: "In our judgment the NASA FY 1970 budget is deficient by $198 million from the amount required."[26]

Mayo slapped this down in a hurry. It took him only a few days to reply with a letter to Paine: "I am not prepared at this time to recommend to the President approval of your requested budget increase." He also wrote to Nixon: "Our first look at the agency recommendations...shows many more increases than decreases. In total these requests, if granted, would make precarious if not impossible the attainment of the surplus forecast by the previous Administration." He advised the President to "make no statements endorsing future space objectives" pending extensive further review, which would include "the total budget context."[27]

Mayo's staff went on to request a $90 million cut in the NASA budget, where Paine had sought a $198 million increase. While Paine succeeded in winning some relief, the final BoB cut still came to $45 million. Part of a $5.5

---

26. Logsdon, *Apollo*, Chapter 3, pp. 2-3; letter, Paine to Mayo, February 24, 1969.
27. Memo, Mayo to Nixon, March 3, 1969; letter, Mayo to Paine, March 3, 1969. Quoted in Logsdon, *Apollo*, p. III-7.

billion cut in an overall federal budget of $200 billion for FY 1970, this represented no more than NASA's fair share. This, however, was no way to build a post-Apollo future.[28]

Nevertheless, the Nixon administration was preparing to chart its own course toward that future, with Paine having a hand in the planning. In December 1968, during the transition between administrations, Nixon had selected A. Lee DuBridge, the president of Caltech, to be his new science advisor. At Caltech, DuBridge's purview had included the Jet Propulsion Laboratory, the nation's principal center for automated exploration of the Moon and planets operated by this university under contract to NASA. Like many of his fellow scientists, however, DuBridge was skeptical of the value of piloted space flight. He favored paying greater attention to automated missions, and he knew them well.[29]

The Townes panel had cast its net broadly, offering recommendations that dealt with international cooperation in space, issues of NASA's internal organization and of its relations with the Defense Department, and even matters that could require new legislation in Congress.[30] The economist Arthur Burns, a Nixon advisor, proposed that DuBridge should direct a new study that would further address these issues. On February 4, Nixon responded with a memo to DuBridge:

> *There is general agreement that our space efforts should continue, although there are notable differences of opinion in regard to specific projects and the amount of annual funding.*
>
> *The report from Arthur Burns' group...proposes the establishment of an interagency committee which would include you, the Administrator of NASA, and a senior official from the Department of Defense. The primary function of this committee would be to furnish recommendations to me on the scope and direction of our Post Apollo Space Program.*[31]

As the president's science advisor, DuBridge served as director of the Office of Science and Technology (OST), one of a myriad of special-purpose

---

28. Logsdon, *Apollo*, p. III-10.
29. *Ibid.*, p. II-23; NASA SP-4102, p. 15.
30. Townes, chairman, *Report*, pp. 22-28. Reprinted in NASA SP-4407, vol. I, pp. 509-511.
31. Memo, Nixon to DuBridge, February 4, 1969.

## THE SPACE SHUTTLE DECISION

bureaus within the Executive Branch. OST staff members proceeded to draft a directive which represented an attempt to give DuBridge the leadership of this review of space policy. No NASA officials took part in drafting this directive, for in the words of one OST staffer, Russell Drew, "there was a concern that we would be called upon to rubber-stamp a NASA document, which we did not want to do."[32]

Learning of this ploy, Paine protested strongly. He disliked the idea that DuBridge and the OST might present him with plans sanctioned by Nixon that he then would have to execute; as he later put it, "You never want one bunch of guys to do the planning and another bunch to carry it out." Nevertheless, Paine's position was quite weak at that moment. He was no more than Acting Administrator, and because he had sent in his resignation, he was merely serving from day to day at Nixon's pleasure, pending appointment of a replacement.

But Paine, a Democrat in a Republican administration, was not about to play the patsy. As he later recalled,

> *I was the person directly reporting to the President, responsible for the space program even though I was only the acting rather than the full Administrator. I nevertheless took the view that I was acting for the new Administrator whoever he might be, and that it was very important that I not give away any of the authority and responsibility...to an advisory staff function [the OST] even though that staff function might reside in the White House....*
>
> *I took the very early and the very strong view...that this must recite the fact that NASA would be responsible for setting up the NASA portion of this, that the Air Force and the DoD would be responsible for the military portion of the space program; and that we then review it with other responsible people...to make it reflect a broad Administration-wide consensus.*

Paine won his argument, in what he called "a rather typical Washington power struggle."[33]

If not the OST and Lee DuBridge, who would direct this interagency study? The answer lay in the original Space Act of 1958 that had created a policymak-

---

32. Logsdon, *Apollo*, p. IV-4.
33. Interview, Thomas Paine, September 3, 1970, pp. 5-6.

ing body, the National Aeronautics and Space Council (NASC). Its members included representatives from all federal agencies with an important interest in space: NASA, the DoD, the AEC and the State Department, the latter because the space program required overseas tracking stations and featured cooperation with scientists of other nations. Though by law the Vice President chaired this body, it had never done much. The recent Townes panel described it as "not very effective," noting that Nixon might ask Congress to abolish it.

The ineffectiveness of the NASC matched that of its new chairman, Vice President Spiro Agnew. He was in a position familiar to new vice presidents, for he had little to do and was looking for ways to make himself useful. In Paine's words, "at that time, he hadn't figured out what his role was going to be in the administration." He willingly agreed to chair the new interagency review, which would go forward within a committee called the Space Task Group (STG).[34]

On February 13, Nixon issued a new memo confirming this arrangement. DuBridge was to join the STG and provide its staff. Paine would also be a member. Nixon initially designated Secretary of Defense Melvin Laird as a third member. Laird, however, chose instead to appoint a representative: Robert Seamans, lately the number two man in NASA and now recalled from MIT to become the new Air Force Secretary.

The group held an initial meeting on March 7, with Agnew in the chair. This meeting served to organize the group's activities. Nixon had directed the STG "to prepare for me a coordinated program and budget proposal"; hence it was appropriate to invite Robert Mayo, director of the Budget Bureau, to sit with the group as an observer. The STG issued similar invitations to two others: Glenn Seaborg, chairman of the AEC, and U. Alexis Johnson, Undersecretary of State for Political Affairs.[35]

Already Paine was wooing Agnew vigorously, for Agnew had no background in space—he had been governor of Maryland—and proved amenable to Paine's bold planning. The Apollo 9 mission flew during that month, atop a Saturn V, which carried a lunar module into Earth orbit along with three astronauts in an Apollo spacecraft. This represented the first test flight of a

---

34. *Ibid.*, p. 7; NASA SP-4102, p. 15; Townes, chairman, *Report*, pp. 7, 27. Reprinted in NASA SP-4407, vol. I, pp. 502, 511.
35. Logsdon, *Apollo*, Chapter 4, pp. 5-7; memo, Nixon to Agnew et al., February 13, 1969. Reprinted in NASA SP-4407, vol. I, pp. 512-513.

## THE SPACE SHUTTLE DECISION

complete moonship. Paine invited Agnew to come to Cape Canaveral on the day of the launch, as an honored guest. While DuBridge and Seamans were there, it was Agnew who received special treatment. He was given a tour of the moonport, with astronauts as his escorts. These included Frank Borman, who had commanded Apollo 8.[36]

The STG held a second meeting two weeks later. Seamans, seeing "considerable military interest and potential use" for a shuttle, won agreement that a joint NASA/DoD panel would study it anew. Paine called for "a new banner to be hoisted," as daring as Apollo, around which the nation might rally. Agnew pursued this thought: Where was the Apollo of the 1970s? Could it be that the United States should undertake a human expedition to Mars? The issue flickered only momentarily for it was too early for such plans to catch fire. Agnew and Paine, however, would return to this topic anew in subsequent months.[37]

The Townes study, the setting-up of the STG, and Paine's attempt to boost his budget all took place between December 1968 and the following March. For Paine, the record was mixed. He sat and watched while Townes' panelists proposed to do little in space that was new. He had lost in his opening encounter with Mayo. He had succeeded, however, in shaping the STG to his liking and had reason to think he would have Agnew as an ally, with Seamans as another highly knowledgeable participant.

The second STG meeting was the last such full-dress meeting until August. During the intervening months, participants would call on planning groups to develop specific proposals. Paine was already doing this, for while he might have little clout at the Budget Bureau, on his own turf he was king. He wanted a space station and a great deal more, and was already working with colleagues to determine the designs.

### *Paine Seeks a Space Station*

There was little fundamentally new in the realm of space shuttle design; the same was true of space stations. The studies of the 1960s had emphasized concepts such as the Manned Orbiting Research Laboratory of Douglas

---

36. Thompson, ed., *Space Log*, vol. 27 (1991), p. 107; "Visit by the Vice President, Kennedy Space Center," March 3, 1969. Thomas Paine papers, Library of Congress.
37. Logsdon, *Apollo*, Chapter 4, pp. 8, 9.

## Mars and Other Dream Worlds

Aircraft, carrying crews of six to nine people and using the Saturn I-B for logistics. This approach had continued to win attention, with a recent version including a component called the Manned Orbiting Module. The name had been chosen with care, for as one NASA official put it, "What congressman would dare vote against anything called MOM?"[38]

Paine wanted more. He not only wanted new and more detailed studies; he wanted new thinking, and he was not shy about giving pep talks. Thus, in a letter to the director of NASA's Lewis Research Center in mid-January, he noted

> *the need to outline bold objectives for the Space Station program. Modest goals, which tax neither our own creativity nor the potential advances of our industrial technology, are not worthy successors to those of Apollo.... Please review this draft work statement thoroughly, and submit a revised document which proposes a substantially stronger and bolder U.S. Space Station Program.*[39]

A month later, addressing a symposium on space stations at NASA Langley, Paine stated openly that the inspiration for his agenda came from the *Collier's* series of the early 1950s:

> *Seventeen years ago a group of forward-looking scientists and engineers proposed that the United States undertake the construction of a large space station over a ten-year period at a cost of $4 billion.... The space station, a 250-foot-diameter ring with artificial gravity and solar power, was to be put into a 1,075-mile altitude high-inclination orbit....*
>
> *The scientists who worked with Collier's Magazine on this proposal included Dr. Wernher von Braun, Dr. Fred L. Whipple, Dr. Joseph Kaplan, Dr. Heinz Haber and Mr. Willy Ley.... Their timetable for space exploration included an orbiting space station by 1967 and a possible first lunar landing by 1977. Five years before Sputnik the scientists warned: "What you will read here is not science fiction. It is serious fact. Moreover, it is an urgent warning that the U.S. must immediately embark on a long-range development program to secure for the West 'space superiority.' If we do not, somebody else will. That somebody else very probably would be the Soviet Union."...*

---

38. *Aviation Week*, October 21, 1968, pp. 25, 26.
39. Letter, Paine to Silverstein, January 14, 1969.

# THE SPACE SHUTTLE DECISION

> *As we meet here today the United States stands at the end of the first decade in space looking forward to the second.... But we are still looking forward to the establishment of that projected major research laboratory in the sky, the permanent U.S. space station accessed by a low-cost space shuttle.*[40]

Yet though Paine was sounding a clarion call, in a vital respect he was blowing an uncertain trumpet, for his center directors lacked clear direction as to what they were to produce. They were quite ready to boldly go where no one had gone before, but their engineers needed more than *2001: A Space Odyssey* when drawing up their specifications. In particular, they needed a well-drawn Statement of Work to direct the new space station studies. Paine, addressing this issue as well, had recently convened a meeting of those directors for this purpose.

The draft Statement of Work of that moment was the seventh in a series, and everyone agreed it left much to be desired. Abe Silverstein of NASA-Lewis, who had received Paine's letter of mid-January, cited three criteria in evaluating the draft: "Would contractors receiving this in the mail know what to do? Could we evaluate their responses? Could the project be completed to meet our specs within time and money?"

He added that "the number-seven draft document fails to meet these criteria and *cannot* be edited to meet these criteria. NASA is asking—in mushy language—for something we should know ourselves before going out." He then gave a list of "things we need to do more homework on": size, weight, orbits, programs and experiments, power, logistic support, and communications. "We need to define all these factors for the contractors," he concluded. "In essence, tell them clearly what NASA wants. We need a document that defines the basics."

With even the basics left undefined, Robert Gilruth of the Manned Spacecraft Center was willing to fill in the blanks in a very expansive spirit:

> *This work statement doesn't set NASA's sights high enough for the future. We should now be looking at a step more comparable in challenge to that of Apollo after Mercury. The space station size should be modular and based on our Saturn V lift capability into 200-mile orbit. Three launches would give us*

---

40. Paine, speech, February 11, 1969.

## Mars and Other Dream Worlds

*one million pounds in orbit, including spent stages. That is the number we should be planning for the core size.*

Gilruth predicted a need for "nuclear power of several hundred kilowatts. The design should also emphasize the utility of the space base as a way-station to the moon and Mars."

Wernher von Braun, representing NASA Marshall, proposed a specific approach:

*Tell the contractors what we want in the long run, what we foresee as the ultimate, the long-range, the dream-station program. Then NASA should define a 1975 station as a core facility in orbit from which the ultimate "space campus" or "space base" can grow in an efficient orderly evolution through 1985. We should start in 1975 to launch the basic core of the space station or space base that we want to be operating in 1980, providing planned orderly growth capability.*[41]

This meeting took place late in January. Over the next several weeks Charles Mathews, George Mueller's deputy at the OMSF in Washington, developed new guidelines that now called for *two* space stations: an initial concept for the short term and a blue-sky version as a follow-on project. However, the latter would show a close relation to the former, for the initial space station would serve as a module or building block. By launching several of them and linking them together in orbit, NASA could assemble a true space base, with accommodation for as many as a hundred people.

This brought a highly compelling concept for a space station program that could extend through the 1970s and beyond. The basic space station would not resemble MORL, with its crew of six. Instead, it was to provide room for as many as 12 people. It would have a diameter of 33 feet, compared to the 22 feet of MORL, and would fly on the Saturn V. NASA would resupply this station using the Saturn I-B or the Titan III-M, with the chosen launch vehicle carrying a modified Gemini spacecraft called Big G. The standard Gemini had carried two people; Big G would retain that spacecraft as a cockpit or flight deck, while adding a passenger section with 7 to 10 additional seats.

---

41. Paine, notes, January 27, 1969.

# THE SPACE SHUTTLE DECISION

This station would be in orbit as early as 1975. Modules of similar size, brought up on their own Saturn V vehicles, would then build it into the final space base, with enough people to fill an office building. After 1975, a space shuttle would become available, and would replace Big G along with its expendable booster.[42]

Here indeed was the boldness Paine wanted. Part of what made this scenario exciting was that, in its essence, it called for no more than a modest extension of concepts with which everyone was familiar. The new station would represent something of a stretch when compared with MORL, but its design would rest solidly on the foundation of earlier studies, and it would use the Saturn V. Its logistics vehicle was old hat; the Big G concept had been around since 1967. The space base would grow from the basic station in a natural way, with a space shuttle—another well-known concept—complementing it strongly.

With these plans evolving rapidly, Paine received a request from the White House dated February 17, in which Nixon solicited his views on issues of policy in space and aeronautics. Paine responded by pouring out his heart in a nine-page letter, again single-spaced. He wanted his space station, naturally, and he not only wanted a presidential commitment; he wanted it quickly. This letter represented an attempt to bypass the deliberations of the Space Task Group, even though Paine was one of its members. It also bypassed the normal budget process.

While that was awkward, Paine and Nixon were both aware that the STG would require the entire spring and summer to carry through with its work, and Paine wrote that the matter was too urgent to wait. He brashly played the Soviet card, warning that Moscow was "pushing toward a dominant position in large-scale long-duration space station operations in Earth orbit.... Their moving clearly ahead of the U.S. in this field would have a continuing impact on the rest of the world." The Soviets might make their move as early as that summer. This would "take the edge off your announcement of a similar U.S. objective in the fall."[43]

Paine knew his man. Nixon had lost the presidency to Kennedy by a whisker-thin margin in 1960 partly because Kennedy had warned that the

---

42. *Aviation Week*: June 19, 1967, pp. 20-21; February 24, 1969, pp. 16-17; Report H321 (McDonnell Douglas); Logsdon, *Apollo*, chapter 3, pp. 23-24.
43. Memo, Paine to Nixon, February 26, 1969. Reprinted in NASA SP-4407, vol. I, pp. 513-519.

Russians were ahead of us in the space race. Nixon, however, refused to bite. He replied to Paine with a courteous note that put the space station issue firmly in the hands of the STG. The PSAC also declined to support NASA, as Lewis Branscomb, chairman of its panel on space science and technology, wrote that

> *if one does not accept the argument that potential Soviet competition in this area compels establishment of the space station at the earliest date, there does not seem to be a compelling operational requirement for a specific target date.*[44]

The work of Mathews and his colleagues now gave a firm basis for a suitable Statement of Work that could guide the new round of space station studies. On April 19, this document went out to prospective contractors. There would be two study contracts of $2.9 million each, one managed by NASA Marshall and the other by the Manned Spacecraft Center, with NASA Headquarters providing coordination. The studies were to put more than half their efforts into defining the basic station with its crew of 12. These studies were also to address issues involving logistic systems, along with concept definition of the eventual space base.[45]

## *Space Shuttles Receive New Attention*

Money talks, and the initial funding of the 1969 Space Shuttle studies showed that these held lower priority than those of the space station. While the station studies came to $5.8 million, there was only $1.2 million at first for shuttle work. This would be divided among four contractors who were to pursue their studies for no more than six months.[46]

Shuttles would now win new attention, for these studies would feed into the work of the STG. George Mueller then set up a program office almost overnight. To direct it, he picked LeRoy E. Day, an Apollo manager who was two levels down from him on the organization chart. It was April; the Apollo 10 mission was only weeks away, and Day was deeply immersed in preparation for a key pre-launch review.

---

44. Memo, Nixon to Paine, March 7, 1969; Logsdon, *Apollo*, p. III-8.
45. Logsdon, *Apollo*, chapter 3, pp. 24-26.
46. *Aviation Week*, February 10, 1969, p. 17.

# THE SPACE SHUTTLE DECISION

Day went into Mueller's office, expecting to discuss this upcoming flight, and saw him covering a blackboard with notes on the Space Shuttle. Day knew that this would not concern him; it was completely foreign to him in any case, so he waited politely for Mueller to finish. The following discussion ensued:

> Mueller: *I want you to really get going on the space shuttle. We've got a whole series of things to be done. We have to complete reports in about sixty days here, and then we have to negotiate with the DOD.*
> Day: *George, what does all this have to do with me?*
> Mueller: *Well, this is what I want you to do.*
> Day: *But you haven't said anything about me leaving my job in Apollo. You know we're just getting ready for the flight readiness review on Apollo 10.*
> Mueller: *I understand all that, but I want you to work on the shuttle. I need somebody to really head up this stuff, and I want you to do it.*
> Day: *Well, gee, I guess—let me go back and kind of timeline how I can get disengaged and come back with some dates to you, and then we'll talk about it, and maybe—I guess I can get out in a couple of weeks, right after this flight readiness review.*
> Mueller: *No. You don't understand. I want you over here now to begin work on the shuttle.*
> Day: *What does "now" mean?*
> Mueller: *Tomorrow morning.*

It was past four in the afternoon, and Day felt that he was completely over his head. He felt even more intimidated when Mueller said that they were to write a proposal for the President's Space Task Group, which would go on to the White House. Mueller assured him that he would not be alone: "You'll have practically a blank check. You'll be able to get people—whoever you need."[47]

Paine was also taking action, in concert with the new Air Force Secretary, Robert Seamans. Their staffs had discussed the formal Terms of Reference for a new joint study of space transportation. Less than three years earlier, a similar joint study had found no design concept that could satisfy the needs of both NASA and the Pentagon. The STG, however, had called for this new

---

47. John Mauer interview, LeRoy Day, October 17, 1983, pp. 1-3.

study, and Paine and Seamans agreed that it would go forward in two parts. At first the DoD and NASA would work separately, each defining the Space Shuttle concepts that would suit its own needs. The two agencies then would work together, seeking to meld their approaches, and would "recommend a preferred concept." A joint committee would manage both phases of the effort, with the co-chairmen being Mueller and Grant Hansen, Assistant Secretary of the Air Force for Research and Development. The study was to be brief; by mid-June, a joint report was to be ready for the STG.[48]

Early in May, Mueller hosted a meeting of the NASA Space Shuttle study contractors and presented them with some new rules. The mission of this shuttle had emphasized space station logistics, with a modest payload capacity of 25,000 pounds to orbit, 3,000 cubic feet, 2,500 pounds returned from orbit. This rationale now was broadening to include the launching of spacecraft, many of which would require upper stages to reach high orbits. Mueller told the attendees that "the principal carload capacity that we would have would probably be liquid hydrogen. So that dictates a fairly low density volume."

The new rules called for a payload of 50,000 pounds carried both up and down, a volume of 10,000 cubic feet. The payload bay could be 15 or even 22 feet in diameter, the latter accommodating craft that would fit atop an S-IVB. This doubling of the payload weight would bring a doubling in the shuttle's takeoff weight. This would suit the Air Force, which had a strong interest in large payloads and had built the Titan III-C to launch them.

Mueller had been a professor of electrical engineering during part of the 1950s. Drawing on this background, he proceeded to lay out a new concept for the use of computers to achieve rapid onboard checkout, and to present flight crews with the information they would use during a mission. This concept would make it possible for a small ground crew to carry out the preflight checks, achieving true aircraft-like simplicity.

Mueller called for designers to equip individual shuttle components, such as rocket engines, with sensors that would monitor their condition. Each component would carry a black box that would keep track of measured parameters. At any moment, some parameters would stand within acceptable limits, while others would lie outside such limits. Still other parameters would

---

48. Ames, chairman, *Report*, p. 1; Logsdon, *Apollo*, p. IV-9; "Terms of Reference for Joint NASA/DOD Study of Space Transportation Systems," April 11, 1969. Reprinted in NASA SP-4407, vol. II, pp. 364-365.

## THE SPACE SHUTTLE DECISION

be within bounds at that moment, but would be drifting toward unfavorable values. In response to a query from the computer, each black box could answer in turn: "I am well," with all parameters within limits. A particular box might answer, "I am sick," with one or more parameters outside the safe zone. Similarly, a box might respond, "I am about to get sick," with a parameter drifting toward danger. Further queries from the computer then could identify the bad parameters and permit cures.

Mueller also wanted the onboard computer to take the initiative in presenting data to the flight crew. While the data would be available on flight-deck instruments, a pilot might easily miss something because there would be many such instrument displays. Mueller preferred to "have the computer sweeping the cabin and looking at the end points of the gauges, and when one is going off from where it ought to be it can flash and show you what the reading is and what it ought to be and tell you what is wrong." With the computer running a display, "it provides you with the information you need when you need it, but it does not spread that information out over so many instruments."[49]

On May 19, two weeks after this meeting, LeRoy Day's task group submitted an initial report. It represented a milestone in presenting the Space Shuttle concept as one that might win serious support, for it broadened the rationale while narrowing the range of acceptable design approaches.

Since the early 1960s, shuttle advocates had been bedeviled by a multiplicity of reusable launch vehicle concepts, all of which could claim the name of a shuttle. In their day these had included boosters powered by scramjets or by LACE, horizontal-takeoff vehicles employing a rocket sled, and behemoths such as the Nexus that matched the weight of an ocean liner. These had fallen by the wayside, but the range of concepts had remained uncomfortably broad: expendable boosters with reusable upper stages, stage-and-a-half partially-reusable configurations such as Lockheed's Star-Clipper, two-stage fully-reusables such as General Dynamics's Triamese. This was somewhat like having the Air Force propose to build a new military airplane, without specifying whether it would be a fighter, bomber, or transport.

The May 19 report now rejected the use of an expendable booster. To meet Mueller's new requirements, such a launch vehicle would have to be

---

49. Mueller, *Briefing*, May 5, 1969.

larger than a Saturn V. It might use a big solid-propellant first stage or rely on low-cost liquid-fueled engines, making it cheaper to buy and fly. This vehicle, however, would splash into the Atlantic every time one of them flew—something that NASA was not about to do. After all, the goal from the start had been to move well beyond the Saturn V, not to develop it anew in a less costly version. In the words of the report, "Fully reusable or near fully reusable systems offer the maximum potential for an economic and versatile space shuttle system."

The report also broadened the rationale. To NASA, though not to the Air Force, a shuttle had primarily held the promise of low-cost logistic support for a space station. That made it a speculation nested within a speculation, for the station existed only at the level of designs and dreams. The report now added several attractive types of missions that reflected current practice or that built on current activities in plausible ways. This broadened rationale also made it more likely that NASA could come up with a configuration that would win Air Force support, thus further widening its usefulness.

A shuttle might not only place satellites in orbit; it could service them. Standard practice amounted to shooting and hoping, as ground crews launched their rockets and trusted their spacecraft to work. A shuttle crew, however, could check out a satellite after carrying it to orbit, ensuring that it was functioning properly. If a costly spacecraft failed in orbit, a shuttle might fly up to fix it. A shuttle might also carry it to a repair facility within a space station, or return it to Earth for rebuilding.

This reusable launch vehicle could also be large enough to carry the highly capable Centaur upper stage and powerful enough to carry communications satellites or planetary spacecraft of considerable size. This would lower the cost of such launches, by taking advantage of the shuttle's reusability, without compromising the demands of spacecraft designers by limiting this service to payloads of only modest weight.

A shuttle could also serve as an interim space station, by carrying an instrumented and crew-tended module within its payload bay. Such a mission might fly for up to thirty days. It could be far less costly than Skylab, while offering duration, internal volume, and onboard power considerably beyond that of Apollo.

Here was a new form of boldness: not a warmed-over version of the *Collier's* agenda, but a well-grounded concept of a completely new approach to the space activities that were already under way or approved. The report continued to list the prime shuttle mission as "space station/base logistics support." But its breadth of rationale for the first time raised the possibility that a shuttle program might take on a life of its own, serving the nation even in the complete absence of a station. On these terms, the shuttle could indeed go forward.[50]

## Space Task Group Members Prepare Plans

Within the STG, Paine represented NASA, Seamans represented the DoD, and DuBridge spoke for the scientific community. Though these leaders and their constituencies held distinctly different views on the future of space flight, the STG would have to reach a consensus if it was to speak with one voice. Much of the work of the STG took place outside its infrequent formal meetings, as these members commissioned studies that would define their positions.

NASA already had a well-established planning procedure in place; Homer E. Newell, the Associate Administrator, had set it up the previous year to develop program options that the Administrator could present to the BoB. This procedure featured a dozen working groups that drew broadly on specialists serving the entire range of NASA activities, with one committee providing coordination and a second committee—the Planning Steering Group—choosing the options that would reach Paine's desk. Newell took on the task of using this machinery to prepare the planning document that Paine would take to the STG.

Though he now held NASA-wide responsibility, Newell had headed the Office of Space Science and Applications (OSSA) that dealt with automated spacecraft. OSSA had operated in the shadow of the far more powerful Office of Manned Space Flight, and Mueller, its director, was not about to defer to Newell when it came to planning. Mueller had a planning group of his own at Bellcomm, a branch of AT&T with close ties to Bell Labs. This group had furnished NASA with planning analyses during the Apollo program; it too would readily serve the needs of the STG.[51]

Newell and Mueller initiated their planning exercises in December 1968, with the work of the Townes panel under way. Though the STG still lay two

---

50. Day, manager, *Summary Report*, May 19, 1969.
51. NASA SP-4102, pp. 256-257; Logsdon, *Apollo*, chapter 4, pp. 17-19, 26-27.

months in the future, Newell hoped to influence the FY 1971 budget that would become the subject of serious negotiation during 1969. During that same December, Mueller convened another of his planning groups, the Science and Technology Advisory Committee (STAC). In the course of that month, Mueller told von Braun that Nixon would go all-out and pick the Space Shuttle as his big program. STAC reviewed the prospects for piloted space flight and showed similar ebullience, calling for "extensive exploration and initial colonization of the Moon," along with planetary missions that would include "a manned expedition to the surface of Mars." STAC also strongly endorsed the Shuttle, giving it highest priority and calling it "the keystone to future development and large-scale practical application of the space program."[52]

Armed with this study, Mueller approached Bellcomm. With two pages of handwritten notes, he joined a Bellcomm staffer on a flight to Cape Canaveral late in March. The two men discussed the outline of what Mueller wanted. It amounted to an extension of the space station approach that was taking shape, wherein a single module would serve as a building block for later construction of a space base. Similarly, a minimum number of major new systems were to serve as many roles as possible.[53]

Spurred on by Paine's desire for boldness, Newell's planners proceeded to develop a scenario calling for a fast-paced effort that would emphasize space stations. The program called for an initial station in 1975, with a crew of 12, followed quickly by additional stations in polar orbit and in geosynchronous orbit, the latter at an altitude of 22,300 miles. A space shuttle would enter service during 1977. At the same time, other stations would be operating in lunar orbit and on the Moon's surface, while NASA would begin to build an Earth-orbiting space base for a crew of 50.

Newell did not ignore thoughts of piloted flight to Mars. His working groups included a task force on planetary exploration. During April he told its chairman, Donald Hearth, to write a position paper on "a mid-'70s decision leading to a manned Mars landing in the mid-1980s." Hearth's target date of 1986 gained influence as it made its way upward through the planning process. In July, a draft report for the STG, which reflected Newell's work, proposed "that the United States begin preparing for a manned expe-

---

52. NASA SP-196, pp. iii, 5, 12; Logsdon, *Apollo*, chapter 4, pp. 16-17, 18, 26.
53. Logsdon, *Apollo*, Chapter 4, pp. 27-28.

## THE SPACE SHUTTLE DECISION

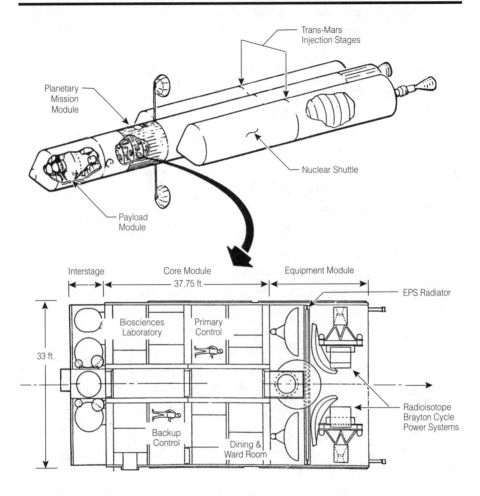

*Design for a nuclear-powered spaceship for an expedition to Mars. (North American Rockwell)*

dition to Mars at an early date." In turn, such a goal could provide a focus for decisions on future programs.[54]

Though Mueller also was interested in Mars, he disliked the overall approach of Newell's planning. In sowing the Earth-Moon system thickly with space stations, Newell was repeating the basic theme of the ambitious Apollo Applications wish list of a few years earlier. Mueller knew from sad experience that this would not work. He had tried it, and all it had produced was Skylab.

---

54. *Ibid.*, pp. 19-20, 23-24, 48.

Newell's approach smacked of building space stations for their own sake, as if NASA was the Bureau of Reclamation with its penchant for dams and water projects. The stations also failed to point a clear path toward Mars.

Mueller, working with the planners at Bellcomm, was devising both an integrated program and an integrated set of projects that could carry it out. This plan resembled LeRoy Day's for the shuttle. Day's plan held a space base as a long-term goal but asserted that the shuttle would pay its way in the nearer future by launching and servicing payloads. Similarly, while Mueller and Bellcomm aimed specifically at Mars, they expected to get there by using rockets and spacecraft that would serve a broad range of activities between the Earth and the Moon. The key was breadth of application. A version of the plan presented to STAC that July stated that "the program of developments and flight activities that comprise the integrated space program will expand this nation's capacity for space flight as far as foreseeable development in technology will permit."

In addition to the Space Shuttle and a space station module, the plan called for three new program elements:

*Space Tug:* This would serve as a general-purpose vehicle that would be based in space, returning to Earth only at rare intervals, if at all. It would draw on the ability of the shuttle to carry propellants in substantial quantities. The tug would operate as a "utility propulsion module capable of transporting men, spacecraft and equipment throughout cislunar space." It also would provide a ferry from a lunar-orbiting space station to the Moon's surface.

*Astronaut-Tended Spacecraft:* These would include large automated telescopes in orbit as well as automated applications satellites. They would be designed to take advantage of the shuttle's capacity for revisit and on-orbit maintenance, including the installation of upgraded instruments by the flight crews.

*Nuclear Shuttle:* This reusable rocket would rely on the NERVA nuclear engine. It would operate between low Earth orbit, lunar orbit, and geosynchronous orbit, with its exceptionally high performance enabling it to carry heavy payloads and to do considerable amounts of work with limited stores of liquid-hydrogen propellant. In turn, the nuclear shuttle would receive this propellant from the Space Shuttle.

The Space Shuttle, space tug, and nuclear shuttle together would constitute a complete reusable space transportation system, with the tug and nuclear

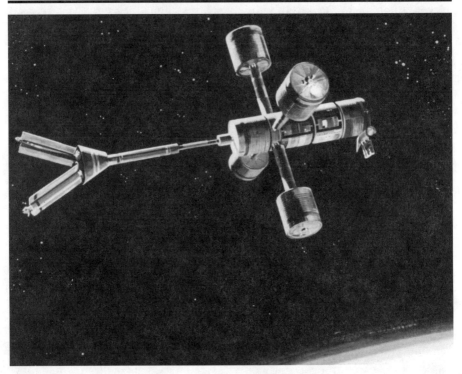
*Space base as envisioned in the late 1960s. (North American Rockwell)*

shuttle both based in space. In turn, this system would serve an array of programs that would focus on applications of a space station module. Variants of this module would operate as components of a large space base, as stations in lunar orbit and on the Moon's surface, and as a geosynchronous station.

What of Mars? A piloted mission could use this same equipment, with the nuclear shuttle providing propulsion for a spaceship that would draw again on the basic space station module. Mueller, like Newell, looked toward such a mission in 1986, defining it also as an ultimate goal. Unlike Newell, however, he expected to get to Mars with equipment that could find plenty of uses closer to home. Just as Day's space shuttle might earn its keep even in the absence of a space station, Mueller's integrated plan would serve the nation even if NASA never received permission to send astronauts to the Red Planet.[55]

---

55. *Ibid.*, pp. 25-26, 27, 30-32, 48-49.

Mueller's integrated plan came along a little too late to serve as the basis for a July draft of NASA's report to the STG, setting forth the agency's proposals. That draft drew principally on Newell's work, adding a discussion of Mueller's plan as nothing more than an appendix. Within a week, however, the appendix became the core of the report. The reason for this, as one might expect, was Tom Paine.

Newell's Planning Steering Group (PSG) had briefed Paine on its activities at the end of May. A dissatisfied Paine saw the report as tantamount to stapling together the contributions of its task groups, for the PSG at that moment had nothing so succinct as a clear emphasis on space stations. Paine was not pleased "with the level of imagination and the level of innovation and the level of forward thrust"; he described the recommendations as "good, workmanlike, but stodgy and unimaginative."

But he reacted quite differently when he received a briefing from Mueller. Mueller's plan proved easy to sell; he recalls that he had to "spend more time making sure my facts were right than convincing Tom." At a June 24 planning review meeting, Paine personally directed Newell to accept Mueller's plan and to use as much of it as possible in preparing NASA's report to the STG. By then there was not much difference between the two plans; both featured space stations galore along with lunar stations and a trip to Mars. Mueller, however, had the more convincing scenario as to how it might happen, for he made Mars appear to grow out of ongoing future activity in a natural way. By contrast, Newell made Mars appear more *ad hoc*.[56]

Yet while Mueller's plan offered exciting theater, it flew in the face of the demand by the Townes panel for redress of the imbalance between the piloted and automated elements of NASA's program. These corresponded respectively to OMSF and OSSA. With Newell having come out of OSSA to direct his NASA-wide planning activity, there had been at least a chance that NASA might respond to Townes' call. When Mueller, head of OMSF, took over the planning with a set of proposals that grew out of his ties to Bellcomm and STAC, it was clear that Paine would go to the STG with an agenda that would be virtually all OMSF. Paine then would learn that there were plenty of people, in both the Air Force and in the scientific community, who opposed a

---

56. *Ibid.*, pp. 22-25, 32-33; interview, Thomas Paine, September 3, 1970, p. 3.

single-minded emphasis on piloted flight, and who certainly were in no hurry to get to Mars.

Robert Seamans was a fellow member of the STG, co-equal with Paine in standing. His report carried a summary that rejected virtually all of Mueller's work. He willingly endorsed and even emphasized the value of a space shuttle:

*I recommend that we embark on a program to study by experimental means including orbital tests the possibility of a Space Transportation System that would permit the cost per pound in orbit to be reduced by a substantial factor (10 or more).*

But he rejected building even one space station, let alone several:

*Even though the development of a large manned space station appears to be a logical step leading to further use and understanding of the space environment, I do not believe we should commit ourselves to the development of such a space station at this time.*

He had similar hard words about flight to Mars: "I don't believe we should commit this Nation to a manned planetary mission, at least until the feasibility and need are more firmly established."[57]

Lee DuBridge, the third member of the STG, developed his own view as well. As chairman of the President's Science Advisory Committee (PSAC), he commissioned a report from the PSAC's Space Science and Technology Panel. He also issued invitations to other institutions to present their own positions. The respondees included the American Institute of Aeronautics and Astronautics (AIAA), the nation's principal professional society in this field.

The AIAA might readily have done the expected by acting as a standard Washington lobby, pleasing its corporate members by asking for the Moon. It did no such thing; it issued a report that was remarkable for its moderation.

Flight to Mars? "This program is the next major step after unmanned exploration of the planets. While it is technically possible to commit to development of vehicles for this program at this time, it would not be reasonable to do so."

---

57. Letter, Seamans to Agnew, August 4, 1969. Reprinted in NASA SP-4407, vol. I, pp. 519-522.

## Mars and Other Dream Worlds

Space stations? "Again, there has been a constant flow of studies for 10 years of small and large space stations, most of which assumed that man was to find an abundance of tasks to perform continually in space." The AIAA declined to endorse any such proposals, opting instead for additional activity within Apollo Applications. Its report noted that such missions "will provide a means of retaining a manned orbital capability until flight experience produces desirable specifications for new orbital-station hardware."

However, concerning space shuttles, the AIAA had a different view:

*We encourage early steps be taken to commit to flight demonstration a partially reusable low-cost space-transportation system which could start initial operations in the 1974-76 period.... For example, most versions of low-cost space-transportation systems can effectively compete with present expendable boosters in delivering medium to large unmanned payloads to orbit....* We consider that commitment to an entirely new space station is less urgent than commitment to a new logistics system [emphasis in original].[58]

One of a number of papers that reached the PSAC, the AIAA report's moderate tone gave it weight. The PSAC also gave attention to Russell Drew, a staffer at the Office of Science and Technology. Drew became convinced that the Shuttle was the key element in the long-range program. In May, the PSAC's Space Science and Technology Panel met with the full PSAC at Cape Canaveral, with the Space Shuttle as a prime topic of discussion. Drew wrote a background paper that emphasized its use in launching automated spacecraft. Significantly, he downplayed its uses in supporting a space station.

The PSAC report made recommendations that closely resembled those of Seamans. Seamans had proposed an expansion of Apollo Applications; the PSAC did so as well, asserting that such a program could provide much of the data on long-duration human space flight "for which a more ambitious space station has been proposed." PSAC rejected such a station, and proposed merely to "keep open the option of manned planetary exploration in the 1980's, but without immediate commitment to this goal."

The PSAC, however, had kind words for the Space Shuttle:

---

58. *Astronautics & Aeronautics*, July 1969, pp. 39-46.

# THE SPACE SHUTTLE DECISION

> *Study, with a view to early development, a reusable space transportation system with an early goal of replacing all existing launch vehicles...with a system permitting satellite recovery and orbital assembly and ultimately radical reduction in unit cost of space transportation.*[59]

It is worth noting that PSAC, AIAA, NASA, and the Air Force all endorsed a shuttle. NASA had developed this concept for use in space station logistics, but the other three institutions rejected such a station out of hand. Nevertheless, they liked the Shuttle because it seemed to promise lower cost and greater effectiveness in supporting automated spacecraft. This unanimity gave critical political support to NASA during 1970 and afterward.

In their overall views, however, the members of the STG were badly split. There was good agreement between the views of Seamans and of DuBridge, but Paine seemed out in the blue sky. This raised the possibility that DuBridge and Seamans might collaborate on a majority report, leaving Paine to tack on his views as an appendix, as Francis Clauser had done with his letter to the Townes panel. The STG, however, had a fourth member: its chairman, Spiro Agnew, who had been developing his own views as well.

## *Agnew Leads a Push Toward Mars*

Like many other children of immigrants, Agnew had advanced in the world largely through his own talent and effort. On the eve of World War II, he had been a claims adjuster for the Lumbermen's Mutual Casualty Company in Baltimore. After returning from service in the Tenth Armored Division, he became a manager at Schreiber Food Stores. He acquired a law degree and launched his political career by winning the presidency of his local PTA. He rose in politics through the next 20 years, and ran for governor of Maryland in 1966. It was a time of racial upheaval, and his Democratic opponent courted the votes of white people with thinly-disguised racial appeals. Agnew built a reputation as a moderate, won support among black voters, and took the election.

He maintained his role as a political moderate, becoming a leading supporter of Nelson Rockefeller, governor of New York, who competed with Nixon for the 1968 GOP presidential nomination. Nixon picked him as his

---

59. Logsdon, *Apollo*, chapter 4, pp. 33-38; DuBridge, chairman, *Post-Apollo*.

running mate largely because he had not been around long enough to draw opposition from any of the party's factions. Agnew himself admitted that his name was "not a household word."

He became better known during the campaign due in part to his personal coarseness. He called Poles "Polacks," referred to a Japanese news reporter as "the fat Jap," and declared that Hubert Humphrey, the Democratic presidential nominee, was "soft on communism." Campaigning in Detroit— certainly a place where discretion was advisable—he remarked that "if you've seen one city slum, you've seen them all." As his reputation spread, protesters began to greet him with signs such as one that read "Apologize Spiro, It Will Save Time Later."[60]

This insensitivity was part of a larger artlessness, for in dealing with the space program he quickly showed that despite having no background in this area, he would rush in where leaders such as Robert Seamans and Lee DuBridge would fear to tread. At Cape Canaveral, six weeks after the inauguration, he stated at a luncheon that he was "all-out for space." Less than three weeks later, at a meeting of the STG, he suggested that Mars could be an Apollo-like goal for the 1970s. He then raised the prospect of Mars repeatedly in subsequent discussions.

In mid-May, the STG met with members of the House and Senate who served on the congressional space committees. In the words of an observer from the Budget Bureau, "a promotional motive...ran virtually unchecked" at the meeting. Agnew declared that the nation could "prove its greatness" with the space program. He added that he "might be all alone," but he "favored a great achievement represented by planetary travel as a way of invigorating the American public."

Early in July, at a meeting with space planners from outside the government, he stated that "a manned spaceflight to Mars" could be the "overture to a new era of civilization." He compared this mission to the early voyages of exploration: "Would we want to answer through eternity for turning back a Columbus or a Magellan? Would we be denying the people of the world the enlightenment and evolution which accompany every great age of discovery?"

Thus far he had restricted his comments to audiences at STG meetings. On July 16, however, as he joined thousands of people at Cape Canaveral for

---

60. Manchester, *Glory*; White, *1968*; see index references.

## THE SPACE SHUTTLE DECISION

the liftoff of Apollo 11, he went public. He stated in an interview that he had the "individual feeling that we should articulate a simple, ambitious, optimistic goal of a manned flight to Mars by the end of this century."[61] This statement brought quick reaction within NASA, as Paine directed his planners to come up with a "very strong, very far-out, but down-to-earth presentation" that would "substantially shake up the STG." Specifically, these planners were to prepare a proposal for a Mars mission at a date well before 1986, a date that Mueller and Newell had previously endorsed in their planning.

Mueller's planners at Bellcomm, who had already devised a concept for a Mars mission based on Mueller's integrated plan, briefed Paine on their scenario on July 19. Characteristically, Paine wanted more, and decided to "wheel up NASA's big gun": Wernher von Braun. He told von Braun to prepare a presentation for the STG on the feasibility of a Mars mission that would resemble Bellcomm's, but that would fly at the earliest possible date.

Such studies had been a specialty of the house at NASA Marshall, which von Braun headed. In his words, "it was an effort of a very few weeks to put a very consistent and good and plausible story together as to how we would use these new elements to go to Mars." Paine's directive suited him personally as well, for as he said a year later,

> *I have been a space man ever since I was a child, and I think I would be betraying my profession if I were to tell you that we should not send men to Mars. I think we should and we will, and I am all for the finest and the most energetic space program we can imagine.*[62]

What brought this sudden focus on Mars? Paine would describe it as a matter of timing:

> *Had we done this in the first meeting of the Space Task Group, we would really have shot our wad too early. Had we waited until too late, the darn thing would have been cast in concrete and we wouldn't have had the opportunity. I felt that this was the right time. Everyone had listened to various proposals. We had listened to everybody; the time had come for us to come*

---

61. Logsdon, *Apollo*, chapter 4, pp. 12-13, 39-40.
62. *Ibid.*, pp. 40-43, 48-50; interview, Thomas Paine, September 3, 1970, pp. 9-10; John Logsdon interview, Wernher von Braun, pp. 11, 13.

## Mars and Other Dream Worlds

*out with a powerful forward look of our own and say, "We really haven't seen the proposal that we can carry to the President for the future NASA program, but we now feel that we have a specific one to lay before you."*[63]

Timing was important in other ways as well, for the early months of the Nixon administration coincided with the brilliant successes of Apollo 8 and Apollo 11. Less than a month after Apollo 8 had carried the first astronauts to orbit the Moon, Nixon had spoken of them in his inaugural address:

*Only a few short weeks ago, we shared the glory of man's first sight of the world as God sees it, as a single sphere reflecting light in the darkness.*

*As the Apollo astronauts flew over the moon's gray surface on Christmas Eve, they spoke to us of the beauty of earth—and in that voice so clear across the lunar distance, we heard them invoke God's blessing on its goodness.*

*In that moment, their view from the moon moved poet Archibald MacLeish to write: "To see the earth as it truly is, small and blue and beautiful in that eternal silence where it floats, is to see ourselves as riders on the earth together, brothers on that bright loveliness in the eternal color—brothers who know now that they are truly brothers."*[64]

Amid the glow of that triumph, Paine had tried to win a budget increase and to gain Nixon's support for a space station. Now in July, the Apollo 11 lunar landing encouraged even more far-reaching thoughts. In addition to this, the success of Apollo 11 had the highly practical consequence of freeing up the first Saturn V for other duty, as Paine committed one of them to launch the Skylab dry workshop.[65] Such broadened use of the Saturn V would be a keystone of an effort that would aim at Mars.

The members of the STG—Paine, DuBridge, Seamans, and Agnew—had not held a full-dress meeting since March 1969, for their staffs had been busy preparing proposals and working papers. On August 4, however, two weeks after the Moon landing, these principals met anew. Paine declared that "Apollo 11 started a movement that will never end, a new outward movement in which man will go to the planets, first to explore, and then to occupy and

---

63. Interview, Thomas Paine, September 3, 1970, p. 11.
64. Nixon, Inaugural Address, January 20, 1969.
65. NASA SP-4208, pp. 109-110.

utilize them." He then turned the meeting over to von Braun, who proceeded to describe a "typical manned Mars mission."

Carrying 12 astronauts, it would leave Earth orbit in two spaceships on November 12, 1981. It would arrive in Mars orbit on August 9, 1982 and would spend 80 days in this orbit, with six crew members descending to the Martian surface and spending up to two months exploring it. The expedition would leave Mars late in October and would swing past Venus four months later. It would arrive back in Earth orbit on August 14, 1983, ending a flight of 640 days.

Yet despite its boldness, this mission would rely mostly on the equipment of Mueller's integrated plan. Nuclear rockets, essential for propulsion, would duplicate Mueller's nuclear shuttle used for flights between Earth orbit and lunar orbit. The Mars ships would be variants of a standard space station module. The only major new item would be a Mars Excursion Vehicle, to carry crews from orbit to the surface of that planet.[66]

This proposal left the STG split right down the middle: Agnew and Paine supporting a strong push toward Mars, DuBridge and Seamans recommending much less. There was a fifth man at that meeting: Robert Mayo, director of the Budget Bureau. Though he was an observer rather than a full member of the STG, his views would carry weight. His staff had been considering proposals as well. They emphasized the need not for a single program, but for alternative programs with budgets at different levels. Mayo now found an ally in Seamans, who had been highly skeptical of Paine's ambitious plans.

DuBridge did not emphasize his own point of view, but tried to encourage a compromise. None was within reach; indeed, while the STG had hoped to recommend a single program to the White House, there was no chance it could agree on one. The disagreements ran deep; Seamans later said he was "sort of like a skunk at a garden party" for opposing Paine, while Agnew told Mayo that he was "nothing but a cheapskate." Nevertheless, the group could follow Mayo's recommendation, which was in line with a standard Washington practice. Rather than continue to seek the elusive single set of recommendations, the STG decided to prepare three program options, which the staffer Russell Drew described:

---

66. Logsdon, *Apollo*, chapter 4, pp. 51-52.

## Mars and Other Dream Worlds

1. "Austere": Level funding at $4 billion per year, with no commitment to Mars but with an option for such a mission retained.
2. "Intermediate": Funding increasing over the next five years to $5-6 billion per year, with a commitment to Mars. This commitment would carry no fixed date, but the mission would probably fly in the mid- to late-1980s.
3. "Vigorous or all-out": Funding increasing to $7 billion per year in the mid-1970s and possibly to $8-10 billion in the latter half of the decade, with a commitment to an early Mars mission.[67]

This was a major victory for Paine. This three-option package again reflected standard practice, with the one in the middle as the one for the President to choose. The other two choices then would appear as too much and too little. The STG's "intermediate" program specifically envisioned a commitment to Mars, with all that would entail: a space shuttle, space station, space tug, and nuclear shuttle. The only questions would involve the pace, schedule, and budget.

Paine, quickly following up this victory with another, won the assignment of preparing the details of the three options. This played to a long-established art whereby the officials chosen to write a White House report can often shape it to suit their preferences. Rather than provide three different programs, Paine's staffers proceeded to develop three different schedules for the *same* plan. That plan took Mueller's integrated scenario as its point of departure, with all three aiming at a piloted Mars mission sometime during the 1980s.

Plan A offered the "maximum progress technically feasible." NASA described it as "comparable to the 1961 Apollo decision to go to the Moon." Closely resembling the plan that Paine and von Braun had presented to the STG on August 4, it called for a mission to Mars in 1981. Plan B, offering "maximum returns from an economical program," was the one Paine hoped to have approved; it differed from Plan A largely in slipping the Mars mission to "1983 or 1986." Plan C offered "minimum investment consistent with continuing technological advance." It also retained the full Mueller program, delaying the Mars mission only to 1986 or 1989. The three alternatives featured dates as follows:

---

67. *Ibid.*, pp. 55-60; John Logsdon interview, Robert Seamans, Washington, September 2, 1970, p. 13.

## THE SPACE SHUTTLE DECISION

|  | Plan A | Plan B | Plan C |
|---|---|---|---|
| **Space Transportation** | | | |
| Space shuttle | 1975 | 1976 | 1977 |
| Space tug | 1976 | 1978 | 1981 |
| Nuclear shuttle | 1978 | 1978 | 1981 |
| **Piloted Space Flight** | | | |
| Space station, crew of 12 | 1975 | 1976 | 1977 |
| Space base, crew of 50 | 1980 | 1980 | 1984 |
| Space base, crew of 100 | 1985 | 1985 | 1990 |
| Lunar orbiting station | 1976 | 1978 | 1981 |
| Lunar surface base | 1978 | 1980 | 1983 |
| First expeditions to Mars | 1981 | 1983 | 1986 |

These plans were ready in mid-August. NASA then added a fourth option, Plan D; it excluded developments related specifically to the Mars expedition. In other respects it was identical to Plan C. Plan D, the least ambitious, called for simultaneous development of a space station and a space shuttle, with both becoming operational in 1977.[68]

Significantly, amid the deliberations of the STG, the members with the most experience—Seamans and DuBridge—favored the most modest initiatives. Paine, the man for Mars, had held his posts within NASA for barely a year and a half. Agnew, most enthusiastic of all, had never dealt with space at any serious level; he hardly knew a rocket from a sprocket. Both Agnew and Paine were living in a dream world.

The nation had changed since 1961. The circumstances that had led to Apollo no longer applied. America now faced new issues and new concerns, and to such a degree as to make even the Moon landing appear merely as an irrelevant distraction. In no way would Nixon endorse a mission to Mars. Indeed, within this new climate, even Plan D would prove to be out of reach.

---

68. Logsdon, *Apollo*, chapter 4, pp. 60-63; Newell, chairman, *America's Next Decades*, pp. 59-60.

# CHAPTER FOUR

# Winter of Discontent

On an afternoon in July 1969, while the Apollo 11 mission stood poised for a flight to the Moon, Tom Paine found himself confronted by a group of civil rights demonstrators. Their leader was Reverend Ralph Abernathy, president of the Southern Christian Leadership Conference. Abernathy had succeeded Martin Luther King in that post, following the death of King a year earlier. Abernathy now came to Cape Canaveral on the eve of NASA's triumph.

A light mist of rain fell intermittently, as thunder rumbled in the distance. Paine stood coatless under a cloudy sky, accompanied only by NASA's press officer, as Abernathy approached with his party, marching slowly and singing "We Shall Overcome." Several mules were in the lead, as symbols of rural poverty. Abernathy then gave a short speech. He deplored the condition of the nation's poor, declaring that one-fifth of the nation lacked adequate food, clothing, shelter, and medical care. In the face of such suffering, he asserted that space flight represented an inhuman priority. He urged that its funds be spent to feed the hungry, clothe the naked, tend the sick, and house the homeless.

Paine replied that "if we could solve the problems of poverty by not pushing the button to launch men to the Moon tomorrow, then we would not push that button." He added that NASA's technical advances were "child's play" compared to "the tremendously difficult human problems" that concerned the SCLC. He offered the hope that NASA indeed might contribute to addressing these problems, and then asked Abernathy, a minister, to pray for the safety of the astronauts. Abernathy answered with emotion that he

151

# THE SPACE SHUTTLE DECISION

would certainly do this, and they ended this impromptu meeting by shaking hands all around.[1]

Their brief conversation brought no lasting consequence. Yet it was heavy with history, for Paine and Abernathy stood as representatives of two deep themes that had marked the nation's experience before America even existed.

Paine was the technologist, heir to a record of splendid accomplishment. His forebears had built ships, constructed transcontinental railroads, dug the Panama Canal, captured water to allow cities to grow in the arid West, flung power and telephone lines from coast to coast. They had built highways and factories, had put the nation on wheels, had mastered the art of flight. At that very moment, others were winning achievement in the realm of computers.

There was, however, another and far more somber side to America's history, for the nation had been conceived in the original sin of slavery. Abraham Lincoln had proposed that "every drop of blood drawn with the lash shall be paid by another drawn with the sword"; yet the stain ran so deep that not even the Civil War could expunge it. Like Lincoln, Martin Luther King had grappled with this sin, had sought the moral authority to sway a deeply divided people; and like Lincoln, he had paid with his life, with his goal only partly won.

## *The Sixties*

"The legacy of Apollo has spoiled the people at NASA," Wernher von Braun remarked in the wake of the Moon landing. "They believe that we are entitled to this kind of a thing forever, which I gravely doubt. I believe that there may be too many people in NASA who at the moment are waiting for a miracle, just waiting for another man on a white horse to come and offer us another planet, like President Kennedy."[2]

In 1969, NASA still lived in the shadow of Kennedy, both in its immediate concern with Apollo and in its institutional hopes. Apollo had taken form as an initiative in foreign policy. It could hardly have been otherwise; Kennedy was very much a cold warrior, who had devoted his inaugural address entirely to foreign affairs. There was a reason for this overriding con-

---

1. Paine, Memo for Record, July 17, 1969.
2. John Logsdon interview, Wernher von Braun, Washington, pp. 18-19.

cern: Kennedy, like his party, carried a heavy burden. The party governed under its own shadow, for they had held both Congress and the White House when China fell to communism in 1949.

It is difficult to overstate the dismay with which America faced the communist threat of the postwar years. It was almost as if to say that our victory in the war was meaningless, that we had defeated Japan and Germany only to face the far greater power of Stalin and Chairman Mao. Less than a year after Mao proclaimed the People's Republic of China, the U.S. was at war in Korea, a war that President Truman would find himself neither able to win nor to end. In turn, this war drove him from office. At home, fear of communism encouraged the excesses of Senator Joseph McCarthy and his allies, who recklessly smeared the reputations of good and decent people because of their political beliefs and activities, real or alleged.

It was the proud boast of Eisenhower's Republicans that while Truman had lost not only China but Eastern Europe, they had held the line. They had ended the Korean War, and had preserved peace amid subsequent dangers in a perilous world. Kennedy's main challenge was to continue to hold this line, to deny Moscow and Beijing any further victories. Under the shadow of China, however, he would not proceed with the calm confidence that had marked Eisenhower and his policies. Living in that shadow, Kennedy's Democrats would find themselves driven to become more anti-communist than the Republicans. In conducting foreign policy, they worked amid gnawing concern that they might prove to be weak, and would compensate by becoming overly bold.[3]

The most important consequence was the war in Vietnam. When the French faced defeat in their struggle against Ho Chi Minh in 1954, Ike had had his chance to intervene massively in that country. Declining to do this, he had left the French to their fate. But Vietnam was adjacent to China, in the one area of the world where further communist advance was both most likely and most unacceptable. Kennedy and his advisors accepted the domino theory, which viewed South Vietnam as a linchpin: if it fell, the whole of Southeast Asia would soon go as well. In 1961, General Lyman Lemnitzer, chairman of the Joint Chiefs of Staff, warned that if Saigon were to fall, "we would lose

---

3. Heppenheimer, *Countdown*, pp. 177-188 treats the background to Kennedy's commitment to Apollo. See also Logsdon, *Decision*.

## THE SPACE SHUTTLE DECISION

Asia all the way to Singapore." Kennedy, accepting this view, made it a basis for policy.[4]

Waging total cold war, Kennedy believed that it was essential to deny Moscow propaganda victories as well as military ones. A prime topic for propaganda was space flight, and in no way would Kennedy concede that the Soviets might concentrate resources into this area while failing their citizens in a host of ways that were far more important. The issue was one of national prestige, what in earlier times had been known as national honor: if the world viewed space as important and saw that the Soviets were ahead, then America would have to meet this challenge and take the lead. Time and again, during the campaign of 1960, Kennedy spoke of other nations and emphasized that leadership in space was essential if America was not to forfeit their support:

> *The people of the world respect achievement. For most of the twentieth century they admired American science and American education, which was second to none. But now they are not at all certain about which way the future lies. The first vehicle in outer space was called Sputnik, not Vanguard. The first country to place its national emblem on the moon was the Soviet Union, not the United States.*
>
> *If the Soviet Union was first in outer space, that is the most serious defeat the United States has suffered in many, many years. Because we failed to recognize the impact that being first in outer space would have, the impression began to move around the world that the Soviet Union was on the march, that it had definite goals, that it knew how to accomplish them, that it was moving and we were standing still. That is what we have to overcome, that psychological feeling in the world that the United States has reached maturity, that maybe our high noon has passed and that now we are going into the long, slow afternoon.*[5]

Ike had refused to be drawn into war in Vietnam, leaving that commitment to Kennedy. At a cabinet meeting in December 1960, Ike had also declined a commitment to the Moon, turning down a specific plan that closely resembled the eventual Apollo.[6] When Kennedy accepted that challenge, only

---

4. Fall, *Hell*, pp. 293-313; Manchester, *Glory*, pp. 915-923.
5. McDougall, *Heavens*, pp. 221-222.
6. Logsdon, *Decision*, pp. 34-35.

five months later, the Moon held a threefold significance. It represented a simple and dramatic goal that everyone could understand. It appeared reachable during that decade, and would not impose a prolonged effort that might lose public interest. In addition to this, the Moon was demanding enough to call for an entirely new array of launch vehicles and spacecraft, requiring far more power than the Soviet rockets of the day could provide. The Soviet lead in rocketry would not help them; like the Americans, they would have to start afresh. Kennedy believed, correctly, that in the resulting competition the U.S. would prove more capable in coming up with the enormous sums of money that would be necessary to reach the Moon.

As the decade of the 1960s progressed, the cold war lost its sense of imminent threat. In 1961, Nikita Khrushchev had provoked a crisis in Germany, and had built the Berlin Wall. By 1968, however, the Democrats could say that they too had held the line. By then nearly 20 years had elapsed since the fall of China had given communism its last major territorial advance. The Soviets had been stymied in Europe; America and its NATO allies had protected West Berlin, even though that city was entirely surrounded by communist territory. Though Fidel Castro ruled Cuba, he had failed to spread his revolution elsewhere in the Caribbean or in Latin America. In addition to this, communism had received a severe setback in Southeast Asia in 1965, for General Suharto of Indonesia broke an attempted communist takeover and went on to crush his country's communist party.[7]

In 1968, the nation was at war in Vietnam. During February, amid the new year celebrations known as Tet, that country's communist forces launched a massive and widespread series of attacks. Battles raged in Saigon, where they penetrated the grounds of the American embassy. They captured the city of Hue, an ancient capital, and held it for several weeks. They laid siege to a Marine base, Khe Sanh, pounding it with mortars and artillery. Dozens of cities came under assault.

As a military engagement, this Tet Offensive failed. Powerful counterattacks routed the communists, retaking Hue, while the Marines held Khe Sanh. As a political exercise, however, the offensive succeeded brilliantly. It drove home the fact that North Vietnam was in the war to stay and would not be defeated by any means short of additional massive escalation. In 1961,

---

7. Heppenheimer, *Countdown*, pp. 196-197; Johnson, *Modern*, pp. 479-480.

## THE SPACE SHUTTLE DECISION

Kennedy had declared that America would "pay any price, bear any burden" to prevail. By 1968, it was clear that the nation would do nothing of the sort, at least not in Vietnam. In the wake of that offensive, the question facing America was not how to win, but how to withdraw. In turn, this reflected the waning of foreign affairs as a paramount concern, for withdrawal clearly meant that the nation would leave the battlefield on terms short of victory.[8]

While foreign affairs lost their life-and-death character, the public turned to domestic concerns with considerable passion. Now these issues that had languished since the late 1930s, amid wars and military preparations, would have their day. Foremost among them was race.

We remember the 1960s for the civil rights revolution. Its roots, however, went back an additional decade, and embraced all three branches of the federal government. In 1954, the Supreme Court showed that it would rule unanimously in upholding the rights of black America, as Chief Justice Earl Warren led his associate justices in handing down the landmark ruling, *Brown v. Board of Education*, that struck down the segregation of schools. Three years later, President Eisenhower showed that he would enforce a desegregation order using federal troops, as he sent elements of the 101st Airborne Division to quell a dangerous mob in Little Rock, Arkansas. Also in 1957, Senate Majority Leader Lyndon Johnson rallied two-thirds of his fellow senators to break a filibuster and enact a civil rights bill. Though the bill was weak, its significance was great; it was the first such measure enacted since Reconstruction.[9]

In the lives of most black and white people, however, nothing had changed. Though the Supreme Court ruling represented binding precedent as case law, it lacked the force of a federal statute. Federal civil rights law remained so weak that the Justice Department lacked the legal standing to initiate lawsuits aimed at achieving desegregation. The civil rights movement had an episodic character; when Ike sent troops to Little Rock, for instance, that city's crisis ended as quickly as if the Seventh Cavalry had come riding to the rescue in a John Wayne movie. Similarly, when Kennedy sent a federal force against armed white rioters at the University of Mississippi in 1962, this news story blazed up and died in a matter of days. Such events made it easy

---

8. *Time*, February 9, 1968, pp. 15-16, 22-33; *Newsweek*, February 12, 1968, pp. 23-33; Manchester, *Glory*, pp. 1124-1126; White, *1968*, pp. 3-5, 10-13; Tuchman, *Folly*, pp. 348-352.
9. Manchester, *Glory*, pp. 734-737, 799-809; Branch, *Parting*, pp. 220-222.

to believe that all was well, that federal marshals would preserve order, and that America could continue without fundamental change.

Then in April 1963, Martin Luther King took his movement to Birmingham, Alabama, which he described as "the largest segregated city in the United States." Opposing him was the city's powerful police commissioner, Eugene "Bull" Connor, an ardent racist. King launched a succession of protest marches and demonstrations that grew in size as the month progressed; Connor struck back by arresting and jailing the demonstrators. King himself became a prisoner; still the protests continued to grow. By early May, Connor had literally run out of jail cells, and when the demonstrations continued, he lashed at them with police dogs and with fire hoses forceful enough to peel bark from a tree.

Television networks had been covering Birmingham as an ongoing news story, and now they showed their power. When viewers saw nonviolent protesters under attack by vicious dogs and equally vicious police, the nation shuddered in dismay. This marked a breakthrough in the cause of civil rights, for that movement now held America's full attention, and would not let it go. A month later, Kennedy himself addressed the nation, calling for a sweeping law that would protect the rights of black citizens. Kennedy took this stand before the election of 1964 and not after, for he expected to win a second term. In turn, his reelection was to vindicate his leadership on this most controversial of issues.[10]

The historian Bruce Catton writes that during the Civil War, newly-freed blacks "were men coming up out of Egypt, trailing the shreds of a long night from their shoulders." For many of their descendants, the passage of a century had brought little change. Thus in 1964, a black woman named Fannie Lou Hamer told of her attempt to register to vote as a resident of Mississippi:

> *I was carried to the county jail. I was placed in a cell. After I was placed in the cell I began to hear sounds of licks and screams. I could hear the sounds of licks and horrible screams, and I could hear somebody say, "Can you say, 'Yes sir,' nigger? Can you say 'Yessir'?"*
>
> *They beat her, I don't know how long, and after awhile she began to pray and asked God to have mercy on these people.*
>
> *And it wasn't too long before three white men came to my cell.*

---

10. Manchester, *Glory*, pp. 943-952, 976-978; White, *1964*, pp. 199-215.

> *I was carried out of the cell into another cell where they had two Negro prisoners. The State Highway Patrolman ordered the first Negro to take the blackjack.*
>
> *The first Negro prisoner ordered me, by orders from the State Highway Patrolman, for me to lay down on a bunk bed on my face, and I laid on my face.*
>
> *The first Negro began to beat, and I was beat until he was exhausted. The State Highway Patrolman ordered the second Negro to take the blackjack. The second Negro began to beat and I began to work my feet. I began to scream, and one white man got up and began to beat me on the head and tell me to "hush."*
>
> *All this is on account we want to register, to become first-class citizens.*[11]

Yet if federal legislation could extirpate such evils, the nation now would certainly make the attempt. The Civil Rights Act, which became law in mid-1964, proved to be only the beginning. A year later, Congress complemented it with a far-reaching Voting Rights Act. In turn, these laws were part of a surge of domestic legislation that was virtually unparalleled. Trust in government was at a peak, and President Johnson, supported by powerful majorities within a willing House and Senate, would make the most of this.

Aid to education topped his list of priorities; over 40 bills dealt with this topic. Congress enacted a law establishing Medicare, which complemented Social Security in addressing the needs of retirees. Johnson had declared war on poverty; Congress responded with a law that set up a new Office of Economic Opportunity, with the rural poor of Appalachia as a particular concern. Other bills established a National Foundation for the Arts and Humanities and a Cabinet-level Department of Housing and Urban Development. Still others fought heart disease, stroke, and cancer. A new immigration law opened the door to newcomers from Asia, heralding a change in the centuries-old predominance of immigration from Europe. To pay for it all, Johnson won a major tax cut that would stimulate economic growth.[12]

Johnson was not about to promote these new programs at the expense of existing ones; hence NASA and Apollo would receive their due. As the nation turned its attention toward these domestic concerns, however, it became

---

11. Catton, *Stillness*, p. 259; White, *1964*, pp. 332-333.
12. Manchester, *Glory*, pp. 1041-1044; White, *1964*, pp. 470-476.

increasingly clear that Apollo represented a response to a Soviet challenge that was about to run its course. Apollo was a creation of its time, and by decade's end that time had come and gone. Events soon demonstrated that Apollo was a program that the nation would neither renew nor long continue. In turn, these events weighed heavily upon Paine's pursuit of Mars. They took the form of budget cuts, imposed within the BoB.

## *Mars: The Advance*

At Gettysburg in 1863, General George Pickett led a charge that reached the top of Cemetery Ridge, only to be driven back by superior strength. NASA's pursuit of Mars would show a similar character, with the contested ground being the budget allocation for FY 1971. NASA accounted for some two percent of the federal budget. While this was far below the allocations of the Pentagon or Health, Education, and Welfare, it was enough to justify the continuing attention of small groups of staffers within both the White House and the BoB.

Peter Flanigan, Assistant to the President, served as the White House link to NASA. He reported directly to Nixon and was one of the more powerful of the presidential assistants. Flanigan had been a Wall Street investment banker; his father had been chairman of Manufacturers Hanover Trust. Following Nixon's election, Flanigan had drawn on his broad social and professional acquaintances and had recruited some 300 appointees for high-level administration positions. His White House responsibilities were correspondingly broad, and he relied on five staff assistants. These included Clay Whitehead, a graduate of MIT, who dealt with the space program as part of his day-to-day concerns. Whitehead had worked on Apollo at the Rand Corp. and helped to plug gaps in Flanigan's experience, for Flanigan had no prior background in space.

Within the BoB, the director Robert Mayo and his deputy, James Schlesinger, were the only political appointees; the rest of the Bureau consisted of permanent Civil Service staff. Schlesinger was also a Rand Corp. alumnus; he worked closely with Whitehead during 1969 in reviewing the NASA budget. This budget fell within the purview of BoB's Economics, Science, and Technology Programs Division, where a small professional group specialized in the pertinent issues.[13]

---

13. Logsdon, *Apollo*, chapter 5, pp. 14-15; *National Journal*, February 28, 1970, pp. 422-425.

## THE SPACE SHUTTLE DECISION

The Space Task Group (STG) was to submit its report to Nixon in September 1969, in time for its recommendations to influence the FY 1971 budget that Nixon would send to Capitol Hill the following February. However, initial exchanges concerning this budget were under way as early as April 1969, barely two months after the inauguration. On April 4, Mayo sent a letter to Paine that asked: "Should the U.S. undertake the development of a long duration manned orbital space station in the FY 1971-73 period?" Attached to this letter was a full page of questions. Paine had recently tried to bypass the budget process by seeking Nixon's approval for a space station in his memo of February 26, but Mayo's letter showed that Paine could still hope to win approval by working within this process. The list of questions amounted to an invitation to justify such a project in detail, with an understanding that when NASA made its case, Mayo's staff would give it close scrutiny. The BoB would give particular attention to its cost.[14]

Though the work of the STG was separate from the budget process, the two activities went forward in parallel. On June 11, Nixon sent a memo to Mayo that made his own attitude perfectly clear:

*Substantively, the continuation of a restrictive fiscal policy to combat the critical problem of inflation will be controlling in formulation of the 1971 budget, and this policy should be applied to the budget requests of* all *departments and agencies. I want it made clear to all departments and agencies that the budget going to Congress will be my budget and that it should reflect the goals and objectives of my Administration.*[15]

Two weeks later, Whitehead sent a memo to Flanigan:

*As you know, I have expressed in the past some uneasiness about the review of the future of our space program. My main concern is that NASA and others will use the enthusiasm generated by a success of Apollo 11 to create very strong pressures on the President to commit him and the Nation prematurely to a large and continuing space budget.*

*The immediate problem is that the space task group chaired by the Vice President appears to be homing in on a single recommended space program*

---

14. Letter, Mayo to Paine, April 4, 1969.
15. Memo, Nixon to Mayo, June 11, 1969.

*that will involve immediate commitments to high levels of lunar exploration simultaneously with a large manned space station program. This may be appropriate and may be the President's ultimate choice. However, a strong case can be made for constraining the NASA budget to its present level or slightly lower....*

*The President should be informed that NASA is making strong public statements about future commitments in space and that there is significant danger that he may find himself in a very difficult situation in the next few months unless he asserts an interest in assessing the desirability of alternative space programs in a considered way without unnecessary pressure being generated by NASA in the press and on the Hill.*[16]

The NASA appropriation for FY 1970 was $3.7 billion. Whitehead noted that "the President is personally interested in a serious evaluation of several alternative NASA budget levels, including one in the vicinity of $2.5 to $3 billion." He proposed that "you or I call Bob Mayo to emphasize the importance" of treating such a level as a formal budget option. He also suggested that Flanigan send a memo to Nixon recommending "that NASA be calmed down during the enthusiasm of Apollo 11, pending a systematic review this fall."[17]

Mayo was not about to chop NASA down to $2.5 billion, at least not at the moment. However, his staff would certainly consider what it would mean to impose cuts to that level, and to even lower levels. Late in August the director of the BoB's Energy, Science, and Technology Programs Division learned of a conversation between Whitehead and the BoB's deputy director, James Schlesinger:

*Mr. Whitehead expressed the view that the President was not eager to proceed with an expanded space program and in fact would like to see it significantly reduced in the near future. Mr. Whitehead had discussed this view with other White House people...and found none of them to be advocates of increased space spending and none who indicated any real problem with significant reductions in the space program....*

---

16. Memo, Whitehead to Flanigan, June 25, 1969.
17. Ibid.; NASA SP-4102, p. 188.

## THE SPACE SHUTTLE DECISION

> Mr. Flanigan claimed to have telephoned Dr. Paine and instructed him to stop public advocacy of early manned Mars activity because it was causing trouble in Congress and restricting Presidential options. According to Dr. Schlesinger, Mr. Flanigan believes the President would like options even lower than $2.5 billion. Also according to Dr. Schlesinger, Mr. Flanigan is basing his comments on personal conversation with the President. In the light of these events, Dr. Schlesinger asked me to define a $1.5 billion per year space program.[18]

His staff set forth budget options in an internal BoB paper. The options would bear comparison with those favored by Paine; but whereas Paine started with the current budget and hoped to go upward, the BoB staff started at the FY 1970 level and considered the consequences of tilting sharply downward.

One alternative, at $3.5 billion per year, eliminated NERVA and stopped production of Saturn V and Apollo spacecraft. This option, however, would maintain a vigorous program in piloted flight, featuring Skylab with three visits as well as six additional Apollo lunar missions. Better yet, such a budget would accommodate "Space Transportation System and Space Station module development with launch of both in 1979."

Two other options, at $2.5 billion, also permitted flight of Skylab with its three visits, along with the six Apollos. There could even be a space station in 1980, with Titan III-Gemini for logistics. However, there would be no Space Shuttle. NASA Marshall would close, while activity at the Manned Spacecraft Center would fall substantially.

At $1.5 billion, the piloted space program would shut down entirely: "All manned space flight ceases with Apollo 14 in July 1970." Not only NASA Marshall but the Manned Spacecraft Center would close, with the Saturn launch facilities at Cape Canaveral shutting down as well. Yet NASA would continue to maintain a vigorous program of automated space flight. Even at $1.5 billion, the agency could send six Viking landers to Mars, and could take advantage of a rare alignment of the outer planets to send spacecraft to Jupiter, Saturn, Uranus, Neptune, and Pluto. NASA would conduct "at least one planetary launch each year in the decade," and would pursue "a relatively ambitious science and applications program with 95 launches in the decade."[19]

---

18. Logsdon, *Apollo*, chapter 5, pp. 15-16.
19. Budget Bureau, "NASA Issues Paper," undated; late August 1969.

Here, in stark contrast, were two visions for NASA's future: Paine's, who hoped for as much as $10 billion and an early expedition to Mars, versus Mayo's, who would consider cuts to one-seventh of that level and a total shutdown of piloted flight. Yet while such options might represent the shape of things to come, Mayo, at least for the moment, would give Paine considerable leeway to argue for his preferred budget. If Paine's arguments proved inadequate then Mayo could lower the boom. However, he would not hasten to do this.

On July 28, Mayo sent a letter to Paine that carried a decidedly mixed set of messages:

*The inflationary outlook, combined with the budgetary momentum of prior commitments and existing laws, make it imperative that we adopt a very restrictive fiscal policy in the 1971 budget.*

*Federal spending plans for 1971 must conform to the President's declared intention to eliminate the income tax surcharge. The resulting loss in revenue will make a balanced budget impossible unless we apply a firm brake on the growth of expenditures. Since a balanced budget is essential to our effort to cope effectively with continuing inflationary pressures, we must maintain a tight rein on budget outlays.*

*Accordingly, a stringent and frugal approach must characterize our 1971 budget proposals. Very few program expansions and new starts can be accommodated.*

An attached sheet gave recommended budget figures. Mayo presented "budget authority," or funds to be appropriated by Congress; he also gave "outlays," which could tap unspent funds from prior years or lay aside such funds for use in the future. He cited an "official target": "the maximum amount that would be available for NASA under the current fiscal outlook for 1971." He also proposed an "alternative target" that represented "a higher resource level, in case subsequent events enable changes in current plans":

|  | *Official Target* | | *Alternative* | |
|---|---|---|---|---|
|  | Budget Authority | Outlays | Budget Authority | Outlays |
| Funding in millions: | $3,470 | $3,500 | $4,500 | $4,200 |

163

## THE SPACE SHUTTLE DECISION

The official target assumed that both budget categories would remain constant at $3.5 billion per year from 1972 to 1978. This would impose a new cut, because the FY 1970 budget stood at $3.7 billion. The alternative target, however, assumed a gradual rise to $6 billion in 1978 that would allow Paine to get a head start toward Mars. Moreover, Mayo suggested in his letter that he might be even more generous: "If you feel that you must request 1971 budget authority or outlays greater than either of these planning figures, you may, of course, do so."[20]

Given an inch, Paine would willingly take enough miles to reach the planets. He proceeded to disregard both Mayo's opening paragraphs, with their words of caution, and his official target of $3.5 billion. Instead, Paine instructed his associates to prepare their final FY 1971 budget proposals in accordance with Program B within his position paper for the STG. This plan aimed to reach Mars as early as 1983, and represented the option that he hoped Nixon would approve.[21]

Paine also faced the issue of having the STG accept his position paper as the basis for the official report that would go to Nixon. The staffer Russell Drew prepared a draft of this report; it was ready on August 27. The members of the STG—Paine, Agnew, DuBridge, Seamans—met anew on September 3, and reached agreement on several basic principles, with all members concurring. This had the important consequence that the STG would not present majority and minority views, but would stand united behind their final report.

They agreed that any program they might recommend was not to include merely the use of existing capability such as Skylab, Titan III, and Saturn V; it was to include the development of new capability. In particular, the STG accepted the eventual development of both the space station and Space Shuttle. This represented a defeat for Seamans, who had rejected the station and had accepted the shuttle only with misgivings. Nevertheless, Seamans agreed not to press his objections.

The members also accepted the concept of an eventual expedition to Mars as the focus for development of the new capability. However, they did not specify the meaning of "eventual," other than to say that it would be prior to the year 2000. This brought DuBridge into the fold, as he too accepted the goal of Mars.

---

20. Letter, Mayo to Paine, July 28, 1969.
21. Logsdon, *Apollo*, chapter 4, pp. 60-62; chapter 5, p. 8.

## Winter of Discontent

Mayo, sitting with the STG as an observer, insisted that the report present a low-cost option that would reflect James Schlesinger's suggestions. DuBridge agreed with Mayo, and Paine agreed to add another alternative, Plan E. It resembled the BoB options at the $2.5 billion level, protecting Apollo and Skylab but shutting down piloted flight. This option offered neither the station nor the shuttle. It did, however, include a strong program of automated spacecraft, with emphasis on planetary missions.

Within these options, now numbering five, Paine and Agnew still hoped to have the report include a strong recommendation for Plan B. The full STG finessed this issue by agreeing not to recommend any particular program to Nixon. This allowed each member to maintain his own views of appropriate budgets, schedules, and pace, without requiring anyone to yield to others.[22]

The next move came directly from the White House. John Ehrlichman, one of Nixon's closest advisors, describes what happened in his memoirs:

*One morning in early September 1969 I had to leave the senior staff meeting early to go see the Vice President. Peter Flanigan had alerted me that Agnew's Space Advisory Committee [sic] was about to make some recommendations to the President that Flanigan knew Nixon could not live with. Peter had been unsuccessful in dissuading the President's science advisor, Lee DuBridge, from agreeing with the staff of Agnew's Advisory Committee that there should be a very costly manned mission to the planet Mars in 1981. So Flanigan had asked for a meeting with Agnew, the ex-officio chairman of the committee, in the hope that we could persuade him to kill it.*

*I had read a briefing paper on the question the evening before, and it seemed obvious to me that Agnew and DuBridge owed it to the President not to include a proposal our budget couldn't pay for. A Mars space shot would be very popular with many people. If the committee proposed it and Nixon had to say no, he would be criticized as the President who kept us from finding life on Mars. On the other hand, if the committee didn't recommend it, we avoided the problem altogether.*

*DuBridge was perhaps to be forgiven for failing to understand such a political argument, but I saw no excuse for Agnew's insistence that the Mars shot be recommended. At our meeting I was surprised at his obtuseness. It*

---

22. Ibid., chapter 4, pp. 63-65.

was, he argued, a reasonable, feasible option. That was what his committee was supposed to come up with, and that was what they intended to do.

*I had been wooed by NASA, the Space Administration, but not to the degree to which they had made love to Agnew. He had been their guest of honor at space launchings, tours and dinners, and it seemed to me they had done a superb job of recruiting him to lead this fight to vastly expand their empire and budget.*

*I finally took off the kid gloves: "Look, Mr. Vice President, we have to be practical. There is no money for a Mars trip. The President has already decided that. So the President does not want such a trip in the Space Advisory Committee's recommendations. It is your job, with Lee DuBridge's help, to make absolutely certain that the Mars trip is not in there."*

*Mr. Agnew was not happy to be told what to do by me. He demanded a personal meeting with the President. This was a matter for Constitutional Officers to discuss.*

*I overlooked the obvious innuendo that I was lying to Agnew about what the President had decided. "Fine," I said. "I'll arrange it at once, and someone will call you."*

*Flanigan and I left Agnew about 9:45 a.m. At 10:00 a.m. the Vice President called me. He had decided to move the Mars shot from the list of "recommendations" to another category headed "Technically Feasible."*

*When I saw President Nixon later that day I told him about our session with Agnew and his telephone call.*

*"Good," Nixon said. "That's just the way to handle him; use that technique on him anytime." Nixon looked at me vaguely. "Is Agnew insubordinate, do you think?"*[23]

The STG staff proceeded to modify the draft of the final report, but only slightly. NASA's Plan A, with its mission to Mars in 1981, lost the status of a formal option. Plan E, which excluded new programs in piloted space flight, also was downgraded. This left Plans B, C, and D, which were redesignated as Options I, II, and III. Because the middle option would remain the one for Nixon to choose if he wished, this reshuffle amounted to delaying the Mars mission from 1983 to 1986—but retaining this expedition as the centerpiece.[24]

---

23. Ehrlichman, *Witness*, pp. 144-145.
24. Logsdon, *Apollo*, p. IV-66.

The STG's final report thus showed a close similarity to NASA's position paper of a month earlier. Plan A, with Mars in 1981, appeared with the designation "Maximum Pace." The STG rejected it with regret, presenting it "only to demonstrate the upper bound of technological achievement." Plan E, described as "Low Level," was one with which "the interests of this Nation would not be served."

With these caveats, the report presented Mueller's integrated plan in full. It described the major elements: space shuttle, space tug, nuclear shuttle, space station module. In turn, these would represent "development of new capabilities for operating in space." The three main options would lift NASA's budget from its 1970 level, $3.7 billion, respectively to $5.5, $7.65 and $9.4 billion, a decade later.

Graphs, published with the report, presented curves of funding for all five plans, giving particular attention to the three main options. Separate curves traced funding levels through 1979 for Plan C; they showed clearly that the shuttle and station, pursued concurrently, would dominate expenditures for new starts through 1976. Their costs would then diminish, while spending for additional new starts—space base, space tug, nuclear shuttle, lunar orbiting station—would rise rapidly to prominence. Spending for a 1986 Mars expedition would also increase sharply beginning in 1978. The report concluded:

> As a focus for the development of new capability, we recommend the United States accept the long-range option or goal of manned planetary exploration with a manned Mars mission before the end of this century as the first target.[25]

Agnew decided that Russell Drew, who had drafted the report, would brief Nixon on its contents. This briefing took place on September 15; Nixon listened attentively, and met as well with STG members and observers, giving them opportunities to comment. These panelists stated that they had rejected the "extreme options" of Mars in 1981 and of eliminating plans for post-Apollo piloted programs. Nixon's press secretary, Ronald Ziegler, then reported that the President "had concurred wholeheartedly in the panel's rejection of the two extremes."[26]

---

25. Newell, chairman, *America's Next Decades*; Agnew, chairman, *Post-Apollo*. Reprinted in NASA SP-4407, vol. I, pp. 522-543.
26. Logsdon, *Apollo*, p. IV-66; *New York Times*, September 16, 1969, pp. 1, 21.

# THE SPACE SHUTTLE DECISION

*Three levels of space activity studied by the Space Task Group in 1969. (NASA)*

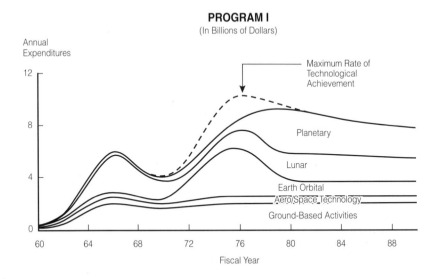

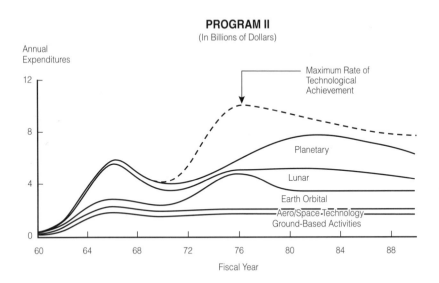

## Winter of Discontent

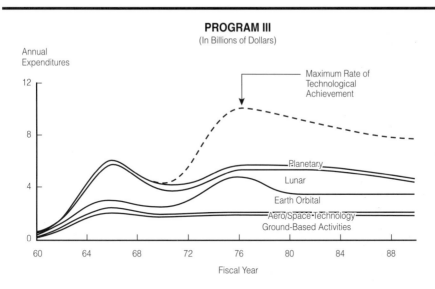

Below, detail of Program II, calling for simultaneous development of a space staion and shuttle, followed by a buildup for a Mars expedition in 1986. (NASA)

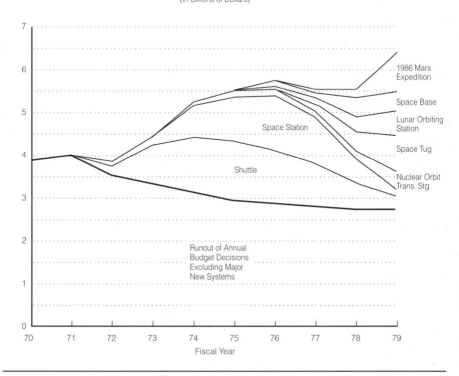

THE SPACE SHUTTLE DECISION

While Nixon's response fell well short of a Kennedy-type commitment to Mars, even as an option for future presidents, it did represent a significant straw in the wind. By endorsing the STG's rejection of Plan E, with its phase-out of piloted flight, Nixon hinted for the first time that he would want more than the Apollo and Skylab missions that he had inherited from previous administrations. He would want a piloted program of his own, and Agnew, as brash as Paine in these matters, promptly sent a letter to Nixon that strongly recommended Plan C (designated Option II in the report), which anticipated Mars in 1986. This letter amounted to an endorsement of Mueller's integrated plan in its original version, which had also called for Mars in 1986. Paine indeed had received the bolder thinking for which he had called. It was clear, however, that this boldness had merely given him leeway to back off to the far-reaching plan that Mueller had proposed in the first place.[27]

## *Mars: The Retreat*

When Nixon met with the STG, Robert Mayo was among those present. He did not need to say much. Everyone knew he had the authority to deal with NASA in his own good way. He already had a staff report that came close to asserting that NASA should follow Plan E, or something very similar. This report had outlined the consequences of holding NASA to future budgets as low as $1.5 billion.

This staff report treated the Space Shuttle at some length, comparing it with upgrades of the Titan III as an alternative. It concluded that even with an active flight schedule of 55 flights per year, the Titan III would represent the less costly way to proceed, with its advantage growing markedly at lower flight rates. The reason for this was that while the Shuttle would reduce the cost of space flight, it would take time and cost money to develop. To the BoB, dollars in future years held less value than present dollars. This was not due to inflation, but rather it reflected the fact that those future dollars would have to earn interest to match the worth of present ones.

NASA had proposed that the Shuttle replace most of the expendable boosters that were currently in use, excluding only the Saturn V. Mayo's staff doubted that NASA and the Pentagon in fact would do this, even if a shuttle became

---

27. Letter, Agnew to Nixon, September 15, 1969.

available. They noted "the existence of strong vested interests and established working relationships in the existing boosters and facilities." Their report stated:

> **Recommendation:** *We recommend against Presidential endorsement of the Space Transportation System at this time.*

Other conclusions were similar: "We recommend against endorsement of a space station now—at least until the orbital workshop [Skylab] is further along in development—perhaps until it has flown." "We recommend against endorsement of the manned planetary expedition (Mars) goal either with or without a target date. In summary, we believe the Mars goal to be much more beneficial to the space program than to the nation as a whole."

The BoB staff showed a similar iconoclasm in its overall view of piloted space flight:

> *The crucial problem with manned space flight is that no one is really prepared to stop manned space flight activity, and yet no defined manned project can compete on a cost-return basis with unmanned space flight systems. In addition, missions that are designed around man's unique capabilities appear to have little demonstrable economic or social return to atone for their high cost. Their principal contribution is that each manned flight paves the way for more manned flight....*
>
> *NASA equates progress in manned space capability with increased time in space, increased size of spacecraft, and increased rate of activity. The agency also insists upon continuity of operational flight programs, which means we must continue producing and using current equipment concurrently with development of next generation systems. Therefore, by definition, there can be no progress in manned space flight without significantly increased annual cost.*

Staff members also reviewed the STG report in draft form. Their comments were scathing, virtually dismissing it out of hand:

> **The report is inadequate** as
> – a basis for Presidential decision,
> – a published justification of Administration decision....

***What are we asking the President to decide?*** *This is not clear from reading the report. For example, does Presidential acceptance of the objective "Developing new capabilities for operating in space" amount to go-ahead decisions on a large earth-orbiting manned Space Station and a Space Transportation System involving three major new systems development for manned and automated systems with both chemical and nuclear engines? The report is susceptible to both "yes" and "no" interpretations.*

***The central issue****—"What is the future of civilian manned space flight activities" is not directly addressed.*

***A good catalogue of technical possibilities*** *for the future is provided. However, in our view these are very optimistic possibilities. For example, ESTP Division staff believe it highly unlikely that a manned Mars mission could in fact be undertaken in 1981 or that a space shuttle...could in fact be developed in five years....*

***The report is lacking in identified outputs*** *for the large-scale manned program recommended. There is therefore little on which to base value judgments.*

***Justification for large-scale manned space effort is only loosely derived.*** *It is based on*

*– challenge to our spirit of adventure*

*– challenge to our national competence in engineering*

*The view then is that a space program supported by national acceptance of these challenges can be used to enhance our national*

*– welfare*

*– security*

*– enlightenment*

*In our view, an unmanned flight program, because of its demonstrated output and lower costs, can be justified directly on the basis of returns to our security, economy, and advancement of science.*

*It is the costly, large-scale manned flight program that requires some overriding decisive force to keep it going....*

***No low-cost options.*** *The report does not contain any program options with annual costs less than current levels. In our view, such options should be identified in the report, and evaluated in terms of returns to the nation—not in terms of entrancing opportunities passed up.*[28]

---

28. Budget Bureau, "NASA Issues Paper," undated; late August 1969.

Armed with this staff review, Mayo wrote a letter to Nixon on September 25, presenting the BoB's assessment of the final STG report. He described it as having "several shortcomings" that "impair its completeness as a vehicle for your *final* decision."

Mayo noted an excessively narrow scope that ignored "the relative standing of the space program in our full range of national priorities" as well as "the future economic context within which the recommended space expenditure increases would have to be considered." He suggested that Nixon have the report reviewed by the Cabinet and perhaps the National Security Council as well. Such reviews would take time, and would give Nixon excellent reason to avoid rushing into any hasty commitments.

Mayo then warned that the report's estimates of the costs of future programs appeared to be "significantly underestimated." He also had other words of caution:

> *The report does not clearly differentiate between the values of the manned space flight program versus a much less costly unmanned program with its greater emphasis on scientific achievement and potential economic returns....*
>
> *The report is written in such a way that your endorsement of any of the recommended program options implies endorsement of major new long-term development projects, which are included in all three of the program options. Therefore, in a practical sense, the report gives you little flexibility except as to timing (and therefore annual costs).... All the defined options involve significant budget increases over current levels....*
>
> *Because the Space Task Group report has now been published, your endorsement now of any specific option will commit us to annual budget increases of at least the magnitudes specified in the report. Therefore, you could lose effective fiscal control of the program.*
>
> *I am convinced that a forward-looking manned space program can be developed for you that does not involve commitments to significant near-term budget increases.*[29]

This letter, circulated within the White House, drew a succinct response from Ehrlichman: "I concur with the Director's recommendations." It also won support from Henry Kissinger, the national security advisor and head of

---

29. Memo, Mayo to Nixon, September 25, 1969. Reprinted in NASA SP-4407, Vol. I, pp. 544-656.

## THE SPACE SHUTTLE DECISION

the National Security Council.[30] With this, Mayo was ready to receive Paine's budget request for FY 1971.

Paine had begun by assembling his associates' estimates totalling $5.4 billion and including $1.0 billion in new starts. This was too much even for him; he responded that their requests were "not consistent with the recommendations made to the President" by the STG, and "far exceed the dollar level that can be reasonably expected." He met with his colleagues, cut their dollar amounts, and presented his proposed budget to Mayo in a letter dated October 8. Paine requested $4.2 billion in outlays and $4.497 billion in new budget authority, with these levels matching those of the "alternative budget" in Mayo's letter of late July.[31]

Mayo and his staff, however, had no intention of granting such largesse. In a staff paper dated November 13, the BoB gave NASA a tentative allowance of $3.349 billion in budget authority and $3.515 billion in outlays. The first of these would require congressional appropriation; it represented a cut of over a billion dollars or more than 25 percent in Paine's request.

Such a budget meant that, at least in FY 1971, NASA would receive no commitment to either a space station or a shuttle. It would cut the launch rate for Apollo missions to as low as one flight per year, and would slam the door on continued production of the Saturn V. It would so restrict NASA that it would prohibit any new starts even in automated spacecraft.[32]

Paine hit the roof. In a letter to Mayo on November 18, he declared that "the allowance and rationale are both unacceptable." He then followed standard procedure by filing a "reclama," a request for review. This too was part of the budget process; it was far from unusual for a department or agency head to receive a cut in a proposed budget. Rather than compromise, however, Paine stuck to his guns, and to his requested budget levels. He got nowhere in a November 21 meeting with Mayo. One participant states that the meeting "broke fairly quickly because we couldn't accommodate anything." Another participant adds that Paine "went away angry."[33]

One should not see this as a personal fight between Paine and Mayo. Paine later noted that "Bob Mayo's son has his wall plastered with NASA

---

30. Memo, Ehrlichman to Staff Secretary, October 7, 1969; memo, Kissinger to Whitaker, November 17, 1969.
31. Letter, Paine to Mayo, October 8, 1969; Logsdon, *Apollo*, chapter 5, pp. 19-20.
32. Logsdon, *Apollo*, chapter 5, pp. 20-22.
33. Ibid., pp. 22-23; letter, Paine to Mayo, November 18, 1969.

posters," adding that while Mayo was "a little hard-headed about things," he was "an easy person to get to know. I was always very comfortable going over and talking to Bob." Rather than keep matters at an impasse, they now agreed that NASA and BoB staffers were to work together to try to narrow their differences.

Mayo proceeded to raise NASA's allowance to $3.7 billion, matching the appropriation for FY 1970. Paine's staff developed alternative budgets that ran as low as $3.91 billion, though he insisted to Mayo that an appropriation of $4.25 billion "is the lowest level you and I can responsibly recommend to the President." This left a gap of over half a billion dollars between their positions.[34]

The reclama procedure called for Mayo to meet personally with Nixon to present the BoB's budget recommendation, and then to inform Nixon of areas of disagreement between BoB and the agency. Paine was not to be present; Nixon did not wish to act as a referee. The meeting took place on December 5. Three days later, Paine talked by telephone with Flanigan, who presented Nixon's decision: "The President says that he doesn't have enough money within the next couple of years and must accept limitation of activity, doing the best he can within the $3.7 limitation." Nixon had come down strongly on the side of Mayo.[35]

Paine still had one more card to play, as he wrote to Nixon directly, urging a "curtailed and spartan" level of $4.075 billion that would keep the Saturn V in production, or a level of $3.935 billion that would suspend Saturn V production but provide startup funds for a space station and shuttle. The two men met just before Christmas, and again Nixon stood firm. Paine would have to accept the BoB figures of $3.7 billion in budget authority and $3.825 billion in outlays. These were the numbers that would go to Congress in the President's budget.[36]

Ordinarily that would have been the end of the matter, with NASA absorbing this cut and making the best of it. In fact, the cuts for FY 1971 were only beginning, and the first new one came from Flanigan. He had tried to develop an independent White House view of an appropriate NASA budget, with his

---

34. Letter, Paine to Mayo, December 5, 1969; Logsdon, *Apollo*, chapter 5, pp. 23-24; E. M. Emme interview, Thomas Paine, August 3, 1970, p. 30.
35. Covert (secretary to Paine), Memo for Record, December 8, 1969; Logsdon, *Apollo*, chapter 5, pp. 24, 28-29.
36. Letter, Paine to Nixon, December 17, 1969; Logsdon, *Apollo*, chapter 5, pp. 29-30.

staff member Clay Whitehead digging into details of this agency's projects. In a letter to Nixon, Paine had warned that at $3.7 billion, "U.S. manned flight activity would end in 1972 with an uncertain date for resumption many years in the future." Flanigan and Whitehead wondered if things were really that serious.

As they pursued their investigations, they became convinced that NASA indeed could live with $3.7 billion, could even receive a budget below that level and still avoid dire consequences. Flanigan advised Ehrlichman of this. Ehrlichman also received counsel from another presidential advisor, Bryce Harlow, liaison with Congress, who warned that a $3.7 billion figure would not win support on Capitol Hill. Ehrlichman discussed the matter with Nixon, and they agreed to seek further cuts.

Amid a flurry of activity within the White House and BoB, Paine soon learned that the $3.7 billion figure that he could not live with now stood at a level higher than what he would have to accept. Early in January 1970, Flanigan presented the news: $3.53 billion in budget authority, $3.6 billion in outlays. The latter figure represented a cut of $225 million from an earlier estimate of $3.825 billion in outlays. Flanigan's memo also stated that "there is no commitment, implied or otherwise, for development starts for either the space station or the shuttle in FY 72. That is a matter to be discussed when the '72 budget is developed."[37]

Paine's initial response was to order the closing of the Electronics Research Center, a NASA facility in Cambridge, Massachusetts. Though it was not a center on a par with the likes of NASA Marshall, it had a staff of 800 and would be missed. Paine then held a press conference on January 13, 1971. He stated that total employment, within NASA and its contractors, would fall from 190,000 to 140,000 during 1971. (As recently as 1966, this total had approached 400,000.) Production of the Saturn V would cease, Apollo lunar missions would fly only at six-month intervals, and Viking missions to Mars would fly in 1975 rather than in 1973, as earlier planned.[38]

Meanwhile, back at the White House, a Cabinet meeting was reaching decisions that would lead to further cuts. The economist Arthur Burns, a presidential counselor, had urged Nixon to bring the overall federal budget into

---

37. Letter, Paine to Nixon, December 17, 1969; memo, Flanigan to Paine and Mayo, January 6, 1970; Logsdon, *Apollo*, chapter 5, pp. 27-28, 30-32.
38. NASA press release no. 69-171, December 29, 1969; Paine, statement, January 13, 1970; Logsdon, *Apollo*, p. V-33.

line with new and lower estimates of revenue. He had won support from George Romney, Secretary of Housing and Urban Development. Romney now called for a uniform reduction of 2.5 percent in all department budgets, along with restrictions on salaries and pay raises. On January 13, as Paine was meeting the press, Nixon met with his cabinet officers and directed them to make such cuts. He put Burns in charge of this effort, which they called Operation Paring Knife.

Nixon directed Mayo to inform Paine that NASA would have to reduce its budget by another $200 million. Paine received the news just as he was arriving at a banquet. He later recalled that

> *while I grandly entered this big ballroom for this event the loudspeaker boomed out that I was to call the White House. And I went with sinking heart knowing damned well that they weren't calling to say that we had more money.*[39]

Paine tried to get by with a cut of only $51 million; Mayo agreed to present this to Nixon. Paine told Flanigan of this, and Flanigan responded angrily, "You mean Mayo capitulated?" But Paine's ploy collapsed within hours, as Nixon rejected his compromise. Paine now had no choice but to take the full reduction of $200 million.

This left NASA with $3.333 billion in budget authority and $3.4 billion in outlays. As recently as October, Paine had requested $4.497 billion and $4.2 billion, respectively. This budget authority represented a cut of 10 percent from the FY 1970 appropriation of $3.697 billion, with inflation eroding its value further. This was merely Nixon's requested budget; Congress was free to make further cuts.[40]

## *The Turn of Congress*

The Budget Bureau was part of the permanent Washington bureaucracy, staffed by members of the Civil Service who took pride in a tradition of nonpartisan concern for the national interest. By contrast, Congress was as

---

39. Logsdon, *Apollo*, chapter 5, pp. 33-34; interview, Thomas Paine, September 3, 1970, pp. 15-16.
40. Letter, Paine to Nixon, January 15, 1970; letter, Paine to Mayo, January 16, 1970; Logsdon, *Apollo*, chapter 5, pp. 34-35.

partisan an institution as that city could offer. Its members paid keen attention to public opinion. When Agnew showed up at the Apollo 11 launch and called for flight to Mars, key senators were quick to respond.

Mike Mansfield, the Senate Majority Leader, declared that he would rule out such efforts "until problems here on Earth are solved." Following the safe return of the Apollo 11 astronauts, Clinton Anderson, chairman of the Senate space committee, stated that "now is not the time to commit ourselves to the goal of a manned mission to Mars." Senator Margaret Chase Smith, a Republican member of that committee, added that the government "should avoid making long-range plans during this emotional period," following the first Moon landing. She warned against becoming involved "in a crash program without the justification we had for Apollo."[41]

There was similar sentiment in the House. Congressman George Miller, chairman of that chamber's space committee, warned against decisions that would "commit ourselves to a specific time period for setting sail for Mars" and proposed that such decisions might be deferred until "five, perhaps ten years from now." Joseph Karth, a space subcommittee chairman, asserted that the success of Apollo would not "translate directly into an urgent mandate to put a man on Mars by 1980 or, for that matter, any other magical date." He declared that NASA was showing "complete lack of consideration for the taxpayer." Congressman Olin Teague, chairman of the powerful Subcommittee on Manned Space Flight, said a year later that "the easiest thing on Earth to vote against in Congress is the space program. You can vote to kill the whole space program tomorrow, and you won't get one letter."[42]

These people were members and leaders of the congressional space committees. If they were willing to take such candid views, what would the rank and file do within the House and Senate? Certainly they would pay close attention to public opinion polls—which were strongly adverse to NASA. Following Apollo 11, a Gallup Poll took a nationwide survey of views concerning flight to Mars. Fifty-three percent of the respondents were opposed to such a program; 39 percent were in favor. A few weeks later, a *Newsweek* poll

---

41. *New York Times*, July 16, 1969, p. 22; *Congressional Record*, July 29, 1969, p. S8739; Logsdon, *Apollo*, p. IV-53.
42. Logsdon, *Apollo*, p. IV-54; *Aviation Week*, August 18, 1969, pp. 16-17; John Logsdon interview, Olin Teague, Washington, August 15, 1970, p. 5.

found that 56 percent of the public wanted Nixon to spend less on space. Only 10 percent wanted him to spend more.[43]

While Paine did what he could to plead his case, he faced entrenched opposition. He met with Senator Edward Kennedy, brother of the late president, and suggested that Apollo astronauts might carry some memento of JFK to the Moon. He quickly learned that the senator had no interest "in identifying Jack Kennedy at all with this landing. He more or less gave me the impression that he felt that this was one of President Kennedy's aberrations."

Unable to sway his critics, Paine soon was dismissing them out of hand:

*One of the games that some people on the Hill might play would be to say, gee, let's hit the space program and wipe it out, and keep the sewers and so forth in. The idea was that, well, the reason the country was so crummy was because we went to the moon, and by God, if we had only spent that money on all these other things that we needed to do, then we would have a great country and a crummy space program. Wouldn't it be better than a great space program and a crummy country. This was the line of reasoning they slipped into.*[44]

Nixon sent his budget for FY 1971 to Capitol Hill on February 2, 1970. The first step was for the space committees to hold hearings, where NASA's officials included a new Associate Administrator for Manned Space Flight. George Mueller, who had held that post since 1963, resigned from NASA in December 1969 and left government service to become a vice president at General Dynamics. His replacement, Dale D. Myers of North American Rockwell, had managed the Navaho missile program in the long ago. Myers had been a vice president in the Space Division and had been general manager of Apollo. He also had directed his company's studies of the Space Shuttle.

Now, in congressional testimony, he spoke of a "shuttle/station" and described it as a single integrated program, offering "the first elements of a transportation system." The shuttle would "transport a crew of two, and

---

43. *Congressional Record*, August 13, 1969, p. H7361; *Newsweek*, October 6, 1969, p. 46. See also NASA SP-4407, Vol. I, p. 546.
44. Paine, Memo for Record, July 1, 1969; interview, Thomas Paine, August 12, 1970, pp. 14, 16; E. M. Emme interview, Thomas Paine, Washington, September 3, 1970, pp. 6-7.

## THE SPACE SHUTTLE DECISION

twelve passengers, into low orbit." In addition to supporting the station, it would accomplish "propellant delivery, satellite repair, short-duration orbital missions, deployment of satellites," and the launch of automated "planetary probes."

The space station would have a crew of 12, "seven men working and five men operating the vehicle itself," and would have "an operational life of ten years, with resupply." It would fly atop a Saturn V, with both the shuttle and station entering service by 1978. Significantly, Myers noted that the FY 1971 budget held no funds for even preliminary studies of a piloted mission to Mars. NASA officials understood that such studies and plans could only hurt the agency.[45]

NASA was requesting $110 million for the shuttle/station, up from $18.5 million in FY 1970. These funds would pay for extensive design work on both projects, including early work on a new engine for the Shuttle. In its original proposal to the BOB in October 1969, NASA had requested over $250 million for these projects. Olin Teague, the most powerful of the space subcommittee chairmen and a power within the full committee as well, was in an expansive mood and was far from willing to accept the BoB's cuts. Proposing to add $80 million for the shuttle/station, he asked Myers what NASA would do if it had more money for piloted space flight. "I don't think we have to rubber-stamp something the Bureau of the Budget does," he argued.

> *We are going along with the people halfway, going along with the people who are supposed to know something. That was the President's Task Group. What should we do, just sit back on our cans and let the Bureau of the Budget dictate every damn thing we do? We are right, you know we are right, and we know more about it than they do, and I bet you this subcommittee of mine knows more about this program than the Bureau of the Budget does.*[46]

The structure of the House space committee paralleled that of NASA. NASA had a powerful Office of Manned Space Flight and a much less influential Office of Space Science and Applications (OSSA) that dealt with automated spacecraft. These offices had counterparts among the House subcommittees, with Teague chairing the one on piloted space flight. A separate

---

45. Logsdon, *Apollo*, chapter 5, pp. 41, 43-44.
46. Ibid., pp. 43-44; AAS History Series, Vol. 4, pp. 245, 247-248.

subcommittee dealt with the concerns of OSSA; its chairman was Joseph Karth. Karth lacked the clout of Teague, much as OSSA had to defer to OMSF. Nevertheless, he was ready to confront Teague, and NASA, when he felt this was necessary.

With their automated orbiters and landers, the Viking missions to Mars fell within Karth's purview. However, he strongly opposed piloted flight to that planet. In 1967, he had been working to win support for Voyager, with its even more ambitious orbiters and landers, when he learned that NASA was requesting proposals for studies of piloted missions to Mars and Venus. He stated that this act left him "absolutely astounded. Very bluntly, a manned mission to Mars or Venus by 1975 or 1977 is now and always has been out of the question—and anyone who persists in this kind of misallocation of resources at this time is going to be stopped."[47]

He responded similarly to the work of the STG. In March 1970, addressing a meeting of the American Institute of Aeronautics and Astronautics, he described its plans as

> *totally unrealistic. Based on my experience with Ranger, Centaur, Surveyor, Mariner, Viking and even Explorer, NASA's projected cost estimates are asinine. NASA must consider the members of Congress a bunch of stupid idiots. Worse yet, they may believe their own estimates—and then we really are in bad shape.*

He opposed Teague's motion in committee, and when Teague prevailed, nailing the $80 million increase to the authorization bill, Karth took his opposition to the floor of the House. Teague viewed this as an unprecedented breach of congressional practice, for Karth, who chaired a subcommittee that did not deal with piloted space flight and who had not participated in the hearings of Teague's own subcommittee, was taking a strong stand against the recommendations of that subcommittee of which he was not a member. Teague became so angry that he vowed that, although Karth was among the most senior members of the full committee, he would personally see to it that Karth would never become its chair.

Karth's amendment called not only for the elimination of Teague's $80 million increase; it demanded elimination of all funds for the shuttle/station,

---

47. Logsdon, *Apollo*, chapter 1, pp. 13-17; chapter 4, pp. 47-48; *Aviation Week*, September 11, 1967, pp. 26-27.

and chopped another $50 million from piloted space programs as well. The entire House took up this amendment on April 23, with Karth insisting that NASA's plans were premature: "Before the Space Shuttle can be a reality, many difficult technological advances must be made in such areas as configuration and aerodynamics, heat protection, guidance and control, and propulsion." Then he dropped a bombshell, suggesting that approval of the shuttle/station would necessarily imply much more: "This in my judgment at least—and there is a great deal of evidence to support the theory—is the beginning of a manned Mars landing program." He spoke of a "back door" to that planet, adding that a decision to "embark upon a $50 billion to $100 billion manned space flight landing program to Mars is something I think we ought to debate loud and clear."

The House had no love for Mars; indeed, even the automated Viking program was controversial. Congressman Edward Koch, a member of Karth's subcommittee and a future mayor of New York City, had stated, "I just can't for the life of me see voting for monies to find out whether or not there is some microbe on Mars, when in fact I know there are rats in the Harlem apartments." In the floor debate, however, the BoB's budget cuts now worked ironically in NASA's favor, for these cuts had eliminated all funds directed toward such a piloted expedition.

"There is no money in here for a manned trip to Mars," countered Don Fuqua, a member of Teague's subcommittee. A Republican member, Richard Roudebush, added: "I am puzzled by the statement that the Shuttle is in some way mixed up with the Mars landing, when nothing is further from the truth." George Miller, chairman of the full committee, also stated authoritatively that there was no relation between the shuttle/station and a Mars expedition.

These reassurances helped to defeat Karth's amendment, but only by the narrowest of margins. Only about one-fourth of the 435 members of the House were present and voting, and the final tally was a tie: 53 for, 53 against. Under House rules, this meant it had failed to pass. Other amendments followed, along with other votes, but the opponents of NASA went down to defeat more handily. The full $190 million for the shuttle/station survived—to face new opposition in the Senate.[48]

---

48. AAS History Series, Vol. 4, pp. 246-251; Logsdon, *Apollo*, Chapter 5, pp. 45-48; *Congressional Record*, April 23, 1970, pp. H3384-H3423; *Aviation Week*, May 25, 1970, p. 27.

## Winter of Discontent

Like Karth, Senator Walter Mondale was a Democrat of Minnesota, with the two men being close colleagues. The Senate had no counterpart of Olin Teague, no one who would push successfully to add funds for the shuttle/station in the authorization bill; the bill that reached the Senate floor contained only the administration request of $110 million. Mondale nevertheless moved to strike this entire amount, and offered an impassioned plea:

*This item involves a fundamental and profound decision about the future direction of the manned space flight era. This is, in fact, the next moon-type program. I believe it would be unconscionable to embark on a project of such staggering cost when many of our citizens are malnourished, when our rivers and lakes are polluted, and when our cities and rural areas are dying. What are our values? What do we think is more important?*[49]

The Senate debated Mondale's amendment for four hours, then sent it to defeat by a vote of 29 to 56. Mondale tried anew in July, when this chamber turned to the appropriations bill. This bill totalled nearly $18 billion and included funds not only for NASA but for the Department of Housing and Urban Development. Senators thus faced a potentially irresistible opportunity to add funds to meet the needs of the nation's cities, and to subtract funds from the space program.

Mondale's colleagues quickly did the former, adding $400 million for urban renewal and for sewer and water projects. Mondale then offered his amendment again, as he sought to delete the $110 million for the shuttle/station as an appropriation. After several hours of debate, his amendment lost—by a margin of only 28 to 32.

With debate resuming the next day, Paine knew that he faced an imminent threat from similar amendments. He discussed the situation at a meeting on Capitol Hill with Senator Hugh Scott, the Republican leader. In Paine's words,

*we decided that the best chance of defeating that would be to offer the people who would be on the floor who had to more or less vote against increases in space, a bill to vote against. And once they had voted against*

---

49. Logsdon, *Apollo*, chapter 5, pp. 48-49; *Congressional Record*, May 6, 1970, pp. S6768-S6817. For Mondale quote see also Chaikin, *Man*, p. 336.

## THE SPACE SHUTTLE DECISION

*NASA, then when the other bill came up proposing to cut us, that they might feel free and a little easier in not voting to cut us, since they had already voted against us once.*

The Senate appropriations bill called for a level of spending slightly below the president's budget, and they decided to seek a member of the Senate space committee who would introduce an amendment to bring it back up to Nixon's request of $3.333 billion. They quickly settled on Barry Goldwater, an active space proponent, as the man they wanted. At that moment, Goldwater was on the Senate floor; they sent him a note, and he met Paine and Scott in the latter's office. He agreed with the strategy and invited them to prepare an amendment that he would introduce.

"We went into the outer office," Paine recalls. "We got the girl to put an amendment form into the typewriter there, and she banged out an amendment. Barry folded it up, put it in his pocket, and walked out." Scott then invoked Senate procedure and arranged for Goldwater to introduce his measure after everyone was back from lunch. It met resounding defeat, 15 to 58—as Paine had expected.

Following this vote, NASA's opponents launched their onslaught. William Proxmire, an ally of Mondale and a strong critic of NASA in his own right, noted that the House had approved a NASA budget that was $136 million below the administration request. This cut had not been aimed specifically at the shuttle/station, but had been spread among a variety of programs. Proxmire now introduced his own amendment, calling for a cut to this level in the Senate appropriation. He asserted that the money saved could "provide a subsidy for the building of some 125,000 to 150,000 new low- and moderate-income housing units." This amendment also failed, 34 to 39.

Senator William Fulbright then introduced yet another amendment, demanding a cut of $300 million for NASA. Nixon had described the Senate as "spendthrift" for having added $400 million to the bill the previous day, for urban programs. "We should all have an opportunity to help balance the current bill," this senator said, adding that his cut in funding for the space program would do precisely that. He presented an explicit appeal to take money from NASA and spend it on the cities: "We voted for sewers. Certainly sewers are more important than going to the Moon." Again Paine found the support he needed as Fulbright's measure went down, 32 to 37.

*Winter of Discontent*

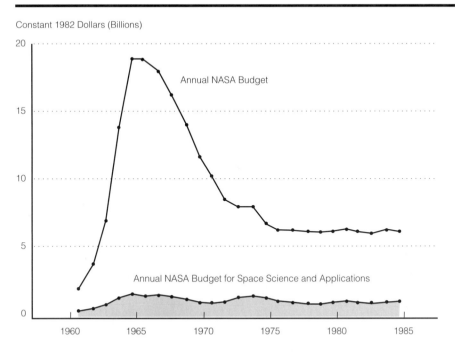

NASA's actual budget in constant dollars, 1960–1985. (Scientific American)

The final Senate appropriation passed easily, 68 to 4. It provided $3.319 billion for NASA, for a cut of only $14 million from the President's budget. This bill went to conference with that of the House; the conferees gave NASA a final appropriation of $3.269 billion. They also granted $110 million for the shuttle/station, eliminating Teague's proposed increase but matching Nixon's original request.[50]

This appropriation represented a drop of $428 million from the level of FY 1970, with inflation reducing the 1971 allocation even further. It marked the fourth year in a row of such cuts, and while no one had a crystal ball, this budget at least would offer the solace that future cuts would be considerably less severe. NASA funding finally hit rock bottom in FY 1974, barely above $3 billion. In constant dollars, this represented only one-third of NASA's peak in the

---

50. Logsdon, *Apollo*, chapter 5, pp. 49-51; *Congressional Record*, July 6, 1970, pp. S10603-S10625; July 7, 1970, pp. S10681-S10700, S10721-S10727; E. M. Emme interview, Thomas Paine, July 9, 1970, pp. 7-9; Low, Personal Notes No. 27, July 18, 1970; Paine, statement, September 2, 1970 (with summary of FY 1971 budget).

mid-1960s. Subsequent budgets stayed close to that level, with adjustment for inflation, and continued at that constant-dollar level through the mid-1980s.[51]

Yet even in 1970 the final cut, from $3.333 billion to $3.269 billion, gave a clear view of congressional attitudes toward Apollo and, by extension, to the challenges that might lie beyond the Moon. This reduction of $64 million represented only two percent of the administration request, but NASA was already so hard-pressed that it had significant consequences. Paine now saved $42 million by canceling two planned Apollo Moon landings. This amounted to a down payment; those two missions would have cost $800 million, spread over several years.

The Apollo program had spent $23.85 billion through mid-1970, and had accomplished the two Moon landings of Apollo 11 and 12. The equipment was in hand for six more. Hence, to save 3.3 percent of the program cost—$800 million out of nearly $24 billion—Paine sacrificed one-third of the remaining missions. The loss in lunar science was greater still. Lunar landings were visiting rugged regions of interest to geologists; indeed, a professional geologist, Harrison Schmitt, flew to the Moon aboard Apollo 17. The final Apollo missions were able to stay longer on the Moon, as astronauts ranged widely by driving a battery-powered vehicle that resembled a dune buggy.

Such waste is inconceivable unless one understands that, to Congress, Apollo was a means to achieve national prestige. By 1970, the nation had reached the Moon and had won whatever prestige it was likely to get from this. Members of Congress could look at the Moon and say "been there, done that." Each Apollo flight cost up to $400 million. Such a sum, following the estimate of Senator Proxmire, could provide housing for as many as a million people. In an era when people looked to Washington to do such things, Apollo would fail totally in any competition.[52]

## Paine Leaves NASA

During the year that followed the landing of Apollo 11 in the Sea of Tranquillity, NASA received a cold bath in the sea of reality. Yet the experi-

---

51. NASA SP-4012, Vol. III, p. 12; *Scientific American*, January 1986, p. 34.
52. Paine, statement, September 2, 1970; Low, Personal Notes No. 30, September 6, 1970; *Aviation Week*, September 7, 1970, pp. 18-19; NASA budget data, February 1970. For later Apollo missions, see Chaikin, *Man*.

ence left Paine unmoved; he remained as ebullient as ever in his hopes. He not only continued to cherish the goals of the STG; he sought to define further goals reaching to the year 2000, three decades in the future. He set up a three-day meeting at which space experts were to brainstorm on such goals; the invitees included Wernher von Braun, Arthur C. Clarke, Robert Gilruth of the Manned Spacecraft Center, and astronaut Neil Armstrong.

Calling for "a completely uninhibited flow of new ideas," Paine offered an "operating manual" for "Spaceship NASA." He wanted new types of engines, "to achieve Wernher's Metaphysical Goals of extending terrestrial life within the solar system and out into the galaxy." He hoped for an "Intercontinental Space Plane," able to fly "anywhere on Earth in an hour." He proposed "Global Telecommunication/Supercomputer Networks" that indeed would take shape as the Internet. Another concept called for "Food Manufacture," synthesizing food from fossil fuels and "freeing man from his 5000 year dependence on agriculture." The list concluded: "Understand Man's Origin and Destiny" and envisioned "the future evolution of terrestrial life to other worlds with eventual communication with other intelligence."

He called for "swashbuckling buccaneering courage" and proposed "fighting ships: both naval and buccaneering" as a model for NASA. Having served personally in the Navy, he proceeded to issue orders:

> *Consider NASA as Nelson's "Band-of-Brothers"—Sea Rovers—combining the best of naval discipline in some areas with freedom of action of bold buccaneers in others—men who are determined to do their individual and collective best to moving the planet into a better 21st Century.*

**INSTRUCTIONS TO CAPTAINS**
*Must be competent and hard working, sensitive but steady nerved, visionary but tough minded, determined and thoughtful. No room for ideology.*

**Scholarship**
*Know the ocean, storms, rocks and shoals you will face. Know your ship, men and fleet commander; keep your watch, quarter and stations bill up to date as casualties and rotation take place. Continuously study your course, position, consumables and destination. Keep a sound man with keen vision and a good glass stationed in the foretop.*

# THE SPACE SHUTTLE DECISION

### Command
*Buccaneer captains with letters of marque and reprisal live dangerously. This danger can be reduced by alert lookouts, fast sailing, superior seamanship, winning the respect and loyalty of the crew. Complete your homework before talking or issuing orders. Be careful of ideology and amateur social science and economics.*

The meeting took place in mid-June of 1970. Paine followed with a letter to Nixon: "The results are exciting and I would like to request an appointment to present to you our best current thinking.... The purpose is...to give you a heretofore unavailable Presidential level long-range view of man's future potential in space."[53]

Then in July, he received an attractive job offer from Xerox. The offer appealed to him, for his government salary was $42,500 per year. As he put it, "with four children in school, I can certainly use a little more money to help support this family and give them a good start in life." He called Jack Parker at General Electric, an old friend and a member of the board of directors, and asked for advice. Parker replied, in Paine's words, "that they would be very anxious to have me come back," and that GE might be able to offer him a very promising position. Paine then talked to the chairman of the board and learned that the position would call for him "to head up all of General Electric's power generation activities including both the conventional steam turbine business and also the nuclear power plants." Paine expressed interest and suggested that he could take the post early in 1971; the chairman replied that GE would need him that summer. There was nothing pressing to keep him at NASA, and on July 28 he sent Nixon his letter of resignation, to become effective on September 15. Having left GE only two and a half years earlier, he now returned, with his tenure at NASA representing merely a brief interlude within a career at GE that spanned nearly three decades.[54]

He had been a liberal Democrat in an administration of Republicans, a Lyndon Johnson appointee held over to serve Nixon's loyalists. In addition to this, he had spent much effort fighting for his own agenda, rather than promoting that of the president. Yet he did not leave Washington under a cloud. Peter

---

53. Memo, Paine to Addressees, May 25, 1970; Memo for the President, Paine to Haldeman, July 9, 1970.
54. Biographical data, Thomas O. Paine papers, Library of Congress; Logsdon, *Apollo*, p. V-53; E. M. Emme interview, Thomas Paine, August 3, 1970, pp. 10-14.

Flanigan described him as a "good soldier" who "accepted decisions after getting a full hearing." Ehrlichman compared NASA's bold proposals to a spring that "had to be stretched in order for it to come back to where it belonged." Nixon, on receiving Paine's letter of resignation, wrote that "the course you have done so much to set will help guide our efforts for years to come."[55]

His push for Mars fell short, but even within the STG report, he expected to defer the serious pursuit of this goal until 1976. Although Congress and the BoB cut his budget, this schedule left time for them to experience a change of heart. In the words of Dale Myers, "Hope springs eternal. After we came off the Apollo peak, it was very difficult to accept that we'd be at a level half or a third of that. We always wanted to think that next year would be better."[56]

With all his swashbuckling, what did Paine accomplish at NASA? Though he did not build this agency in the manner of his predecessor, James Webb, he was the captain on its bridge during the run-up to the Moon landings. He pushed successfully for the dramatic Apollo 8 mission that orbited the Moon at Christmas in 1968; he approved the dry workshop for Skylab. He was at the helm when the landings took place.

Amid his setbacks, Paine followed the lead of George Mueller and steered NASA onto the new course that Nixon noted. Mueller had tried and failed to build a major post-Apollo effort, Apollo Applications, based on use of Saturn-class launchers and Apollo spacecraft. This effort was in tatters by mid-1968. Mueller responded by envisioning a space shuttle as a focus for the future. Paine took this vision, made it his own, encouraged Mueller to strengthen it with bolder thinking, and sold it to the STG.

Though Mars provided a long-term goal, the shuttle/station was to represent the main work of the 1970s. When budget cuts hit home, Paine held to this plan, preserving options for the future by sacrificing those of the past as he shut down Saturn production and canceled Apollo Moon landings. Congress also signed on for the shuttle/station, appropriating $110 million to start the work during FY 1971.

As that fiscal year began, however, in mid-1970, NASA's situation was tenuous in the extreme. Funding for the shuttle/station had survived by votes of 53-53 in the House and 32-28 in the Senate, which left the program vulnerable

---

55. Logsdon, *Apollo*, p. V-55; John Logsdon interview, John Ehrlichman, Santa Fe, New Mexico, May 6, 1983, p. 15; letter, Nixon to Paine, July 28, 1970.
56. Author interview, Dale Myers, Leucadia, California, December 6, 1996.

## THE SPACE SHUTTLE DECISION

to even a slight increase in anti-space sentiment. Similarly, Paine had not won the endorsement of Nixon for this program. Lacking such endorsement, NASA could proceed with detailed studies of the shuttle/station, but could not award the contracts that would build it.

Arthur Cleaver, a leader in British rocket development, quoted the Duke of Wellington in describing the votes in Congress as "a damn close-run thing—the nearest run thing you ever saw in your life."[57] If NASA was to avoid meeting its own Waterloo, it would need new sources of strength. It would find them by abandoning the plans of the STG, dropping the space station, placing all hope in the Space Shuttle as a separate project, and making common cause with the Air Force.

57. *Astronautics & Aeronautics*, October 1970, pp. 70-72.

# CHAPTER FIVE

# Shuttle to the Forefront

"I wouldn't want to be quoted on this," President Johnson told a gathering in 1967.

> *We've spent $35 or $40 billion on the space program. And if nothing else had come out of it except the knowledge that we gained from space photography, it would be worth ten times what the whole program has cost. Because tonight we know how many missiles the enemy has and, it turned out, our guesses were way off. We were doing things we didn't need to do. We were building things we didn't need to build. We were harboring fears we didn't need to harbor.*[1]

Within NASA, Apollo addressed the perception of power that Moscow's highly publicized space spectaculars pointed to communism as the way of the future. The Air Force had a separate space program that dealt with the reality of power. Working closely with the CIA, the Air Force had the task of launching reconnaissance satellites that could determine the Soviet order of battle, counting that nation's bombers and missiles while determining the location of their bases and their operational readiness. In turn, these satellites provided strategic intelligence that shaped America's Cold War policies.

## *The Air Force in Space*

The background to the Air Force program dated to 1953, shortly after the inauguration of President Eisenhower. In August of that year, the Soviets det-

---

1. Richelson, *Secret Eyes*, p. 93.

onated a nuclear weapon with a yield of 400 kilotons. By studying its fallout, American analysts determined that it was not a true hydrogen bomb. However, it did represent a large step upward in Soviet nuclear power. In addition to this, the CIA learned that the Soviets were building a turboprop bomber, the Tu-95, with enough range to strike the United States. An intelligence estimate, issued early in 1954, predicted that Moscow would have 500 such bombers in 1957.[2]

In March 1954, Eisenhower met with a group of advisors and warned them that he feared a surprise attack, a new Pearl Harbor that would destroy cities rather than battleships. Lee DuBridge, the president of Caltech and chair of this advisory group, responded by taking steps to set up a high-level commission, the Technological Capabilities Panel. It would recommend new policies that could meet this danger. To chair it, Ike recruited James Killian, the president of MIT.

A subpanel, Project 3, dealt with the technical means for surveillance. The people who learned of it included Clarence "Kelly" Johnson of Lockheed, one of the country's top aircraft designers. He had already prepared a design for a reconnaissance aircraft and, without success, had tried to win support from the Air Force. Johnson now joined with Trevor Gardner, a special assistant to the Air Force Secretary, and approached Project 3 for a new try. The subpanel's chairman, Edwin Land, had invented the Polaroid camera and was president of the Polaroid Corporation. He and Killian took the proposal to Ike and convinced him to accept it. The plane that resulted was the U-2.[3]

In mid-February 1955, the full Killian Committee issued its report, titled "Meeting the Threat of Surprise Attack." It declared, *"We have an offensive advantage but are vulnerable to surprise attack"* (emphasis in original). "Because of our vulnerability, the Soviets might be tempted to try an attack." In Edwin Land's section of the report, he wrote,

> We must find ways to increase the number of hard facts upon which our intelligence estimates are based, to provide better strategic warning, to minimize surprise in the kind of attack, and to reduce the danger of gross overestimation or gross underestimation of the threat.[4]

---

2. Rhodes, *Dark Sun*, pp. 523-525; Zaloga, *Target*, pp. 85-88.
3. McDougall, *Heavens*, pp. 115-117; Burroughs, *Deep Black*, pp. 69-75; Killian, *Sputnik*, pp. 67-71.
4. Killian, *Sputnik*, pp. 70, 71, 72, 79, 302.

## Shuttle to the Forefront

At the time, the available "hard facts" were often meager. The 1953 Soviet nuclear test had caught everyone by surprise. Then, on May Day of 1954, at a public air show, the Soviets showed off a new jet bomber, the Bison. Here was another surprise—a Soviet jet bomber! It was all the more worrisome because no one in the U.S. had known of it until the Kremlin displayed it openly. A year later, in preparations for the next such air show, American observers saw a formation of 10 of these aircraft in flight. In mid-July came the real surprise. On Aviation Day, Colonel Charles Taylor, the U.S. air attaché in Moscow, counted no fewer than 28 Bisons as they flew past a review in two groups. This bomber now was obviously in mass production. The CIA promptly estimated that up to 800 Bisons would be in service by 1960.

In fact, Taylor had seen an elaborate hoax. The initial group of 10 Bisons had been real enough. They then had flown out of sight, joined eight more, and this combined formation had made the second flyby. Still, as classified estimates leaked to the press, Senator Stuart Symington, a former Air Force Secretary, demanded hearings and warned the nation of a "bomber gap." The flap forced Ike to build more B-52 bombers than he had planned, and to step up production of fighter aircraft in the bargain. Yet even when analysts discovered the Aviation Day hoax, they took little comfort. If Moscow was trying to fool the CIA, it might mean that the Soviets were putting their real effort into missiles rather than bombers.[5]

The U-2 became operational in mid-1956, and proceeded to deliver photos of the highest value. One mission returned with pictures that showed far fewer heavy bombers than expected at Soviet bases. This started a process of downward revision of Moscow's estimated air power. One of Ike's military aides declared that "very quickly we found the Bomber Gap had a tendency to recede. It was something that each year was going to occur. But in fact it did not occur." The U-2 also looked at targets of opportunity, and Richard Bissell, the project manager within the CIA, would recall an example: "He was flying over Turkestan, and off in the distance he saw something that looked quite interesting and that turned out to be the Tyuratam launch site. He came back with the most beautiful photos of this place." It was one of the

---

5. Zaloga, *Target*, pp. 81-85; Burroughs, *Deep Black*, pp. 67-68; Prados, *Soviet*, pp. 41-50; Klass, *Sentries*, pp. 6-9.

## THE SPACE SHUTTLE DECISION

principal bases for missile and space launches; yet the CIA had not known of its existence.[6]

Nevertheless, the U-2 delivered far less than it had promised. When it entered service, Soviet radar promptly picked it up. Following the second overflight, the Foreign Ministry lodged a protest. The protests escalated, and after only six such missions, all during July 1956, Ike ordered a standdown. Subsequent flights required his personal approval; over the next four years, only about 15 took place. Then, in May 1960, a Soviet antiaircraft missile downed a U-2 near the city of Sverdlovsk. With this, the overflights ceased completely.[7]

But now a new concern had arisen: the missile gap. Early in 1960, a debate developed in Washington in response to a new intelligence estimate, which predicted that Moscow would possess up to 450 ICBMs in mid-1963. This would be twice America's anticipated strength in missiles. This was frightening enough; deeper skepticism was raised by the fact that the estimate actually represented a substantial reduction from earlier ones. Senators Symington and Johnson asked whether Ike perhaps was cooking the books, downgrading the perceived threat during an election year. Clearly the nation needed additional strategic reconnaissance, and needed it quickly.

By then, the CIA's Bissell had been working for nearly two years to address this problem. Early in 1958, he had initiated a highly classified program, Corona, that sought to build reconnaissance satellites known as Discoverer. These were to fly to orbit atop Thor-Agena rockets. It took a year and a half, however, to get the system to work successfully. The first attempt, Discoverer 1, did not even reach orbit. Following launch from Vandenberg Air Force Base in February 1959, it wound up near the South Pole. Finally, in August 1960, Discoverer 13 proved the lucky 13 in the series. Though it carried no photo equipment, it successfully demonstrated the release of a capsule from orbit and its recovery in the Pacific. This was the first spacecraft to reenter from orbit and be retrieved following descent by parachute.

With this encouragement, Bissell allowed Discoverer 14 to fly with its camera. Its capsule, too, was recovered successfully, this time in midair, on August 19. The film soon arrived at the CIA's Photographic Interpretation Center, and the photo interpreters gathered in an auditorium. The director,

---

6. Prados, *Soviet*, pp. 46, 47; Ranelagh, *Agency*, pp. 316-317.
7. Richelson, *Espionage*, pp. 142-152; Ruffner, ed., *Corona*, p. 3; Prados, *Soviet*, pp. 33-35; Powers, *Secrets*, pp. 95-97.

Arthur Lundahl, spoke to them about "something new and great we've got here." His deputy then presented a map of the Soviet Union. These maps had previously featured a single narrow line to indicate the coverage along the path of a U-2. This one had eight broad swaths running north to south across the USSR and Eastern Europe, covering over one-fifth of their total area. They represented the regions that this single mission had photographed, and people broke out in cheers. Some photos were fogged by electrostatic discharges, but the resolution was 20 to 30 feet, which analysts described as "good to very good." Clearly, this was a turning point.[8]

During the election campaign that autumn, Kennedy stressed the issue of the missile gap, warning that the Republicans had done too little to counter its threat. After the election, he appointed a deputy defense secretary, Roswell Gilpatric, who believed strongly that this gap was real. On taking office in January 1961, Gilpatric and his boss, Defense Secretary Robert McNamara, went to the Air Force intelligence office on the fourth floor of the Pentagon and spent several days personally studying Discoverer photographs.

The Air Force held the view that Moscow was building large numbers of well-camouflaged missile sites. Sites for the presumed disguised installations included a Crimean War memorial and a medieval tower. McNamara and Gilpatric, however, preferred the view of Army intelligence: that the Soviet ICBM, designated R-7, was very large and unwieldy and could move only by rail or military road. Discoverer satellites had taken photos along the Soviet Union's railroads and principal highways—and had found no missile launchers. In February, at an off-the-record press conference, a newsman asked about the missile gap. McNamara replied that "there were no signs of a Soviet crash effort to build ICBMs." Reporters raced to their phones, newspapers blossomed with the word that no such gap existed, and Kennedy himself had to step in, declaring that it was too early to draw such conclusions.[9]

Then in June and July, Discoverers 25 and 26 flew with nearly complete success. While they were only the third and fourth missions to return photos having intelligence value, together these four flights covered more than half of the regions suitable for ICBM deployment. Within this vast area, photo analysts found no more than two new and previously unsuspected ICBM

---

8. Prados, *Soviet*, pp. 82-83, 86-95; McDougall, *Heavens*, pp. 219-220; *Time*, February 8, 1960, pp. 16-19; Ruffner, ed., *Corona*, pp. 3-24, 119-120; Richelson, *Secret Eyes*, pp. 41-44.
9. Richelson, *Secret Eyes*, pp. 57-58; Prados, *Soviet*, pp. 114-115, 119; *Time*, February 17, 1961, pp. 12-13.

bases. Three others were photographed a second time. By comparing them with one another, and with a known testing complex at Tyuratam, the analysts came away with a clear understanding of just what an ICBM base would look like. That made it possible to eliminate a number of "suspect" launch sites and to give a clear and definitive estimate of Moscow's ICBM strength.

This assessment, National Intelligence Estimate 11-8/1-61, titled "Strength and Deployment of Soviet Long Range Ballistic Missile Forces," came out on September 21. It stated:

> *We now estimate that the present Soviet ICBM strength is in the range of 10-25 launchers from which missiles can be fired against the US, and that this force level will not increase markedly during the months immediately ahead.*
>
> *The low present and near-term ICBM force probably results chiefly from a Soviet decision to deploy only a small force of the cumbersome, first generation ICBMs, and to press the development of a smaller, second generation system. On this basis, we estimate that the force level in mid-1963 will approximate 75-125 operational ICBM launchers.*[10]

There indeed was a missile gap—but it favored the United States, and by a large margin. In 1961, the U.S. was already deploying substantial numbers of its first-generation Atlas, Titan, Thor, and Jupiter missiles. In addition, the first Polaris submarines were on station at sea. Beginning in October 1962, the nation would also have the Minuteman ICBM, which would reach the field in even larger numbers.[11]

Yet it was hardly a secret in Moscow that the Soviet R-7 was clumsy and unwieldy; that nation's planners had known this from the start. Why, then, had they taken the trouble to develop it? An answer was in hand, courtesy of Oleg Penkovskiy, a colonel in the Chief Intelligence Directorate of the Soviet army. He had recently begun working for MI-6, Britain's intelligence agency, and had gone on to help the CIA as well. In May 1961, he delivered rolls of microfilm that included minutes of Kremlin meetings in which officials decided to use the R-7 for space launches, but not as an ICBM.

---

10. Ruffner, ed., *Corona*, pp. 26-27, 129-130, 137-140; Richelson, *Secret Eyes*, p. 56; Richelson, *Espionage*, p. 180.
11. Prados, *Soviet*, pp. 119-122; Neufeld, *Ballistic Missiles*, pp. 226, 234-237.

## Shuttle to the Forefront

At the outset of the R-7 program, during the mid-1950s, Soviet officials had expected to fire it from secret bases. This missile would take up to 20 hours to fuel and prepare for launch, and during that time, it would be highly vulnerable to attack. However, if the U.S. did not know where these bases were located, the R-7 would remain safe. The advent of American strategic reconnaissance upset this plan, by giving America the intelligence needed to strike during pre-launch preparations. The head of the Soviet strategic missile force, Marshal Mitrofan Nedelin, accordingly decided to delay the deployment of a large fleet of ICBMs until he could receive a more advanced version that could be fueled and launched on short notice.[12]

The Corona program had sought to use satellites to assess the Soviet threat. It did more; it markedly reduced this threat, at least for a time, by piercing the secrecy that formed a major element of Moscow's strategic calculations. Then, as the 1960s proceeded, the Air Force and CIA introduced a number of important advances in the satellites and went on to fly them routinely.

Improved resolution was an early goal. The cameras of 1961 were only moderate in resolution, and better versions were in service later in the decade. To make the best use of these increasingly sharp images, the post-1961 Discoverer satellites mounted dual cameras that could photograph a site from different directions. This permitted stereophotography, whereby analysts could study images that appeared three-dimensional. Later versions of this spacecraft also carried more film and stayed up longer. The first Discoverers had mission times of a single day; subsequent models stretched this to three weeks and longer.

Resolution always represented a limit for Discoverer imagery. The Thor-Agena booster, used by the Discoverer, had only a modest payload capacity. Beginning in 1963, however, the Air Force employed the Atlas-Agena and then the Titan III, which could launch larger spacecraft with telescopes of greater acuity. These rockets supported a separate program, Gambit, that achieved much greater resolution. For the closest looks, the Air Force used the closest orbits, with Gambit spacecraft dropping down to perigees as low as 76 miles.[13]

---

12. Prados, *Soviet*, p. 116; Richelson, *Espionage*, pp. 56-65; Zaloga, *Target*, pp. 51-54.
13. Richelson, *Secret Eyes*, pp. 353-360; Ruffner, ed., *Corona*, pp. xiv-xv, 27-37; McDonald, ed., *Corona*, pp. 301-307; *Quest*, Summer 1995, pp. 22-33.

*THE SPACE SHUTTLE DECISION*

What did Corona and Gambit show? They photographed all Soviet ballistic-missile launch complexes, following existing as well as new missiles through development and deployment. In particular, they found and repeatedly observed a major center at Plesetsk, near the northern city of Arkhangelsk. Plesetsk specialized in launching reconnaissance satellites and other military spacecraft. At its height, it accounted for more than half of all space launches in the entire world, with Tyuratam a distant second, and Cape Canaveral and Vandenberg Air Force Base far behind.

Corona also was first to see Severodvinsk, the main construction site for ballistic missile submarines. This made it possible to monitor the launching of new classes of subs, and to follow them through to operational deployment. The CIA also observed the rapid growth of the Soviet surface navy. Coverage of aircraft plants and air bases kept analysts up-to-date on bombers and fighters, while other coverage allowed Army experts to learn the nature of the tank forces that NATO would face if the Soviets were to invade Europe.

Corona photography uncovered the construction of antiballistic missile sites near Moscow and Leningrad, along with the radar installations that supported them. Other photos located antiaircraft batteries and made it possible for the Strategic Air Command to find routes for its bombers that could avoid these missiles. Specialized satellites, conducting geodetic mapping, became the main source of data for the military charts of the Defense Mapping Agency.

As recently as the mid-1950s, the Soviets had been able to fool the Americans concerning their air strength, and to touch off a major Washington flap over a supposed "bomber gap," merely by flying the same aircraft around twice at an air show. By contrast, a 1968 intelligence report contained the unequivocal statement: "No new ICBM complexes have been established in the USSR during the past year." As early as June 1964, Corona had photographed all 25 of the complexes then in existence. If there had been any new ones, the CIA would have seen them.[14]

## *The Air Force and NASA*

In 1494, the Treaty of Tordesillas divided up the New World by drawing a line down the Atlantic, with Spain claiming lands to the west of this line and

---

14. Ruffner, ed., *Corona*, pp. xiv, 37.

## Shuttle to the Forefront

Portugal claiming lands to the east.[15] The activities of NASA and the Air Force lent themselves to similar demarcation. With NASA emphasizing Apollo while the Air Force dealt largely with satellite reconnaissance in low orbit, there was little overlap between their concerns. However, these two agencies did not run independent programs; there was a great deal of cooperation.

This cooperation was particularly strong in the realm of launch vehicles. In launching automated spacecraft, the most important such vehicles were derived from the Thor, Atlas, and Titan ballistic missiles; both NASA and the Air Force used these rockets repeatedly, and procured them from the same contractors. They also shared in ongoing developments that increased their payload capacities.

As early as February 1961, an agreement between NASA's James Webb and the Pentagon's Roswell Gilpatric stipulated that neither agency would initiate the development of a new launch vehicle without first seeking the consent of the other. Then in 1962, a joint NASA-DoD Large Launch Vehicle Planning Group issued a report that contained a recommendation: "The 120-inch diameter solid motor and the Titan III launch vehicle should be developed by the Department of Defense to meet DOD and NASA needs, as appropriate in the payload range of 5000 to 30,000 pounds, low Earth orbit equivalent."[16]

The Titan III brought the prospect of wasteful duplication, for it competed directly with NASA's Saturn I-B. This Saturn carried over 36,000 pounds to low orbit. The Titan III-C, the first operational version, had a rated payload of 23,000 pounds; its immediate successor, the Titan III-D, raised this to 30,000. In addition to this, the projected Titan III-M promised to carry as much as 38,000. Nevertheless, as early as 1967, the President's Science Advisory Committee noted that "the launch costs of the [Saturn I-B] are about double those of the Titan III-M."[17]

Because NASA was accustomed to receiving launch vehicles that the Air Force had developed, it yielded gracefully when the Saturn I-B came under pressure. NASA had conducted the initial flight test of a Saturn-class first stage as early as October 1961, at a time when the Titan III was still at the level of preliminary study. In view of this early start, and because the Saturn

---

15. Durant, *Reformation*, p. 264.
16. NASA SP-4102, p. 218; NASA SP-4407, Vol. II, pp. 318, 323.
17. NASA SP-4012, Vol. III, pp. 27, 39; Thompson, ed., *Space Log*, Vol. 27 (1991), p. 125; *Quest*, Fall 1995, p. 18; Long, chairman, *Space Program*, p. 36.

# THE SPACE SHUTTLE DECISION

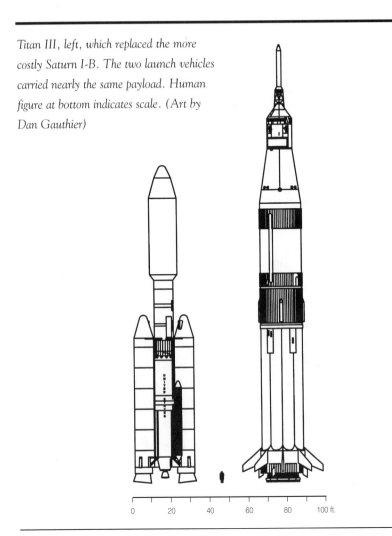

Titan III, left, which replaced the more costly Saturn I-B. The two launch vehicles carried nearly the same payload. Human figure at bottom indicates scale. (Art by Dan Gauthier)

I-B was essential for Apollo, NASA went on to build 14 of them, though George Mueller hoped for more as he pursued Apollo Applications. When budget cuts hit home, however, NASA abandoned the Saturn I-B and turned to the Titan III-E Centaur. It had the energy to launch large payloads on missions to Mars and the outer planets, and did so repeatedly.[18]

In addition to launch vehicles, NASA turned to the Air Force for facilities used for launch and tracking. When NASA's rockets flew from Cape

---

18. NASA SP-4012, Vol. II, pp. 54-57; Vol. III, pp. 40-41; *Aviation Week*, August 3, 1970, p. 45.

Canaveral, they proceeded down the Eastern Test Range—which the Air Force operated. That service provided tracking stations, and when NASA built stations of its own on the islands of Antigua and Ascension, they were co-located near those of the Defense Department.

The Air Force also built up an extensive array of launch facilities at Cape Canaveral. When NASA took over nearly exclusive use of some of them, the Air Force transferred them to NASA outright. These included Launch Complex 12 for Atlas-Agena, LC 36 for Atlas-Centaur, and LC 19 for the Titan II. Other launch pads served both agencies: LC 17 for Delta, LC 41 for Titan III. In addition to this, NASA launched early versions of Saturn, including the Saturn I-B, from LC 34 and 37, which had been built on land owned by the Air Force.

The two agencies also cooperated closely in research. The Air Force had a valuable set of wind tunnels and engine-test facilities at its Arnold Engineering Development Center in Tennessee. This service, however, did not attempt to duplicate the far more extensive facilities of NASA Ames, Langley, and Lewis. In addition to a broad array of supersonic wind tunnels, NASA offered such unique installations as a wind tunnel at Ames Research Center with a 40 by 80-foot cross section, big enough to hold and test full-size fighter aircraft. At NASA Langley, a 60-foot vacuum sphere could accommodate large spacecraft and rocket stages.[19]

In addition to sharing facilities, NASA and the Air Force also pursued joint ventures in research. The X-15 was one; another, the XB-70, involved large aircraft that could fly at Mach 3. The agencies also collaborated in building immense solid-propellant rockets. At Edwards Air Force Base, NASA built test stands for rocket engines used in Apollo. These complemented earlier Air Force test facilities.

Institutional arrangements also bound them closely. Between 1958 and 1964, NASA and the Defense Department executed some 88 major agreements. A joint Aeronautics and Astronautics Coordinating Board (AACB) dealt with such areas as aeronautical research, launch vehicles, spacecraft, and piloted space flight. NASA's Deputy Administrator and the DoD's Director of Defense Research and Engineering co-chaired this board; as early as 1966, an AACB subpanel carried out an important review of concepts for reusable launch vehicles.

---

19. NASA SP-4102, pp. 213, 221, 229, 236; NASA SP-440.

## THE SPACE SHUTTLE DECISION

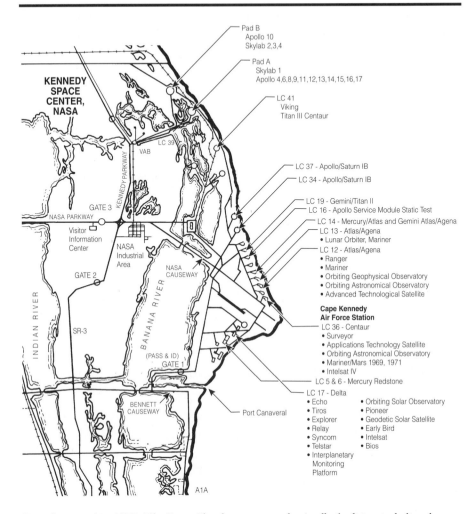

*Cape Canaveral in 1972. The Space Shuttle was to use the Apollo facilities, including the VAB and LC-39. (NASA)*

Within the Defense Department, the Air Force Systems Command (AFSC) held overall responsibility for that service's space and missile programs. In downtown Washington, an AFSC liaison office shared a building with NASA's Office of Manned Space Flight (OMSF). After 1962, NASA had its own Office of Defense Affairs that performed a similar function.[20]

---

20. NASA SP-4102, pp. 213, 217-220, 294 (footnote 17); Ames, chairman, *Report*.

Yet this interagency cooperation would only go so far. In August 1963, Webb and McNamara signed an agreement that sought "to ensure that in the national interest complete coordination is achieved" in pursuing a joint space-station project. Only a month later, McNamara sent Webb a follow-up letter that expressed his reservations. Then in December, McNamara made it clear that at least for the short term, the Air Force would want a piloted orbital facility of its own.[21]

When he canceled Dyna-Soar, on December 10, he handed the Air Force a consolation prize by inviting that service to conduct studies of a new project, the Manned Orbiting Laboratory (MOL). MOL took shape as a cylinder, 10 feet across by 41 feet long, with a Gemini spacecraft at one end; this ensemble was to ride to orbit atop a Titan III. McNamara could not grant formal approval for MOL; that had to come from the White House, and this raised anew the question of what the Air Force might do with such a facility. While that service had failed to provide good justification for Dyna-Soar, this time it came up with a fine reason for MOL: strategic reconnaissance.

The eventual plan called for MOL to carry a telescope with an aperture of six feet, offering resolution of nine inches. Astronauts would avoid photographing cloud-covered regions, but would scan the ground with binoculars, looking for items of interest. The Air Force won support from such key figures as Kermit Gordon, director of the Budget Bureau, and Donald Hornig, the White House science advisor. In August 1965, President Johnson gave MOL his endorsement, which meant it could go forward to contract award and development.

By then George Mueller was nurturing hopes for Apollo Applications, which raised anew the prospect of duplication. NASA officials, unwilling to affront the Air Force, supported MOL and took the view that it was not the national space station contemplated in the 1963 Webb-McNamara agreement. Nevertheless, members of Congress as well as Budget Bureau officials soon were asking whether NASA could adopt a version of MOL for its own use.[22]

In January 1966, Senator Clinton Anderson, chairman of the Senate space committee, sent a letter to Webb that recommended use of MOL. Within the House, the Military Operations Subcommittee criticized Mueller's plan for "unwarranted duplication" and called for MOL to proceed as a joint NASA-

---

21. NASA SP-4407, Vol. II, pp. 356-360.
22. NASA SP-4102, pp. 230-235; Richelson, *Secret Eyes*, pp. 82-83, 90-91; NASA SP-4208, pp. 17-19.

## THE SPACE SHUTTLE DECISION

Air Force program—with the Air Force in charge. Budget Bureau officials also supported a common program. In February 1967, the President's Science Advisory Committee added its own views, calling for "maximum utilization" of MOL and calling on NASA to carefully consider its use "before substantial funds are committed" to Apollo Applications.

NASA responded by having Douglas Aircraft, the MOL prime contractor, evaluate the suitability of MOL for NASA's objectives. The agency also conducted in-house studies. These began by acknowledging that the Saturn I-B was far more costly than the Titan III, and considered whether it might be advantageous to have the latter launch Apollo spacecraft. The OMSF concluded that while this was possible, it would cost $250 million to develop such a Titan-Apollo, which would then require 17 launches before the savings surpassed the initial cost of conversion.

The OMSF also concluded that MOL was too small for NASA's needs. It was no larger than a house trailer, whereas Mueller had described his proposed wet workshop as being the size of "a small ranch house." While the Air Force had a proposal in hand for a larger MOL, this would cost an additional $480 million and would take four years to develop. In comparison, even the Saturn I-B would cost less to use. These arguments mollified the critics, and Apollo Applications went forward, though with a reduced budget.[23]

MOL also went forward, with strong Pentagon support. The Air Force, however, never having carried through the development of a piloted spacecraft, failed to control its cost. Between 1965 and 1969, the projected cost of MOL ballooned from $1.5 billion to $3 billion. During those same years, the escalating Vietnam War placed military programs under severe strain.

The future of MOL came up for discussion at a White House meeting between Nixon, national security advisor Henry Kissinger, and Budget Bureau director Robert Mayo. Though the program carried the strong endorsement of Defense Secretary Melvin Laird and the Joint Chiefs of Staff, it proved to lack support from a key official: Richard Helms, Director of Central Intelligence and head of the CIA. In the words of the analyst Jeffrey Richelson, Helms's advisers "feared that an accident that cost the life of a single astronaut might ground the program for an extended period of time and cripple the reconnaissance program."

---

23. NASA SP-4208, pp. 43, 46-48, 54; NASA SP-4102, p. 232; Long, chairman, *Space Program*, pp. 23-25.

Mayo suggested canceling MOL; Nixon and Kissinger agreed. Only then did Mayo discuss the matter with Laird, who had not even believed that MOL was in trouble. Though Laird appealed directly to Nixon, emphasizing that the Joint Chiefs firmly supported this program, Nixon turned him down. The public announcement of the end of MOL came on June 10, 1969, with its first piloted flight still three years in the future.[24]

During the 1960s, the Air Force pursued two major and separate efforts—Dyna-Soar and MOL—that sought to place military astronauts in orbit. This service ended the decade with both projects canceled and with nothing to show for its efforts. Clearly, if it was to send such astronauts aloft, it would not do so on its own, but would have to work in cooperation with NASA.

In addition to this, the experiences of the Titan III, the Saturn I-B, MOL, and Apollo Applications had shown clearly that these agencies could easily introduce wasteful duplication by pursuing their own programs. This made it plausible that a cooperative NASA-Air Force program, focusing on piloted flight to orbit, would take shape as a national program, a unified effort shaped to serve the needs of both agencies.

Clearly, Air Force involvement would emphasize strategic reconnaissance, which represented the main rationale for that service's activities in space. The experience of MOL, however, showed that it would not do simply to propose that astronauts could operate telescopes and cameras from orbit. Instead, the Air Force would have to use piloted flight to support its work with automated reconnaissance satellites, such as those of Corona and Gambit.

As early as 1963, NASA and the Air Force had executed the Webb-McNamara agreement, which contemplated a joint space station. With MOL now canceled, it was difficult to see how the Pentagon could justify major participation in the space station that NASA's Tom Paine wanted so badly. The Space Shuttle was another matter. By launching, retrieving, and servicing spacecraft, it might significantly enhance the ability of the Air Force to conduct strategic reconnaissance. In turn, by serving Air Force needs, the Shuttle might indeed take shape as a truly national system, carrying military as well as civilian payloads. Beginning in 1969, the evolution of the space shuttle concept took a sharp turn in this direction.

---

24. Richelson, *Secret Eyes*, pp. 101-103.

## A New Shuttle Configuration

When a new round of shuttle design studies got under way, early in 1969, the field had seen no truly new concept since Max Hunter's partially-reusable Star Clipper of several years earlier. While work went forward at the contractors, Max Faget, at the Manned Spacecraft Center, carried through a parallel effort of his own that indeed came up with a new approach. His configuration not only went on to dominate the alternatives; it changed the terms of the ongoing discussions. These discussions had emphasized such issues as full versus partial reusability, with neither approach finding expression in a generally-accepted design concept. Faget now introduced a specific concept: a two-stage fully-reusable shuttle. As it gained acceptance, it spurred debate over its specific features, notably size, payload capacity, and choice of wing design. By focusing the debate, Faget's work thus narrowed the topics that subsequent studies would address, and enabled these studies to achieve greater depth.

Faget was an aerodynamicist who had built his career at NACA's Langley Aeronautical Laboratory. He was a member of the Pilotless Aircraft Research Division, an early nucleus of activity in high speed flight. In 1954, he took part in an initial feasibility study that led to the X-15. He then found a point of departure for his subsequent career in the findings of his fellow NACA aerodynamicists, H. Julian Allen and Alfred Eggers. They had shown that for a reentering nose cone, a blunt shape would provide the best protection against the heat of reentry.

Working with a longtime associate, Caldwell Johnson, Faget proceeded to devise a suitable blunt shape for Project Mercury, which put America's first astronauts in orbit. His Mercury capsule took shape as a cone, with its broad end forward and covered with a thick layer of material to provide thermal protection. (A cutaway view of this concept, elegantly rendered, hangs in Faget's offices to this day.) He came to Houston as a founding member of the Manned Spacecraft Center (MSC), where he became Director of Research and Engineering. He also adapted his basic shape to provide capsules for Gemini and Apollo.[25]

---

25. Author interview, Max Faget, Houston, March 4, 1997; AAS History Series, Vol. 8, p. 299; NASA SP-4307 and -4308, index references under "Faget."

*Maxime Faget. (NASA)*

"My history has always been to take the most conservative approach," he declares. In this frame of mind, he disliked much of the work done to date on Space Shuttle concepts. Lifting-body configurations were popular; Lockheed's Max Hunter had used them in his Star Clipper. Faget acknowledged their merits: "You avoid wing-body interference," which brings problems of aerodynamics. "You have a simple structure. And you avoid the weight of wings." He saw difficulties, however, that were so great as to rule out lifting bodies for a practical shuttle design.

They had low lift and high drag, which meant a dangerously high landing speed. As he put it, "I don't think it's charming to come in at 250 knots." Engineers at McDonnell Douglas, studying their Tip Tank lifting body, had tried to improve the landing characteristics by adding small wings that would extend from the body during the final approach. This appeared as very makeshift.

Because they required a fuselage that would do the work of a wing, lifting bodies also promised serious difficulties in development. It would not be possi-

## THE SPACE SHUTTLE DECISION

*Faget's shuttle concept. (NASA)*

ble to solve aerodynamic problems in straightforward ways; the attempted solutions would ramify throughout the entire design. In his words, "They're very difficult to develop, because when you try to solve one more problem, you're creating another problem somewhere else." His colleague Milton Silveira, who went on to head the MSC Shuttle Engineering Office, held a similar view:

> *If we had a problem with the aerodynamics on the vehicle, where the body was so tightly coupled to the aerodynamics, you couldn't simply go out and change the wing. You had to change the whole damn vehicle, so if you made a mistake, being able to correct it was a very difficult thing to do.*[26]

Instead, Faget proposed to build each of his shuttle's two stages as a winged airplane, with thermal protection on the underside. Before it could fly as an airplane, such a shuttle would first have to reenter, which meant it

---

26. Author interview, Max Faget, Houston, March 4, 1997; Joe Guilmartin and John Mauer interview, Milton Silveira, Washington, November 14, 1984, p. 14.

would need the high drag of a blunt body. "With extremely high drag," he notes, "you throw a big shock wave in front of you, and all the energy goes into that shock." Even with thermal protection, he did not want to fly his shuttle during reentry, in the manner of an airplane: "It's a hell of a lot easier to do a no-lift entry than a lifting entry, from the standpoint of heat protection." With airplane-style reentry, "you are stuck in the atmosphere, going fast for a long time." Rather than lose energy to a shock wave, the airplane would experience drag through friction with the atmosphere which would transfer heat to its surface.

Faget expected to turn his airplane into a blunt body by the simple method of having it reenter at a very high angle of attack, with its broad lower surface facing the direction of flight. In effect, he would take an Apollo capsule, with its large circular heat shield, and trim it to the shape of an airplane with wings. This concept drew on the experience of the X-15 that looked like a fighter plane but reduced its reentry heating by coming in nose-high. It also revived a design approach introduced a decade earlier by NASA's Charles Mathews. He had also proposed to build a winged spacecraft as a glider that would reenter with its bottom side facing forward.

Faget wrote that "the vehicle would remain in this flight attitude throughout the entire descent to approximately 40,000 feet, where the velocity will have dropped to less than 300 feet per second. At this point, the nose gets pushed down, and the vehicle dives until it reaches adequate velocity for level flight." This dive would cost some 15,000 feet of altitude. The craft then would approach a runway and land at a moderate 130 knots, half the landing speed of a lifting body.

Faget wrote that because its only real flying would take place during this landing approach, a wing design "can be selected solely on the basis of optimization for subsonic cruise and landing." The wing best suited to this limited purpose would be straight and unswept, like the wings of fighter planes in World War II. A tail would provide directional stability, again as with a conventional airplane. By moving control surfaces on the horizontal stabilizer, a pilot then could raise the nose slightly just before touching down on a runway, in a maneuver called a flare which adds lift and makes the touchdown gentle.[27]

---

27. *Astronautics & Aeronautics*, January 1970, pp. 52-61; author interview, Max Faget, Houston, March 4, 1997; Faget and Silveira, *Fundamental Design Considerations*, October 1970.

## THE SPACE SHUTTLE DECISION

Faget's concept had the beauty of simplicity and, inevitably, knowledgeable specialists would criticize it as being too simple. The Air Force Flight Dynamics Laboratory (FDL), at Wright-Patterson Air Force Base, quickly emerged as a center of such criticism. The FDL had sponsored space shuttle studies in parallel with those of NASA, and had investigated such concepts as Lockheed's Star Clipper. One of its managers, Charles Cosenza, had been a leader in the development of ASSET. Another FDL scientist, Alfred Draper, was a leader in the field of space systems. Beginning in early 1969, he took the initiative in questioning Faget's approach.[28]

Draper did not accept the idea of building a shuttle as an airplane that would come in nose-high, then dive through 15,000 feet to pick up flying speed. With its nose so high, the plane would be fully stalled, and the Air Force disliked both stalls and dives, regarding them as preludes to an out-of-control crash. Draper preferred to have the shuttle enter its glide while still supersonic, thus maintaining much better control while continuing to avoid aerodynamic heating.

If the shuttle was to glide across a broad Mach range, from supersonic to subsonic, then it would encounter an important aerodynamic problem: a shift in the wing's center of lift. Although a wing generates lift across its entire lower surface, one may regard this lift as concentrated at a point, the center of lift. At supersonic speeds, this center is located midway down the wing's chord (the distance from leading to trailing edge). At subsonic speeds, this center shifts and moves forward, much closer to the leading edge. Keeping an airplane in proper balance requires the application of an aerodynamic force that can compensate for this shift.

The Air Force had extensive experience with supersonic fighters and bombers that had successfully addressed this problem, maintaining good control and handling characteristics from Mach 3 to touchdown. Particularly for large aircraft—the B-58 and XB-70 bombers, and the SR-71—the preferred solution was a delta wing, triangular in shape. Typically, delta wings ran along much of the length of the fuselage, extending nearly to the tail. Such aircraft dispensed with horizontal stabilizers and relied instead on elevons, control surfaces resembling ailerons set at the wing's trailing edge. Small deflections

---

28. Jenkins, *Space Shuttle*, pp. 36, 56; Hallion, ed., *Hypersonics*, p. 459; *Astronautics & Aeronautics*, January 1971, p. 28.

*Shuttle to the Forefront*

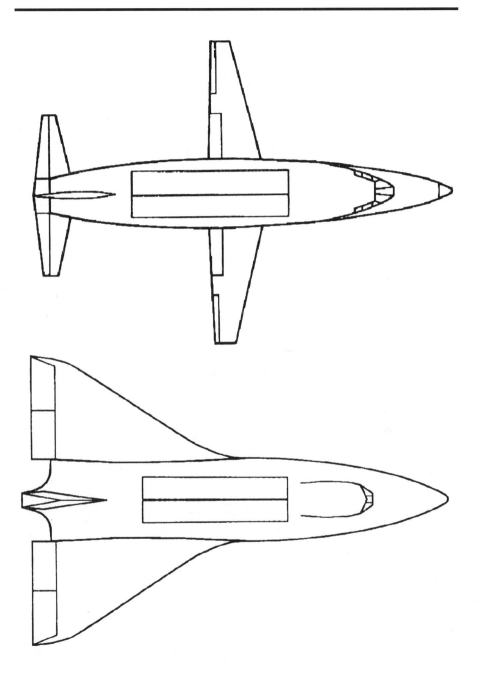

Straight-wing orbiter, top, and delta-wing orbiter. (Art by Dennis Jenkins)

of these elevons then compensated for the shift in the center of lift, maintaining proper trim and balance without imposing excessive drag.[29]

Draper proposed that both stages of Faget's shuttle should feature delta wings, rather than straight ones. Faget would have none of this. Though he acknowledged the center-of-lift problem, he expected to avoid it: "The straight wing never flew at those speeds; it fell at those speeds." A delta wing with elevons promised problems at landing, when executing the flare prior to touchdown. That flare was to add lift, but raising the elevons would increase the drag—with the added lift coming only after the nose had time to come up. This momentary rise in drag would make the landing tricky and possibly dangerous.

To achieve a suitably slow landing speed, Faget argued that the delta wing would need a large wingspan. A straight wing, having narrow chord, would be light and would offer relatively little area demanding thermal protection. A delta of the same span, necessary for a moderate landing speed, would be physically much larger than the straight wing. It would add considerable weight, and would greatly increase the area that would receive thermal protection.

Draper responded with his own viewpoint. For a straight wing to deal with the shift in center of lift, a good engineering solution would call for installation of canards, small wings mounted well forward on the fuselage that would deflect to give the desired control. Canards produce lift, and would tend to push the main wings farther to the back. These wings would be well aft from the beginning, for they would support an airplane that was empty of fuel but that had heavy rocket engines at the tail, placing the airplane's center of gravity far to the rear. The wings' center of lift was to coincide closely with this center of gravity. Draper wrote that the addition of canards "will move the wings aft and tend to close the gap between the tail and the wing." The wing shape that fills this gap is the delta; Draper added that "the swept delta would most likely evolve."

The delta also had other advantages as well. Being thick where it joins the fuselage, it would readily offer room for landing gear. Its sharply-swept leading edge meant that a delta would produce less drag than a straight wing near Mach 1. In addition to this, when decelerating through the sound barrier, a delta would shift its center of lift more slowly. The combination of a sudden drag rise near Mach 1, combined with a rapid center-of-lift shift, would pro-

---

29. Hallion, *Hypersonic*, p. 1032; author interview, Dale Myers, Leucadia, California, December 6, 1996.

duce a sudden and potentially disconcerting change in the stability characteristics of a straight-winged shuttle when slowing through the speed of sound. This change in stability would be much less pronounced with a delta, and would give a pilot more time to react.[30]

The merits of deltas might have remained a matter for specialists except for another important feature of the delta: Compared to the straight wing, it produced considerably more lift at hypersonic speeds. Using this lift, a reentering shuttle could achieve a substantial amount of crossrange, flying large distances to the left or right of an initial direction of flight. The Air Force wanted plenty of crossrange, and the reasons involved its activity in strategic reconnaissance. In particular, these reasons drew on recent experience involving the Six-Day War in the Middle East in 1967, and the Soviet invasion of Czechoslovakia in 1968.

The Six-Day War broke out suddenly, pitting Israel against a coalition led by Egypt whose tanks and aircraft came largely from the Soviet Union. Though America's intelligence community sought to follow the fighting closely, its means proved to be limited. A ship near the Israeli coast, the *USS Liberty*, monitored the communications of the belligerents—until the Israelis bombed it. Though spy planes, such as the U-2 and SR-71, could look down through clear desert skies, experience had shown that the U-2 was vulnerable to antiaircraft missiles. Satellite reconnaissance relied on Gambit and Corona, which had been designed to follow the slow development and deployment of missiles and other strategic weapons. They were not well-suited to the swift battle maneuvers of the 1967 war, and Defense Secretary McNamara does not recall that these spacecraft played any role in U.S. intelligence-gathering during those six days. By the time Corona photography became available, the war was already over.

Then, in August 1968, the Soviets stormed into Prague. A Gambit spacecraft, launched on August 6, performed poorly and was deorbited before the invasion took place. In addition to this, the CIA had a Corona satellite that entered orbit on August 7. It carried two capsules for film return. The first one appeared reassuring; it showed no indications of Soviet preparations for an attack. The second capsule returned photos that clearly showed such

---

30. Author interview, Max Faget, Houston, March 4, 1997; *Astronautics & Aeronautics,* January 1970, pp. 26-35; AIAA Paper 70-1249.

preparations, including massing of troops. By the time that film reached Washington, however, those photos were of historical interest only. The invasion had already taken place.[31]

Clearly, the CIA needed real-time space reconnaissance, and its pursuit of this goal would represent one more instance wherein a task originally thought to require astronauts would be accomplished using automated electronics. The true solution would lie in doing away with photographic film, which took time to expose and return. This film would give way to a new electronic microchip called a charge-coupled device. With an image focused onto this chip, it would convert the image into a rapid series of bits. The data, transmitted to the ground, would give the desired real-time photography, and with very high resolution. In addition to this, by freeing reconnaissance satellites from the need to carry and return film, this invention would allow such spacecraft to remain in orbit and to operate for years.[32]

The charge-coupled device grew out of the work of two specialists at Bell Labs, William Boyle and George Smith. In 1969, such technology still lay in the future. The view in the Air Force was that the CIA would need piloted spacecraft to produce the real-time photos. The late lamented MOL had represented a possible method, for an onboard photointerpreter might take, develop, and analyze photos on short notice. Now, with MOL in its graveyard, attention turned to the Space Shuttle. It might fly into space, execute a single orbit, and return to its base with film exposed less than an hour earlier.

Because much of the Soviet Union lies above the Arctic Circle, the Air Force was accustomed to placing reconnaissance satellites into polar orbits. It could not do this by firing its boosters from Cape Canaveral; geography dictated that these boosters would fly over populated territory. A launch to the north carried the hazard of impact in the Carolinas; a launch to the south would compromise security if the rocket fell on Cuba. Hence, the Air Force maintained its own space center at Vandenberg AFB, on the California coast. It offered a clear shot to the south, across thousands of miles of open ocean.[33]

While a satellite orbit remains fixed in orientation with respect to distant stars, the Earth rotates below this orbit. This permitted single reconnaissance missions to photograph much of the Soviet Union. However, it meant that if

---

31. Richelson, *Secret Eyes*, pp. 94-96, 97-99.
32. Ibid., pp. 124-132, 362; *Quest*, Summer 1995, pp. 31-32.
33. *Time*, December 15, 1958, pp. 15, 41-42.

a shuttle was to execute a one-orbit mission from Vandenberg, it would return to the latitude of that base after 90 minutes in space only to find that, due to the Earth's rotation, this base had moved to the east by 1,100 nautical miles. Air Force officials indeed expected to launch the Shuttle from Vandenberg, and they insisted that the Shuttle had to have enough crossrange to cover that distance and return successfully.

The Air Force had other reasons to want once-around missions. Its planners were intrigued by the idea of using the Shuttle to retrieve satellites in orbit. They hoped to snare Soviet spacecraft in such a fashion—and because Moscow might defend such assets by deploying an antisatellite weapon, the Air Force took the view that if the thing was to be done at all, it was best to do it quickly. A once-around mission could snare such a spacecraft and return safely by the time anyone realized it was missing.

In addition to this, NASA and the Air Force shared a concern that a shuttle might have to abort its mission and come down as quickly as possible after launch. This might require "once-around abort," which again would lead to a flight of a single orbit. A once-around abort on a due-east launch from Cape Canaveral would not be too difficult; the craft might land at any of a number of sites within the United States. In the words of NASA's LeRoy Day, "If you were making a polar-type launch out of Vandenberg, and you had Max's straight-wing vehicle, there was no place you could go. You'd be in the water when you came back. You've got to go crossrange quite a few hundred miles in order to make land."[34]

The Air Force had ample opportunity to emphasize its desire for crossrange by working within the Joint Study Group that Paine and Seamans had set up to seek a mutually-acceptable shuttle design. There were informal discussions as well. George Mueller, who continued to head NASA's OMSF through the whole of 1969, met repeatedly with Air Force representatives at his home in Georgetown, close to downtown Washington. One of his guests was Michael Yarymovych, an Air Force deputy assistant secretary. Another guest, Grant Hansen, was assistant secretary for research and development. He and Mueller also were co-chairmen of the joint study.[35]

---

34. Personal discussions with John Pike, Federation of American Scientists, July 1997; John Mauer interview, LeRoy Day, October 17, 1983, p. 41; Pace, *Engineering*, pp. 146-149.
35. Mueller, *Briefing*, 5 May 1969, p. 2; Pace, *Engineering*, p. 103.

## THE SPACE SHUTTLE DECISION

These Air Force leaders knew that they held the upper hand. They were well aware that NASA needed a shuttle program and therefore needed both the Air Force's payloads and its political support. The payloads represented a tempting prize, for that service was launching over two hundred reconnaissance missions between 1959 and 1970.[36] In addition to this, Air Force support for a shuttle could insulate NASA quite effectively from a charge that the Shuttle was merely a step toward sending astronauts to Mars.

Yet while NASA needed the Air Force, the Air Force did not need NASA. That service was quite content with existing boosters such as the Titan III. "Sure, NASA needs the shuttle for the space station," Hansen said in the spring of 1970. "But for the next 10 years, expendables can handle the Air Force job. We don't consider the Shuttle important enough to set money aside for it."

Yarymovych has a similar recollection:

*NASA needed Air Force support, both for payloads and in Congress. I told Mueller we'd support the Shuttle, but only if he gave us the big payload bay and the crossrange capability, so we could return to Vandenberg after a single orbit. Mueller knew that would mean changing Max Faget's beloved straight-wing design into a delta wing, but he had no choice. He agreed.*[37]

It was not that simple, of course; no impromptu discussion with Mueller would settle such an issue. Rather, it was a matter for the formal protocols of Air Force-NASA cooperation. There was strong conflict between these agencies' wishes, for a NASA baseline document of June 1969, "Desired System Characteristics," emphasized that NASA needed only 250 to 400 nautical miles of crossrange, enough to assure a return to Cape Canaveral at least once every 24 hours. Faget's straight-wing shuttle could achieve 230 nautical miles; straightforward modifications would meet NASA's modest requirements.

To give the Air Force its 1,100 nautical miles of crossrange would impose a serious penalty in design, by requiring considerably more thermal protection. The change to a delta wing, even without crossrange, would add considerably to the wing area demanding such protection. Crossrange then would increase this requirement even further. The Shuttle would achieve its

---

36. *Quest*, Summer 1995, pp. 22-33; Winter 1995, pp. 40-45.
37. Grey, *Enterprise*, pp. 67-68.

## Reference Heating Rate Histories for Orbiter

[Chart: Heating Rate, $Q_{REF}$ BTU/ft² sec vs. Entry Time, Sec. Two curves shown: L/D = 1.7 (higher, prolonged) and L/D = 0.5 (lower, brief).]

*Comparison of types of orbiter. The straight-wing design has low lift, L/D=0.5. It achieves little crossrange, but its reentry is brief and limits the heating rate. The delta-wing orbiter has high lift, L/D=1.7, and achieves large crossrange. Its reentry, however, is prolonged and imposes both a high heating rate and a high total heat. (NASA)*

crossrange by gliding hypersonically, and hence would compromise the simple nose-high mode of reentry that would turn it into a blunt body. This hypersonic glide would produce more lift and less drag. It also would increase both the rate of heating and the duration of heating. Crossrange thus would call for a double dose of additional thermal protection, resulting in a shuttle that would be heavier—and more costly.[38]

Even in its simple straight-wing form, Faget's concept of a two-stage fully-reusable shuttle did not take NASA by storm. It won its pre-eminence only after a process of review and evaluation that extended through 1969 and into 1970. The framework for this process involved the contractors' studies of shuttle configurations that had begun early in 1969. Those studies ruled out expendable boosters for a reusable shuttle, for such boosters were found to exceed the Saturn V in size. Fully- and partially-reusable shuttle concepts

---

38. Day, manager, *Task Group Report*, Vol. II, June 12, 1969, pp. 40-42; *Astronautics & Aeronautics*, January 1970, pp. 57, 59; Faget and Silveira, *Fundamental Design Considerations*, October 1970, pp. 5, 17.

remained in the running, and Faget's concept counted as a new example of the former. NASA proceeded to examine it alongside several alternatives, beginning in mid-1969.

The initial round of studies, during the first half of 1969, had come to $1.2 million, divided equally among four contractors. NASA now extended these studies by giving $150,000 more to each of three contractors, with McDonnell Douglas receiving $225,000. The participating companies also received new instructions that redirected their work.

North American Rockwell had examined expendable boosters. With this approach now out of favor, this firm was free to direct its attention to something new. This proved to be Faget's straight-wing concept, largely in the form he recommended.

McDonnell Douglas, which had examined its Tip Tank stage-and-a-half design, now switched to two-stage fully-reusables. These, however, were not Faget's, but rather continued an earlier line of work. They featured orbiter designs derived from the HL-10 lifting body, with this contractor's engineers considering 13 possible configurations for the complete two-stage vehicle.

At first, the other two contractors saw little change in their assignments. Lockheed was to continue with studies of Star Clipper and of its own version of the Triamese. General Dynamics, home of the initial Triamese concept, was to study variants of this design, and would also apply its background to design a fully-reusable concept having only two elements rather than the three of Triamese.

The orders for this redirection went out on June 20, 1969. Within weeks, the studies brought a flurry of activity that further narrowed the admissible choices. Expendable boosters had already fallen by the wayside. On August 6, a meeting of shuttle managers brought a decision to drop all partially-reusable systems as well. With this, both Lockheed's Star Clipper and McDonnell Douglas's Tip Tank were out. This decision meant that NASA would consider only fully-reusable concepts.[39]

Partially-reusable designs had represented an effort to meet economic goals by seeking a shuttle that would cost less to develop than a fully-reusable system, even while imposing higher costs per flight. This approach had held promise prior to the spring of 1969, when the shuttle had been con-

---

39. Akridge, *Space Shuttle*, pp. 53, 71-72, 90; Jenkins, *Space Shuttle*, pp. 60-64.

*Shuttle to the Forefront*

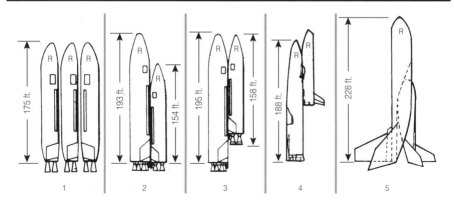

*Fully-reusable shuttle concepts of 1969; R indicates the stage is reusable. 1) Triamese of General Dynamics. 2) Two-stage arrangement with both stages thrusting at launch. 3) Two stages, upper stage ignited at high altitude. 4) Faget's concept. 5) Concept of NASA Langley, with both stages as lifting bodies. (NASA)*

sidered largely as a means of providing space station logistics. Now its intended uses were broadening to include launches of automated spacecraft, which meant it might fly far more often. The low cost per flight of a fully-reusable now made it attractive, and encouraged NASA to accept its higher development cost.[40]

There were at least five ways to build a fully-reusable shuttle, and NASA had appropriate designations and descriptions:

**FR-1:** the Triamese;

**FR-2:** a two-stage vehicle with the engines of both stages ignited at launch;

**FR-3:** a two-stage vehicle with engines in the orbiter ignited only upon staging (Faget's shuttle was an FR-3; so were the concepts of McDonnell Douglas);

**FR-4:** a variant of the Triamese with the core stage not of the same length as the twin booster stages;

**FR-5:** a concept designed to avoid a shift in its center of gravity as its propellant tanks would empty, thus easing problems of stability and control.

---

40. Day, manager, *Summary Report*, December 10, 1969.

## THE SPACE SHUTTLE DECISION

On September 4, another meeting eliminated the Triamese configurations. The initial concept, the FR-1, had called for three elements of common length and structural design. It had proven difficult, however, to have one shape serve both as booster and orbiter; to Silveira, "it gets all screwed up, so you get a lousy orbiter and a lousy booster, but you don't get one that does well." Advocates of the Triamese had turned to the FR-4, with its unequal-length design. This, however, proved heavier than the FR-3, while requiring two booster elements rather than one. It also lost much of the potential cost saving from design commonality between the three elements.

The FR-3 and FR-5 remained. The latter had few advocates; the problem of center-of-gravity shift was not so severe as to call for the design innovations of this class of concepts. The manager Max Akridge writes, "It was felt at this time that clearly, the FR-3 configuration was the forerunner."[41]

These decisions brought a further redirection in the studies, for while North American Rockwell had gotten an early start on Faget's concept, the other three contractors had to change course. McDonnell Douglas, having found no advantage in its lifting-body orbiters, turned to winged orbiters resembling those of Faget.

While Lockheed also turned to the FR-3, it did not embrace Faget's concept wholeheartedly. This company had spent several years studying Star Clipper, which featured a lifting-body orbiter, triangular in shape. This now looked like a good way to meet Air Force crossrange requirements, and Lockheed's new design retained this lifting body, with a broad underside in the shape of a delta.

General Dynamics showed its own individuality. That firm had designed its Triamese with retractable wings, which would fold into the body during flight but swing outward for landing. This eased the problem of providing these wings with thermal protection, because the fuselage would shield them. This feature now reappeared in the company's new FR-3. It drew on more than Triamese; it also reflected company experience with swinging wings. These were part of the F-111 fighter-bomber, which swung its wings to achieve good performance in both subsonic and supersonic flight.[42]

---

41. Akridge, *Space Shuttle*, p. 93; Jenkins, *Space Shuttle*, pp. 66-70; Joe Guilmartin and John Mauer interview, Milton Silveira, Washington, November 14, 1984, pp. 12-13.
42. Reports MDC E0056 (McDonnell Douglas); GDC-DCB69-046 (General Dynamics); SD 69-573-1 (North American Rockwell); LMSC-A959837 (Lockheed); Jenkins, *Space Shuttle*, pp. 63-70.

Hence, by the end of 1969 NASA had settled on the FR-3 as its choice, with Faget's specific concept in the forefront. This raised important questions concerning thermal protection. The booster was to be as large as a Boeing 747, yet was to outperform the X-15, reaching considerably higher speeds. The orbiter would be longer than a Boeing 707. For both, the thermal protection had to be reusable.

Within the industry, a standard engineering solution called for the use of hot structures. This approach had a background that included the X-15, Dyna-Soar, ASSET, as well as the Lockheed SR-71 that was flying routinely above Mach 3. Hot structures typically called for titanium as the basic material, covered with high temperature insulation and an outer skin formed of metallic shingles. The metal was molybdenum or columbium, to withstand extreme temperatures while radiating away the heat. Like the shingles on a roof, those on the surface of a hot structure were loosely attached, to expand and contract freely with temperature change.

Such structures were complex, and the shingles posed difficulties of their own. Columbium and molybdenum oxidize readily when hot, and required coatings to resist this. The Dyna-Soar had been designed to use such thermal protection, and Faget declared that "the least little scratch in the coating, the shingle would be destroyed during re-entry." In turn, lost shingles could bring the loss of a vehicle.

NASA and Lockheed now were developing a new surface material: an insulation made of interlaced fibers of silica that could be applied to the outside of a vehicle. These could withstand temperatures of 2,500 degrees Fahrenheit, making them suitable for all but the hottest areas on a reentering shuttle. The outer surface would radiate away the heat, in the fashion of the shingles. The thickness of the silica then would prevent most of the heat from reaching the vehicle's skin. This material would not oxidize. It also was light, weighing as little as 15 pounds per cubic foot, or one-fourth the density of water.

This material would form the well-known "tiles" of the Shuttle program, being attached to the skin in the form of numerous small shapes somewhat resembling bricks. In 1969, their immediate prospect lay in simplifying the design of hot structures. These might now dispense with their shingles; engineers instead would use titanium to craft an aircraft structure, with skin covering an internal framework, then provide thermal protection by covering the skin with the tiles.[43]

The design studies of 1969 raised another tantalizing prospect: that these tiles might offer enough heat resistance to build the basic structure of aluminum rather than titanium. Titanium was hard to work with; few machine shops had the necessary expertise. Moreover, its principal uses in aerospace had occurred within classified programs such as the SR-71, which meant that much of the pertinent shop-floor experience itself was classified. This metal could withstand higher temperatures than aluminum. Yet, if tiles could protect aluminum, the use of that metal would open the Shuttle to the entire aerospace industry. In Silveira's words, building aluminum airplanes was something that "the industry knew how to do. The industry had, on the floor, standards—things like, 'What are the proper cutting speeds?' They knew how to rivet or machine aluminum."[44]

Hot structures, built of titanium, would continue to represent an important approach in shuttle design. As early as 1969, however, Lockheed took the initiative in designing a shuttle orbiter built of aluminum and protected with tiles. General Dynamics added its own concept, featuring aluminum protected by shingled hot structures that could keep internal temperatures below 200 °F.

At the end of 1969, the contractors' orbiter concepts were as follows, with the boosters being similar:[45]

|  | Configuration | Main structure | Thermal protection |
|---|---|---|---|
| **North American Rockwell** | Faget-type, straight wing | Titanium | Tiles |
| **General Dynamics** | Deployable straight wing | Aluminum | Hot structure |
| **Lockheed** | Delta lifting body | Aluminum | Tiles |
| **McDonnell Douglas** | Faget-type, straight wing | Titanium | Tiles |

---

43. Author interview, Max Faget, Houston, March 4, 1997; proceedings, NASA Space Shuttle Symposium, October 16-17, 1969, pp. 581-591; Report MDC E0056 (McDonnell Douglas), pp. 10-11; Jenkins, *Space Shuttle*, p. 64.
44. Heppenheimer, *Turbulent Skies*, p. 209; Joe Guilmartin and John Mauer interview, Milton Silveira, Washington, November 14, 1984, p. 16.
45. Reports MDC E0056 (McDonnell Douglas); GDC-DCB69-046 (General Dynamics); SD 69-573-1 (North American Rockwell); LMSC-A959837 (Lockheed); Jenkins, *Space Shuttle*, pp. 63-71.

These represented variants of Faget's two-stage concept, which showed that shuttle design had come a long way during that year. Twelve months earlier, the candidate configurations included expendable boosters as well as partially- and fully-reusable concepts. The range of alternatives included the Titan III-M with an enlarged Dyna-Soar, the Star Clipper, and the Triamese. These were as mutually dissimilar as a fighter, a bomber, and a commercial airliner.

People still debated such issues as delta wings vs. straight, aluminum vs. titanium, and hot structures vs. tiles. By year's end, however, everyone agreed that the Shuttle would look much like Faget's. This meant that the most basic issues of configuration had been settled, allowing engineers to advance to deeper levels of detail. The studies of 1970 would pursue such levels, and would lay important groundwork for the eventual evolution of complete engineering designs, explicit in all particulars.

The work of 1969 had given the space station more support than the Shuttle. The studies of that year initially had allocated $5.8 million for the station and only $1.2 million for the Shuttle; the additional funds granted for the latter at midyear, totalling less than $0.7 million, did little to redress this imbalance. In 1970, however, NASA would bring the shuttle to the forefront. In this year, the centerpiece of effort would involve two $8-million contracts for further work on the Shuttle. There would be no significant amount of new funding for the station. These internal NASA decisions would point toward abandonment of the station, at least for a number of years, and elevation of the Shuttle into the sole focus for NASA's future.[46]

## *Station Fades; Shuttle Advances*

In preparation for the work of 1970, NASA and its contractors established a new set of institutional arrangements. Following a proposal of Dale Myers, the agency gave responsibility for managing the upcoming study contracts to NASA Marshall and to the Manned Spacecraft Center (MSC). Each center would hold a complete contract, covering both the booster and orbiter. Marshall, however, would provide technical direction for the booster

---

46. Logsdon, *Apollo*, p. III-26; *Aviation Week*, February 10, 1969, p. 17; Akridge, *Space Shuttle*, p. 71.

portion of both contracts, while MSC would give similar direction for the orbiter portions. This continued the arrangements of Apollo, which had assigned responsibility for the Saturn V to Marshall and for the Apollo moonship to MSC.[47]

Within the industry, competing companies made their own arrangements. There would be only two principal new study contracts, but there were many more than two firms eager for the business, and several of them proceeded to form teams. This reflected the Shuttle's two-stage design, for the complete shuttle was likely to be too large a project for a single contractor to handle. Accordingly, Lockheed teamed up with Boeing, with the two companies proposing respectively to handle the orbiter and booster. North American Rockwell joined with General Dynamics, while McDonnell Douglas associated with Martin Marietta.[48]

Another and highly important set of decisions extended the scope of Air Force-NASA cooperation. The two agencies had collaborated on a joint study of shuttle requirements as part of the work of the STG; this had led to the issuance of a three volume report in June 1969, classified Secret. That collaboration, however, had merely served the immediate needs of the STG and its supporting studies. Now, in February 1970, Paine and Seamans agreed to set up a permanent coordinating committee, with members to be drawn in equal numbers from each agency. As in the 1969 joint study, there again would be two co-chairmen: Dale Myers and Grant Hansen. Hansen had co-chaired the earlier study, while Myers would replace Mueller, on behalf of NASA.[49]

On the matter of crossrange, at least for the moment, they agreed to disagree. Neither agency would seek to impose its will on the other. Rather, each main study contract would conduct two design exercises in parallel: one for an orbiter with crossrange of 200 nautical miles, the second with capability of 1,500 nautical miles. The first was well within the reach of Faget's straight-wing concepts; the second called for more than the Air Force would need. Like a baseball player who swings two bats during warmup and then finds that his single bat feels lighter, the exercise of designing for 1,500 nautical

---

47. Low, *Personal Notes No. 5*, January 17, 1970, p. 4; *Aviation Week*, February 16, 1970, p. 14.
48. *Aviation Week*, January 19, 1970, pp. 17-18; February 9, 1970, p. 27.
49. Akridge, *Space Shuttle*, p. 70; NMI 1052.130, February 17, 1970. Reprinted in NASA SP-4407, Vol. II, pp. 367-368; see also p. 369.

miles would stretch the minds of engineers and make it easier for them to achieve the 1,100 nautical miles that represented the real requirement.

From the outset, however, Paine and Seamans could agree on an important point which Paine described as "a payload volume of 15 feet diameter by 60 feet long." Previous studies had considered payload bays not only of this size but of 22 feet diameter by 30 feet long. The new requirement reflected the needs of both agencies.[50]

The Air Force needed length, for its reconnaissance satellites amounted to orbiting telescopes, and these had to be long to yield the sharpest images. Moreover, such satellites were growing markedly in length. The Corona spacecraft of the 1960s, each with an attached Agena upper stage, had started at 19 feet and quickly grew to 26. The CIA was now readying a new class of satellites that would win the name of Big Bird. With dimensions of 40 by 10 feet, it represented a backup to MOL, whose length and width had been virtually the same. The next generation of satellites, called Kennan, would keep the 10-foot diameter but would grow in length to 64 feet.

Big Bird was in an advanced state of development in 1970; the first of them flew to orbit atop a Titan III-D in June 1971. Their photos gave a resolution of two feet. Kennan was still in initial studies, for it would introduce long-duration operations and the use of charge-coupled devices rather than film. Though it would not fly until December 1976, its images would show resolution as sharp as six inches.

In 1970, the size of Kennan had not been fixed; indeed, very little about this project had been fixed. It was clear to Air Force planners, however, that they would face an increasing need to launch long satellites. Accordingly, they declared that they would need a length of 60 feet for the shuttle's payload bay.

While NASA did not need so much length, its officials wanted a 15-foot diameter to accommodate modules for a space station. This reflected a new approach to the design of such stations. The studies of the 1960s, including those that Paine had initiated in 1969, had envisioned a space station as a single unit that would fly atop a Saturn I-B or Saturn V. As the prospects for Saturns faded while those of the Shuttle seemed to advance, it appeared prudent to envision a class of stations that could be assembled in space as an

---

50. Letter, Paine to Seamans, January 12, 1970; Reports NAS 9-10960 (NASA), p. 2; LMSC-A959837 (Lockheed), p. 2; Low, Memo for Record, January 28, 1970. Reprinted in NASA SP-4407, Vol. II, pp. 366-367.

## THE SPACE SHUTTLE DECISION

array of cylindrical modules, one module per shuttle flight. A shuttle bay with this diameter would accommodate modules 14 feet across, intermediate between the 10 feet of MOL and the 22 feet of Skylab.[51]

In addition, a 15-by-60 foot bay would serve the needs of both agencies by providing room for the space tug and its payloads. Many spacecraft would fly to high orbit, including geosynchronous orbit, and the payload bay had to address such expectations as that future communications satellites would also grow larger. Thus, when Dale Myers asked Grant Hansen to weigh the merits of a reduction to 12 x 40 feet, Hansen replied:

> *The length of the payload bay is the more critical dimension affecting DOD mission needs. If the payload bay length is reduced to 40 feet, then 71 of the 149 payloads forecasted for the 1981 to 1990 time period in Option C and 129 of the 232 payloads forecasted in Option B of the mission model will require launch vehicles of the Titan III family....*
>
> *The 15 foot diameter by 60 foot length payload bay size previously stated as the DOD requirement is based upon payloads presently in the inventory, on the potential use of a reusable upper stage to accomplish our high energy missions, and on a capability to provide limited payload growth. This requirement is still considered valid.*
>
> *In summary, should you elect to develop the shuttle with a 12 ft x 40 ft payload compartment, it will preclude our full use of the potential capability and operational flexibility offered by the shuttle.... Also, if a portion of the present expendable launch vehicle stable must be retained to satisfy some mission requirements, then the potential economic attractiveness and the utility of the shuttle to the DOD is severely diminished.*[52]

On February 20, NASA officials issued the formal Request for Proposals (RFP) that would lead to awards of the Shuttle study contracts of 1970. In addition to studies of the complete two-stage vehicle, a separate RFP solicited proposals for similarly detailed studies of a main engine. Competing companies had 30 days to respond; the submitted proposals would then go before a Source Evaluation Board that would pick the winners. This board included members from the Air Force. On May 12, it chose the teams headed by

---

51. Richelson, *Secret Eyes*, pp. 48, 90, 105-106, 124-130, 361-362; Pace, *Engineering*, pp. 110-116.
52. Letter, Hansen to Myers, June 21, 1971.

## Shuttle to the Forefront

McDonnell Douglas and by North American Rockwell and accepted their bids. These companies now were to proceed with 11-month studies that would carry their shuttle designs to new levels of detail, with each study contract being funded at $8 million.

A year earlier, those same companies had won similar $2.9 million contracts for studies of the space station. This reflected their strength, for they were the only firms to have designed and built both piloted spacecraft and large hydrogen-fueled rocket stages for Apollo. McDonnell had built the Mercury and Gemini spacecraft; Douglas' credits included the S-IVB stage, MOL, and the upcoming Skylab. North American's record included the Apollo spacecraft and the S-II, the second stage of the Saturn V. Other companies had gained strong achievements: General Dynamics's Atlas, Lockheed's Agena stages used with Corona, Martin Marietta's Titan family, Grumman's Apollo lunar module. In their experience, however, McDonnell Douglas and North American Rockwell were in a class by themselves.[53]

They had initiated their current space station studies in September 1969. By the following July, these companies had carried their designs to a good level of detail. Their stations would take shape as a cylinder with diameter of 33 feet, suitable for launch by Saturn V, surmounted by an enormous solar array. Jack Heberlig, a space station manager at MSC, described the internal layout as "basically four decks with a cellar and an attic." The cellar and attic would house spacecraft equipment, including storage tanks as well as noisy fans and blowers, with acoustic insulation to keep their noise from disturbing the crew. The four decks would provide room for living and working.

Mockups, built by the contractors, showed how crew members would live in comfort. Each person would have a stateroom resembling a small college dorm room, with a bunk bed, desk and chair, television and communications equipment, and plenty of storage space in drawers and a closet. Two communal lavatories would each provide a shower stall, urinal, and a zero-gravity toilet called a "dry john." The commander's stateroom would feature a personal lavatory, a small conference table, and a computer terminal—which was rare in 1970.

---

53. *Aviation Week*, September 22, 1969, p. 100; February 16, 1970, p. 14; February 23, 1970, p. 16; May 21, 1970, p. 18; letter, Paine to Seamans, January 12, 1970; letter, Paine to Teague, May 28, 1970.

*Space station concept of 1970, intended for launch with the Saturn V. (North American Rockwell)*

A U-shaped galley, with plenty of countertop space, would include ovens, dishwasher, trash compactor, refrigerator, and storage cabinets. Tables, each with four seats, would stand near a wall of freezer cabinets. Within a recreation area, a similar table would provide room for hobbies and games, with television and movies also being available. A medical room with a treatment table would serve as a dispensary and first-aid station, while also supporting studies of crew members' general health.

Though a Saturn V was to launch this station, its logistics were to depend on use of the Shuttle. Heberlig projected "a minimum of 91 Shuttle flights over a 10-year period." But in July 1970, with the shuttle/station having survived by the narrowest of margins in both the Senate and House, it was clear that to proceed with simultaneous development of both projects would court disaster.[54]

Dale Myers responded by deciding to move ahead with the Shuttle first. This brought a major NASA decision that redirected the ongoing space station studies. While work on 33-foot-diameter versions would continue, the agency now would emphasize investigation of modular versions, which the Shuttle could build in orbit as well as service.

This decision took shape as part of the broader cutbacks that Paine announced in September 1970. Those cutbacks canceled two Apollo missions, freeing their Saturn V rockets for other duty, and Paine noted hopefully that they might find use "in the Skylab, Space Station or other programs where manned operations or a heavy boost capability is required."[55]

Early in 1969, a year and a half previously, it had appeared obvious that Paine's station, with its crew of 12, would fly atop a Saturn V. The space station studies of that decade had all assumed use of Saturn-class launch vehicles, with the Shuttle merely as a handmaiden, a logistics vehicle that was poorly defined. By mid-1970, however, the Shuttle and station had reversed positions. The Shuttle now was ready to stand on its own, justified in large part by the work it could do in supporting the Air Force's *automated* spacecraft. With the Saturn V now representing the rarest and most valuable of commodities, NASA could expect to use it only after the most searching examination of alternatives. These alternatives would particularly include the construction of modular space stations, which could avoid use of Saturn-class boosters entirely.

Now it was the station that retreated to the realm of initial studies. There was no clear understanding, at least in 1970, of how specifically one might proceed with such modular stations, for the work to date had emphasized the design of such stations as large single units. North American Rockwell and McDonnell Douglas carried through an appropriate round of studies during 1971, and issued their reports. Then, with budget cuts continuing to squeeze future goals, NASA largely dropped its plans for such stations pending com-

---

54. *Aviation Week*, September 22, 1969, pp. 100-113; August 3, 1970, pp. 40-45; Report SD-70-536 (North American Rockwell), AIAA Paper 69-1064.
55. Low, Personal Notes No. 27, July 18, 1970, p. 13; Paine, statement, September 2, 1970; letter, Low to Shultz, September 30, 1970, p. 4.

# THE SPACE SHUTTLE DECISION

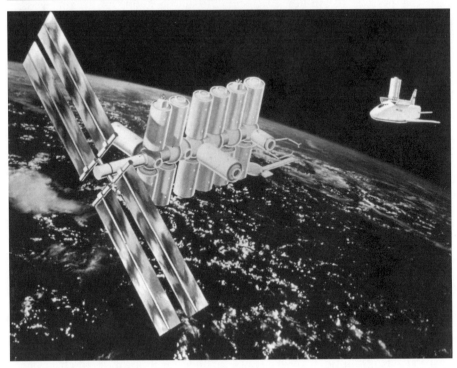

*Modular space station, with modules to be carried aboard the Space Shuttle. (McDonnell Douglas)*

pletion of shuttle development. Not until 1984, with the shuttle operational, would President Reagan finally give the go-ahead for a space station program.[56]

After 1970, with the space station off in the wilderness, its supporters could at least take heart that the Shuttle's capacious payload bay had been sized to accommodate its modules. The size of that bay also served a more immediate purpose, for it helped to nail down Air Force support. During the latter months of 1970, events showed that by working to win this support, and by deferring the space station to a much later time, NASA was well on the way to overcoming its opposition in Congress.

The NASA appropriation for FY 1971 had been part of an omnibus spending bill that had included funds for the Department of Housing and Urban

---

56. Reports MDC 2570, MDC 2727 (McDonnell Douglas); SD 70-153, SD 70-160, SD 71-214, SD 71-217-1, SD 71-576 (North American Rockwell).

Development. In debate during July, the Senate had added $400 million for the programs that Paine described as "sewers." This led Nixon to veto the entire bill, forcing NASA and the other affected agencies to get by for a time through the temporary funding provided by a continuing resolution in Congress. The Senate again took up the appropriations measure on December 7, and once more Senator Mondale introduced his amendment that sought to delete funding for the shuttle/station.

This time it went down to defeat by a comfortable margin, 26 to 50. Several senators switched their votes to support NASA, with one of them being John Pastore, chairman of the Senate Appropriations Committee. Following the vote, he stated, "This matter came before the Senate twice; the matter has been decided. It has been decided by the House and it has been decided by the Senate." Mondale continued to introduce similar amendments on subsequent occasions, but found his quest becoming increasingly lonely as they met defeat by even larger margins: 22 to 64 in 1971, 21 to 61 in 1972. With the Senate vote of December 1970, however, it was already clear that opposition had collapsed.[57]

Also during 1970, the joint NASA-Air Force Space Transportation System Committee emerged as a forum where representatives of the two agencies could hammer out solutions in areas of disagreement. The February 1970 concordat between Paine and Seamans, in establishing this committee, also stated that shuttle development "will be managed by NASA" and "will be generally unclassified." At meetings of this committee, however, Air Force officials proved quite frank in presenting classified material to support that service's point of view.

At a committee meeting on June 29, the Air Force gave a briefing on the size and weight requirements of DoD payloads. This began by disclosing the size and weight of current payloads, and went on to project the specifications of future payloads, eight to ten years ahead. The presentation also reviewed the history of Air Force launch vehicle payload capabilities and the length of payload fairings.

This briefing supported the Air Force demand for a 60-by-15 foot cargo bay. The length would accommodate both current and future payloads; the

---

57. Logsdon, *Apollo*, chapter 5, pp. 51-52; *Congressional Record*, December 7, 1970, pp. S19525-S19552; Low, Personal Notes No. 37, December 20, 1970, p. 4; *Aviation Week*, July 5, 1971, p. 19; *National Journal*, August 12, 1972, p. 1290; May 12, 1973, p. 689.

diameter would provide room for a space tug. It would also alleviate design complications associated with the restrictions of current launch vehicles such as the Titan III-D, which limited spacecraft to diameters of 10 feet.[58]

NASA willingly accepted this requirement, incorporating it explicitly within the Study Control Document that would guide the new round of space shuttle studies. However, this document did not specify a payload weight. At that same meeting on June 29, the Air Force declared that it wanted the Shuttle to carry 40,000 pounds to low polar orbit. This again would provide room for growth; Big Bird, a year away from its first launch, weighed close to 30,000 pounds, and future spacecraft would certainly be heavier.

NASA officials gave their response at another committee meeting, on October 2. They noted that the baseline mission, described in the Study Control Document, had continued to involve logistics resupply of a space station or space base, in an orbit with an inclination of 55 degrees. The Shuttle was to carry 25,000 pounds to this orbit. Following reentry, however, it would not glide to its home base, but would fly with power from turbojet engines. New studies now showed that an operational shuttle could indeed glide safely to this base, dispensing with those jet engines. That would save weight; the Air Force's requirement of 40,000 pounds then would indeed be achievable.[59]

By now it was clear that the Air Force was very much in the pilot's seat when it came to steering the Shuttle program. Max Faget learned this late in 1970, when he wrote a memo to his deputy director at MSC: "The USAF appears not to be nearly as firm on the 15 ft. diameter requirements as they are in length. NASA has no need for 15 ft. diameter either. It is suggested that you attempt to have the payload diameter reduced to 12 ft."

Faget was a power within NASA. He was Director of Engineering and Development at the MSC and reported directly to the head of that center, Robert Gilruth. It took the Air Force only three days to put him in his place, with a reply that read: "The USAF fully supports and stands firm on the present Level I requirement for a payload diameter of 15 feet and a length of 60 feet." This reply came from one Patrick Crotty—whose rank was no higher than major.[60]

---

58. NASA SP-4407, Vol. II, pp. 367-368, 373-374; Pace, *Engineering*, pp. 113-114.
59. NASA SP-4407, Vol. II, p. 374; Pace, *Engineering*, p. 115; Report NAS 9-10960 (NASA), p. 2; Richelson, *Secret Eyes*, p. 106.
60. NASA SP-4307, pp. 170, 212-213; Pace, *Engineering*, pp. 115-116, 130.

## Shuttle to the Forefront

Late in December, NASA formally upgraded the Shuttle's status. It had been managed by the Space Shuttle Task Group, headed by LeRoy Day. The space station had resided within a similar task group. Now the Shuttle received a separate program office, headed by Charles Donlan, Deputy Associate Administrator for Manned Space Flight. Donlan reported directly to Dale Myers, an Associate Administrator, and also kept Day as his own deputy. On NASA's organization chart, this raised the Shuttle to the status of such programs as Mercury, Gemini, and Skylab.

A month later, on January 19 and 20, 1971, NASA hosted a meeting in Williamsburg, Virginia, that included representatives from shuttle study contractors and from the Air Force. This meeting had the purpose of defining shuttle requirements that would guide the work of these contractors. NASA used the occasion to give the Air Force everything it wanted. In particular, the Shuttle would have a delta wing, with crossrange of 1,100 nautical miles. Its payload capacity, 40,000 pounds into polar orbit, would correspond to 65,000 pounds in a due-east launch from Cape Canaveral.[61]

One sometimes hears that when two parties are in a relationship, the one that wants it more is the weaker. NASA certainly had been pursuing support for the Shuttle with unmaidenly eagerness, and the Williamsburg rules were the result. The agency now was promising to build a bigger and heavier shuttle than it had wanted for its own uses, with considerably more thermal protection. It also was prepared to treat the Shuttle as a national asset—which meant the Air Force would not pay for its development or production and yet would receive the equivalent of exclusive use of one or more of these vehicles, entirely gratis. That service would not receive the Shuttle on a silver platter, but would pay for construction of its own launch facilities at Vandenberg AFB. Even so, with the Air Force having by far the larger budget as well as greater political clout, the Williamsburg agreement resembled a treaty between a superpower and a small nation.

Max Faget, for one, was not about to bow to the Air Force's superior wisdom. He was responsible for providing technical direction on the orbiter portions of the shuttle study contracts. His designers duly proceeded to turn out representative designs of delta-wing configurations that could guide the work of the contractors. His heart, however, remained with the straight wing.

---

61. *National Journal*, March 13, 1971, pp. 541, 545; Pace, *Engineering*, p. 116.

## THE SPACE SHUTTLE DECISION

He continued to come forward with new design variants until well into 1971, when as Dale Myers recalls, "I just denied MSC further activities on the straight-wing version."[62]

Having carried through its elaborate courtship (some people would prefer a different word) of the Air Force, NASA was now about to reap its reward. This came in March 1971, when Air Force Secretary Seamans presented testimony before the Senate space committee:

> *Now let me address the Air Force views regarding development of the Space Transportation System. The DOD supports its development if the results of current NASA Phase B studies and our own complementary studies show that such a system is feasible and can offer the desired performance and cost advantages over current systems. Preliminary indications from these studies are that such a system can be developed. If the final study results confirm this, and we think they will, the Air Force will provide a strong recommendation that Shuttle development be authorized. When the operational system is achieved, we would expect to use it to orbit essentially all DOD payloads, "phasing out" our expendable booster inventory with the possible exception of very small boosters such as the SCOUT.*[63]
>
> *The DOD investment over the next two to three years is planned to be small. However, in the future, we will require major funding to equip a DOD fleet and to provide unique DOD hardware, facilities and operational support.*[64]

While Seamans was not ready to give full consent, or to promise to give up the Air Force's cherished Titan III family, he certainly was saying "maybe," and was saying it emphatically.

The Shuttle program was advancing in another respect. With studies of its main engine having been under way for some time, NASA was about to award a contract for its actual development. With this engine as a long-lead item, this would be the first major component of the Shuttle to reach this level of commitment, advancing beyond the level of studies and design exercises to become a true hardware project.

---

62. Jenkins, *Space Shuttle*, pp. 107-117; Hallion, ed., *Hypersonic*, pp. 1006-1030, 1048-1051; author interview, Dale Myers, Leucadia, California, December 6, 1996.
63. A four-stage solid-fuel launch vehicle with payload capacity of some 425 pounds in orbit. NASA SP-4012, vol. III, pp. 28-31.
64. Seamans, Senate testimony, March 30, 1971, pp. 9-10.

## The Space Shuttle Main Engine

The legacies of the 1960s include the use of hydrogen as a rocket fuel, which powered two important engines of the period. The RL-10, developed by Pratt & Whitney, had a rated thrust of 15,000 pounds. Two of them were in the Centaur upper stage. The J-2 of Rocketdyne was an important component of Apollo, with five of these engines in the S-II stage of the Saturn V and a single J-2 in the S-IVB. Its thrust was 230,000 pounds.

The RL-10 and J-2 certainly did not represent the last word in rocketry. As early as 1967, well before the Shuttle began to take shape, the Air Force initiated an advanced propulsion program that led to new work at both Rocketdyne and Pratt & Whitney. These firms selected different approaches toward improving the hydrogen-fueled rocket engine, with the intent of building test hardware.

At Rocketdyne, the point of departure involved an inescapable shortcoming of conventional rocket nozzles, which had the shape of a bell. Within these nozzles, during and after liftoff, atmospheric pressure retarded the free expansion and outward flow of an engine's exhaust. This reduced both its thrust and exhaust velocity—and did so just when the launch vehicle was heavy with fuel and was burning propellant at the most rapid rate, thereby needing all the thrust and performance it could get.

The cure appeared to lie in a new type of engine, the aerospike. It required a ring-shaped combustion chamber surrounding a central body that resembled an upside-down volcano, with inward-sloping flanks and a central vent. Turbine exhaust flowed through the vent; the main engine exhaust expanded against the flanks, with no wall or barrier separating this exhaust from the atmosphere. Atmospheric pressure thus worked freely to shape the exhaust plume as it exited the engine. The aerospike concept offered a compact engine installation that could perform nearly as well at sea level as in a vacuum. However, this performance was somewhat less than could be achieved with a conventional bell nozzle.[65]

Accordingly, Pratt & Whitney accepted the disadvantages of the bell and sought to achieve the highest possible performance by raising the engine's internal pressure. The J-2 had operated at 763 pounds per square inch. Pratt

---

65. Ibid., pp. 23, 27; Sutton, *Rocket*, pp. 60, 62-63; *Technology Week*, March 13, 1967, pp. 18-19; proceedings, NASA Space Shuttle Symposium, October 16-17, 1969, pp. 377-403.

# THE SPACE SHUTTLE DECISION

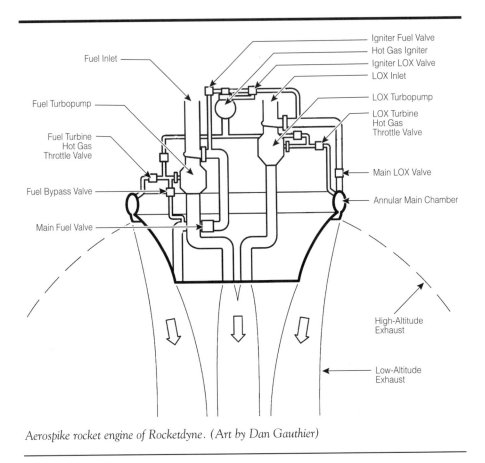

*Aerospike rocket engine of Rocketdyne. (Art by Dan Gauthier)*

& Whitney expected to go much higher. This pressure increased the exhaust velocity. It also allowed an engine to produce high thrust within a compact and lightweight package.

The Air Force awarded a contract to Pratt & Whitney for a new engine, the XLR-129. It aimed at 2,740 psi, nearly four times the pressure of the J-2, and was to deliver 250,000 pounds of thrust. NASA went on to support the work at Rocketdyne, which went forward with its own program for the development of aerospikes. Thus, when concepts of the Shuttle began to jell, during 1969, these programs offered two paths toward development of a shuttle main engine.[66]

Midway through that year, Wernher von Braun, director of NASA Marshall, sent telexes to the nation's rocket-building companies that asked a

---

66. Rocketdyne, *Expendable Launch Vehicle Engines*; *Technology Week*, March 13, 1967, p. 18; *Aviation Week*, August 31, 1970, p. 38; Perkins, *History*, January-June 1970, p. 49.

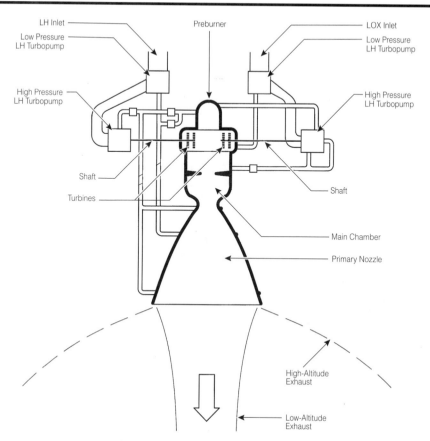

High-pressure rocket engine with bell nozzle, studied at Pratt & Whitney. Note the use of a single preburner. (Art by Dan Gauthier)

number of specific questions. One appreciates the flavor of his queries by noting a few:

> Is industry ready to commit to hard design and development, an engine operating at, say, 800K pounds sea level thrust that will meet the requirements of the space shuttle? Is it technically realistic, and can an orderly development program be accomplished to meet a PFRT [Preliminary Flight Readiness Test] date of mid 1974, with delivery of first flight engines concurrent?
>
> State the top 10 technical problem areas in order of significance that would be expected in achieving the development program.

## THE SPACE SHUTTLE DECISION

> *If a 15 percent to 25 percent thrust uprating became necessary after the engine design is committed, what changes in the design would be required, and what is your assessment of the problems involved?*

Other questions raised searching issues in a host of technical areas: turbine design for high temperatures, high-speed turbopumps, seals and pump bearings, ground-test facilities, and onboard engine checkout using computers. Von Braun also expressed concern that engine materials would become brittle when exposed to hot hydrogen at high pressures. In addition to this, he sought to uncover shortcomings in the aerospike, which faced possible problems of delayed ignition, combustion instability, sources of hot gas to drive the turbines, and the credibility of estimates for component weights and efficiencies.[67]

The aerospike held on through the summer, with shuttle managers not only continuing to consider it on an equal basis with the high-pressure bell, but even looking at shuttle designs offering interchangeability between both types of engine. NASA and its contractors, however, had no real experience with the aerospike, though they had plenty with the bell; indeed, all their rocket engines built and flown to date had been of the bell type. In October, a meeting of shuttle managers brought a decision to use the bell type only. This decision won unanimous support from key technical people at both NASA Marshall and the Manned Spacecraft Center. With this, the aerospike was out in the cold.[68]

This was bad news for Rocketdyne, which had conducted a limited amount of high-pressure work but nevertheless had spent several years placing its money on the wrong horse. It was correspondingly good news at Pratt & Whitney, which had a solid head start. That firm had already built and tested a high-pressure thrust chamber, suitable for the XLR-129, though that chamber had lacked its own turbopumps. High-performance turbopumps, however, were becoming a specialty of the house within this company, which was also pursuing a NASA project. It was building versions suitable for a high-pressure engine of 350,000 pounds of thrust.[69]

Even so, NASA was not about to accept the XLR-129, for that engine promised 250,000 pounds of thrust and NASA wanted 415,000, to reduce the

---

67. Von Braun, NASA MSFC Telex No. P181655Z, August 1969. Contents reprinted in letters: Mulready (Pratt & Whitney) to Weidner (NASA-MSFC), August 20, 1969; Stiff (Aerojet) to Weidner, August 21, 1969.
68. Akridge, *Space Shuttle*, pp. 94-96; Memo for Record, Long (NASA HQ), January 16, 1970, p. 2.
69. Proceedings, NASA Space Shuttle Symposium, October 16-17, 1969, p. 370; *Astronautics & Aeronautics*, June 1971, p. 1 (advertisement).

number and weight of the engines in the Shuttle's booster. In turn, those 415,000 pounds were to come from an entirely new rocket motor, the Space Shuttle Main Engine (SSME). Still, it was highly likely that this SSME would take shape as an enlarged XLR-129.

In February 1970, NASA issued a Request for Proposal that would lead to the award of three contracts, each funded at $6 million, for detailed engineering studies of the SSME. These contracts went to Rocketdyne, Pratt & Whitney, and Aerojet General. Aerojet had its own strong base of experience; it had built the hydrogen-fueled NERVA test engines in collaboration with Westinghouse, and had also built main and upper-stage engines for the Titan family. Though the Air Force now turned the XLR-129 effort over to NASA, Air Force engineers remained active within this program, transferring XLR-129 technology to Rocketdyne and Aerojet and making sure that its lessons were understood.

During 1970, Pratt & Whitney showed anew that it had the team to beat. In its work on the NASA 350,000-pound engine, it built a hydrogen turbopump that produced over 100 horsepower per pound of weight of its turbomachinery. In World War II, nearly 30 years earlier, builders of aircraft piston engines had counted it as a milestone to achieve a single horsepower per pound; the new turbopump thus was 100 times better. It also had five times the power density of Rocketdyne's J-2, which offered only 20 horsepower per pound.

In August, Pratt & Whitney demonstrated a hydrogen turbopump for the XLR-129. This test drove its turbine with a flow of hot gas from a preburner, a high-pressure auxiliary combustion chamber. This work had particular significance because it went forward unusually quickly. In developing turbopumps for the 350,000-pound engine, this company had taken two years to raise the working pressure of the hydrogen pump to 6,000 psi. Its engineers attained 6,700 psi with the new XLR-129 pump in no more than six months.

The company also used XLR-129 hardware as a testbed for the SSME. The latter was to have a combustion chamber pressure of 3,000 psi. While the XLR-129 had a design pressure of 2,740 psi, tests of its thrust chamber repeatedly demonstrated successful operation at and above the level of the SSME. Pratt & Whitney went on to conduct some 200 test firings at these elevated pressures.[70]

---

70. *Aviation Week*, January 19, 1970, p. 17; August 31, 1970, pp. 38-44; March 8, 1971, p. 186; June 14, 1971, pp. 51-57; *Astronautics & Aeronautics*, June 1971, p. 1; proceedings, NASA Space Shuttle Symposium, October 16-17, 1969, pp. 326-341; Nayler, *Aviation*, pp. 148-151; Reports GP 70-35, GP 70-271 (Pratt & Whitney); Executive Summary, PWA FP 71-50 (Pratt & Whitney).

## THE SPACE SHUTTLE DECISION

At Rocketdyne, the head of the SSME effort was Paul Castenholz, a corporate vice president who had previously been project manager on the J-2. While he knew he would need more than paper studies to win against Pratt & Whitney, he also saw an opportunity to go that company one better. The complete SSME would include preburners driving turbopumps, a main combustion chamber fed by an injector, and a nozzle. To build SSME-class turbopumps was out of the question; the work would take too long. The rest of the engine was another matter.

Castenholz saw that he could build a complete thrust chamber with everything but the turbopumps. He could craft it using SSME materials and manufacturing processes. Lacking pumps, he would have to feed its propellants using tanks under high pressure. He then could run tests that would demonstrate essential features of a successful SSME: a thrust of 415,000 pounds, stable combustion, a chamber pressure of 3,000 psi, and cooling of the engine. He could do these things at full scale. Pratt & Whitney had already done most of them repeatedly, but only in an XLR-129 thrust chamber, of 250,000 pounds of thrust. By leaping ahead into the realm of the SSME, Castenholz hoped to put his firm back in the race.

The recent NASA engine study contract provided no funds for this work. He approached the president of Rocketdyne, William Brennan, and asked for up to $3 million in company money. Brennan sent him on to meet with Robert Anderson, president of the parent company of Rockwell International, who approved this expenditure. Castenholz's engineers then set to work, with an important concern being the prevention of combustion instability within the engine.

Good injector design would suppress this instability. Castenholz started with an injector based on that of the J-2, which had shown its stability during this engine's repeated operational use. He then added technical features that experience had shown would promote even more stability. In the words of a close associate, Robert Biggs, "We put two big preventers on an injector that was basically stable to begin with." While this was like wearing both belt and suspenders, it offered a reasonable guarantee that the thrust chamber would work properly on its first try.

The engine testing proceeded at the Nevada Field Laboratory, a rocket facility some 20 miles northeast of Reno, in the Virginia Mountains. The initial work, late in 1970, involved an uncooled thrust chamber that worked as a

## Shuttle to the Forefront

*Rocketdyne SSME thrust chamber under test in 1970 or 1971. (Rocketdyne)*

heat sink, absorbing heat within thick metal rather than using flows of propellant to carry the heat away. While the thickness of the metal made the engine strong and unlikely to explode, it still could run only very briefly before it would burn a hole in its side and blow up. The first tests dealt only with starting the engine, with ignition trials that ran to durations less than five seconds and pressures that stayed well below rated levels.

Early in 1971, the cooled thrust chamber was ready, aiming at NASA's requirements: 415,000 pounds of thrust, 14,670 ft/sec in exhaust velocity, 3,000 psi in pressure. The last test achieved full thrust for only 0.45 seconds. It nevertheless bettered these numbers substantially, delivering 505,700 pounds, 14,990 ft/sec, and 3,172 psi. This was twice the rated thrust of the XLR-129, and 60 ft/sec greater in its exhaust velocity. Though small, this improvement was significant. It promised a shuttle payload nearly 2000 pounds heavier than Pratt & Whitney's engine might carry to orbit, along with a saving in the program cost of close to $50 million.[71]

---

71. Author interviews: Paul Castenholz, Colorado Springs, August 18, 1988; Ventura, California, March 18, 1997; Robert Biggs, Canoga Park, California, January 21, 1997; AIAA Papers 70-044, pp. 2-3; 71-658, p. 4; 71-659; Perkins, *History*, July-December 1967, pp. 36-37; Report RSS-8333-1 (Rocketdyne), chapter IV; *Aviation Week*, June 21, 1971, pp. 60-62; Executive Summary, PWA FP 71-50 (Pratt & Whitney), p. 3.

## THE SPACE SHUTTLE DECISION

No engine, not even the XLR-129, had yet operated as a complete unit, with turbopumps together with its thrust chamber. Pratt & Whitney soon accumulated a total of 2,877 seconds of test operation of its XLR-129 chamber, rated at 250,000 pounds, along with its full-pressure tests of advanced turbopumps for both this engine and the 350,000-pound model. Rocketdyne, in turn, had demonstrated successful starting and stable operation within its SSME thrust chamber, though only briefly. Then, in late January, NASA officials changed the requirements, raising the planned thrust of the SSME from 415,000 to 550,000 pounds. This reflected the growing weight of the Space Shuttle as a launch vehicle, which now was to carry up to 65,000 pounds in payload. Though Rocketdyne had not known that this thrust was to increase, its high-thrust engine test, delivering 505,700 pounds in a chamber built for 415,000, meant that it remained close even to this new requirement. By contrast, the best of Pratt & Whitney's achievements, at 350,000 pounds, lagged well behind.[72]

"The highest risk I've ever taken, in terms of a rocket engine, was to build this full-scale thrust chamber for a proposal," Castenholz recalls. "We worked around the clock. We slept at Rocketdyne in the hospital, every night for a month." Though this did not help his marriage (he and his wife Marilyn later divorced), it did provide what he would need to take to NASA.

High-speed cameras had filmed the tests in Nevada. Those films now were ready to show to officials at NASA Marshall, who were managing work on the SSME. Castenholz arranged to take along the actual thrust chamber: "We thought it was necessary that everyone who would be on the evaluation program should see that we'd actually done it." He wanted them to see his chamber and touch it, not just read about it: "If you can touch something, you feel more comfortable."

Castenholz made a presentation to officials that included Eberhard Rees, the center director. He had succeeded von Braun, who had taken a position at NASA Headquarters in Washington. Robert Biggs, Castenholz's associate, described this as "the best briefing I've ever seen." It included multiple slide projectors and movies with sound, and not all the scenes were of rockets. When Castenholz said that they had done the test in winter, his slides showed the desert covered with snow. Following the briefing, Rees turned to a colleague and said, "Now I really believe it can be done."

---

72. *Aviation Week*, June 14, 1971, pp. 51-57; Executive Summary, PWA FP 71-50 (Pratt & Whitney), p. 3.

## Shuttle to the Forefront

It was March 1971; NASA now was issuing the Request for Proposal that would lead to the award of a contract for actual development of an operational SSME. Rocketdyne's proposal included an executive summary, seven volumes on technical issues, five volumes on management, and 81 more of supporting data. "Thousands of pages, and beautifully done," Castenholz recalls. "A thick document devoted to every detail. Materials. What the materials included. How it was manufactured—pictures of the manufacturing. The tools we used. The test program—how many tests. Each test. Why. Where."

Copies of the proposal went off to NASA Marshall on April 21, with each complete copy filling a bookcase. The denouement came on July 13, 1971. Castenholz recalls that it was rather casual: "I got a call one day. Bill Brennan called me in and said, 'We won.' Bob Anderson came the next day." Then Rocketdyne held a party, inviting all participants. Castenholz recalls feeling "tremendous, joyous that we'd accomplished it. The idea that we'd won was almost mystical." He thought of things that could have gone wrong, but concluded, "I was fortunate to lead a good team. What won? Running that chamber won, and writing a super proposal."[73]

There was no joy in West Palm Beach, Florida, the home of Pratt & Whitney. That firm had taken out advertisements in major aerospace magazines, stating, "We can't wait to start working on the SSME. So we haven't." It had solicited help from the local congressman, Paul Rogers, who had led his state's congressional delegation—including both senators—in writing a letter to Nixon on behalf of this company and its proposal. (Nixon had failed to intervene, and it would not have escaped his attention that California had more electoral votes than Florida.) There was still a possibility, however, that Richard Mulready, Castenholz's counterpart, might yet win the chance to build the SSME. Rocketdyne's margin of victory had been narrow indeed: a score of 711 to 705 in the score of the source evaluation board. It might not take much to tip the balance in the other direction.

Accordingly, the president of Pratt & Whitney, Bruce Torell, lodged a formal protest with the General Accounting Office, with his legal representatives filing a 100-page brief that listed six areas of complaint. The attorneys

---

73. Author interviews: Paul Castenholz, Colorado Springs, August 18, 1988; Robert Biggs, Canoga Park, California, January 21, 1997; *Aviation Week*, March 8, 1971, p. 186; July 19, 1971, p. 12; AAS History Series, Vol. 13, pp. 73-74.

## THE SPACE SHUTTLE DECISION

asserted that Rocketdyne's proposal held technical deficiencies that violated NASA specifications. NASA also had allegedly failed to conduct meaningful negotiations, had treated the Pratt & Whitney proposal in an "arbitrary and capricious" manner, and had accepted procedures that "maximize the risk of cost overruns." Rocketdyne had supposedly obtained an unfair advantage by diverting funds from the Saturn program to support its SSME effort. A sixth cause of complaint was perhaps the most heartfelt: "Selection of Rocketdyne wastes 11 years of knowledge, test-proven design, and government investment in prior Pratt & Whitney programs." The company had been working on high-performance hydrogen-fueled engines since 1960. Yet if the contract award could not be overturned through this appeal, the firm would have nothing but its RL-10 engine for the Centaur, developed a decade prior to 1971.

This appeal put the contract award on hold. The Comptroller General, Elmer Staats, would not negotiate directly with the competing companies; rather, he would deal with NASA, which had reviewed the contracts and had held the legal responsibility to conduct the competition fairly. NASA's case now harmonized with Rocketdyne's: that the agency had followed proper procedures and had conducted a valid assessment of the proposals. Rocketdyne retained its own legal counsel; Castenholz recalls that "we had an attorney in New York who handled our case." Until Staats made his decision, however, the SSME was moribund.[74]

During that summer of 1971, it became clear that the Shuttle program as well was in serious trouble. Not only had it failed to win presidential authorization to proceed; it also was receiving increasingly severe treatment at the hands of critics. While NASA's pact with the Air Force had stilled most of the doubters in Congress, these critics would not be mollified so easily. They were in the Bureau of the Budget.

---

74. *Astronautics & Aeronautics*, June 1971, p. 1 (advertisement); letter, Rogers et al. to Nixon, June 16, 1971; *Aviation Week*, August 9, 1971, p. 23; August 23, 1971, p. 23; author interview, Paul Castenholz, Colorado Springs, August 18, 1988. Points of protest are summarized in letter B-173677, Staats to Fletcher, March 31, 1972.

# CHAPTER SIX

# Economics and the Shuttle

The Space Shuttle effort had a full share of optimists, with one of the more noteworthy being Francis Clauser, chairman of the college of engineering at Caltech. As a member of the Townes panel that had reviewed the space program, immediately following Nixon's election, he had written, "I believe we can place men on Mars before 1980. At the same time, we can develop economical space transportation which will permit extensive exploration of the Moon." His views of the Shuttle were similarly hopeful.

In May 1969, Clauser proposed that the coming decade "will see the cost of space transportation reduced to the point that the average citizen can afford a trip to the Moon." He emphasized that "when I speak of *low-cost* space transportation, I define low to be so low that the *citizenry* can afford to buy tickets for space." To achieve such a goal, he put his trust in single-stage launch vehicles burning hydrogen for high performance, and capable of routine flight to orbit. With such craft, NASA might undertake as many as 40,000 missions "before flight costs would begin to absorb a major share of its minimal budget."

Lockheed's Max Hunter had a similar outlook, as he abandoned his partially-reusable Star Clipper to embrace NASA's two-stage fully-reusable configuration. Speaking at the University of Michigan in mid-1970, he proposed that a schedule of 95 flights per year would bring a cost per flight of some $350,000, or $7 per pound of payload delivered to orbit. He added that Texas Instruments would conduct manufacturing operations in space if the cost went below $50 per pound; at $5 per pound, the Hilton family would build a hotel in orbit.[1]

Was there any basis for such optimism? There was a modest but significant base of experience with existing rocket engines and with the X-15. In addition to this, experience with commercial airliners offered a set of approaches that appeared to be potentially useful. Other approaches reflected the work of design engineers, who expected to meet specifications calling for low cost.

## *Why People Believed in Low-Cost Space Flight*

In October 1969, at a Space Shuttle symposium held in Washington, George Mueller presented opening remarks:

> *The goal we have set for ourselves is the reduction of the present costs of operating in space from the current figure of $1,000 a pound for a payload delivered in orbit by the Saturn V, down to a level of somewhere between $20 and $50 a pound. By so doing we can open up a whole new era of space exploration. Therefore, the challenge before this symposium and before all of us in the Air Force and NASA in the weeks and months ahead is to be sure that we can implement a system that is capable of doing just that.*
>
> *Let me outline three areas which, in my view, are critical to the achievement of these objectives. One is the development of an engine that will provide sufficient specific impulse,[2] with adequate margin to propel its own weight and the desired payload.*
>
> *A second technical problem is the development of the reentry heat shield, so that we can reuse that heat shield time after time with minimal refurbishment and testing.*
>
> *The third general critical development area is a checkout and control system which provides autonomous operation by the crew without major support from the ground and which will allow low cost of maintenance and repair. Of the three, the latter may be a greater challenge than the first two.[3]*

---

1. *Astronautics & Aeronautics*, May 1969, pp. 32-38; seminar, Department of Aerospace Engineering, University of Michigan, June 15, 1970; Townes, chairman, *Report*, letter attached. Reprinted in NASA SP-4407, vol. I, p. 512.
2. A measure of performance, equivalent to exhaust velocity.
3. Proceedings, NASA Space Shuttle Symposium, October 16-17, 1969, pp. 3-8.

## Economics and the Shuttle

At that time, when the 50,000-pound payload was still the standard, Mueller's cost goal represented a cost per flight of from $1 million to $2.5 million. This would not allow ordinary citizens to buy tickets into space, and was somewhat higher than Max Hunter's figure of $350,000. Regardless, if realized, it would be a long leap downward from the $185 million of a Saturn V.

The X-15 had already established itself as a reusable and piloted rocket airplane, with performance approaching at least that of a shuttle booster, though not of an orbiter. As program participants developed experience, they brought the turnaround time to as little as six working days. Individual X-15 aircraft could fly as often as three times a month.

A careful post-flight inspection followed each mission and took about two days. Inspectors examined the aircraft closely, looking for loose fasteners, cracks, hydraulic or propellant leaks, and overheating. Technicians checked the engine system for leaks using pressurized helium. The pilot reported in-flight problems, while other problems became known through study of data from onboard instruments. These post-flight activities guided subsequent work of maintenance and repair.

The engine received particularly close attention. At the start of the X-15 program, an engine run was required before each flight. In subsequent years, an engine still required a pre-flight run after replacement or major maintenance, or after three flights. A test pilot played an essential role during these engine tests, sitting in the cockpit and operating the aircraft systems. These tests disclosed such problems as rough engine operation and faulty operation of a turbine or pump, with the source of the problem being found and fixed.

All aircraft systems received complete tests prior to the next flight. They also received close inspection and overhaul at stated intervals. After every five flights, the landing gear, which was under high stress, was x-rayed for cracks. Because flaps were essential for a safe landing, their gear boxes were checked for wear after every five flights as well. Stability augmentation systems, which helped to maintain control during reentry, were tested for alignment. An engine demanded major maintenance after 30 minutes of operation; it thus had a long life between overhauls, for at full thrust an X-15 would burn a complete load of propellant in less than 90 seconds.

In the X-15 program, the principal maintenance problems centered on structural repairs and on propellant and pneumatic leaks. The latter often resulted from failures of gaskets or O-rings. Most of the structural repair items were minor. Significantly, the hot structure of the X-15, which absorbed the heat of reentry, did not represent an important source of problems. Working at Edwards Air Force Base, a ground crew of modest size successfully handled most issues of maintenance and repair. Three X-15 aircraft thus conducted 199 powered flights between 1959 and 1968, when the program ended.[4]

The X-15 represented one element of experience pertinent to the Space Shuttle. Another element involved the high-performance liquid rocket engines of the 1960s. The Space Shuttle Main Engine (SSME) was initially planned for 100 starts and a 10-hour life, representing a twentyfold improvement over the engine of the X-15. This long life would be essential for a low-cost shuttle, by reducing the number of costly engine overhauls and eliminating downtime due to engine changeouts and major maintenance. Although the engines of the 1960s had not been designed for long life in service, tests had shown that they already were close to achieving the requirement for an SSME.

The RL-10, with 15,000 pounds of thrust, had been the first to show this. As early as 1963, individual engines had been operated for over two and a half hours, with more than 50 restarts. By 1969, the total duration for a single test engine exceeded that of 50 shuttle missions, while a thrust chamber, sans turbopumps, received a series of test firings that totaled more than 11 hours.[5]

The engines of Apollo showed similar life. The F-1 was rated for 20 starts and 2,250 seconds in total duration. Yet by replacing the liquid-oxygen pump impeller and the turbine manifold at 3,500 seconds, test engines achieved as many as 60 starts and total durations of 5,000 to 6,000 seconds. The J-2 did even better, with a test engine running for 103 starts and 6.5 hours, without overhaul.

"We never wore out an engine of the J-2 type," recalls Rocketdyne's Paul Castenholz, who managed its development. "We could run it repeatedly; there was no erosion of the chamber, no damage to the turbine blades. If you looked at a J-2 after a hot firing, you would not see any difference from before that firing. The injectors always looked new; there was no erosion or corrosion on the injectors. We had extensive numbers of tests on individual engines," which demonstrated their reliability.[6]

---

4. NASA TM X-52876, vol. V, pp. 33-44; Miller, *X-Planes*, pp. 106-111.
5. *Aviation Week*, August 31, 1970, p. 38; *Astronautics & Aeronautics*, January 1964, p. 44; proceedings, NASA Space Shuttle Symposium, October 16-17, 1969, p. 360.
6. Proceedings, NASA Space Shuttle Symposium, October 16-17, 1969, p. 401; AIAA Paper 89-2387; author interview, Paul Castenholz, Ventura, California., March 18, 1997.

## Economics and the Shuttle

This experience meant that existing engine-design practice gave a reserve of engine life that engineers could draw on in meeting SSME goals. SSME requirements, however, were far more demanding because it was to operate at much higher pressures. The chamber pressures of the F-1 and J-2 were modest by later standards: 763 and 982 psi, respectively. At full power, that of the SSME would be 3,280 psi. Preburners, which fed the main combustion chamber, were to operate at pressures up to 5,500 psi. In turn, these preburners received propellants from the turbopumps, whose pump discharge pressures had to be higher still: as much as 8,000 psi.

The turbopumps thus would face enormous stresses, produced not only by pressure but by extremes of temperature. These turbopumps would be driven by hot gases and were to pump liquid oxygen and liquid hydrogen at temperatures hundreds of degrees below zero. They had to be built as compact units—which meant that across a distance of no more than two or three feet, a red-hot turbine would be driving a deeply chilled pump. These temperatures would cause the metals and materials of a turbopump to expand and contract every time the engine was fired, and designers had to ensure that the resulting stresses would not produce cracks.

In addition to this, the turbopumps were to operate at extraordinary power levels. The hydrogen turbopump, more powerful than the oxygen pump, was to approach 75,000 horsepower—in a unit the size of an outboard motor. This compared with the 55,000 horsepower that drove the liner *Titanic* early in the century, in an era when engine rooms covered an acre of space below decks. Moreover, its rotating turbomachinery was to spin at over 36,000 rpm. Yet its bearings had to work without lubrication, for the use of oils or greases was out of the question. At the hot end of a shaft, these lubricants would evaporate. At the cold end, they would freeze solid. Within the oxygen turbopump, exposed to liquid oxygen, such substances would explode.[7]

Pratt & Whitney built prototypes of such pumps for both its XLR-129 and its NASA 350,000-pound engine, and Rocketdyne expected to do likewise. To deal with thermal stresses produced by the temperature extremes, designers were accustomed to using high-strength ceramics that expanded and contracted less than metals. Though hydrogen made some alloys brittle, designers

---

7. Rocketdyne, *Expendable Launch Vehicle Engines*; Pocket Data RI/RD87-142 (Rocketdyne), pp. 1-9, 2-15, 2-17, 2-29, 2-31; Lord, *Night*, p. 103.

could protect them with thin coatings of gold, deposited on hot engine parts. This led to talk of "gold-plated engines," but gold was desirable because it would not corrode.

There were a variety of means to design turbopumps. Conventional ball bearings were of stainless steel, but specialized ceramics and glasses offered greater hardness and resistance to wear. It even was possible to dispense with ball bearings altogether and introduce hydrostatic bearings that relied on fluid pressure to maintain a clearance between a shaft and its housing. This avoided having parts in contact that could experience friction and wear. While hydrostatic bearings demanded a great deal of testing to ensure that they would operate properly, Rocketdyne's Robert Biggs noted that when such bearings were used on the SSME, they "worked beautifully."

Although a complete SSME would have 45,000 parts, it was not necessary that all of them last for the rated engine life between overhauls. Engineers expected to design for ease of maintenance, by providing for ready replacement of some parts and components. "Line-replaceable units" could be removed and reinstalled while leaving an engine as a whole attached to its mounts. Through these approaches—design for maintainability, design for relief of thermal stresses, alternate means for building heavily stressed bearings, reliance on the reserve of engine life afforded through existing experience—engineers expected to meet the challenge of developing an SSME with long life.[8]

Other alternatives existed in the area of thermal protection for the booster and orbiter, to guard against the heat of reentry. Hot structures offered a well-established but complex and tricky approach; while tiles of matted silica fiber promised simplicity, they were in an early stage of development in 1970. There also was a third approach: ablative heat shields of light weight and low cost.

In their earliest forms, such heat shields dated to the missile nose cones of the mid-1950s. They had been standard elements of the Mercury, Gemini, and Apollo spacecraft, showing particular merit on the latter, which returned from the Moon by reentering the atmosphere with twice the energy of a return from Earth orbit. Ablative shields carried away the heat of reentry by vaporizing or charring in a controlled manner; hence they were not reusable.

---

8. Author interviews: Robert Biggs, Canoga Park, California, January 21, 1997; Paul Castenholz, Ventura, California, March 18, 1997.

New versions, however, had densities as low as 15 pounds per square foot, matching the low weight of tiles. There also was strong interest in low-cost methods for fabrication of large ablative panels that could be installed and removed readily while covering substantial areas of a shuttle's wings and body. Engineers thus were confident that they could use such panels as an interim method for thermal protection, allowing them to get a shuttle up and flying even if development of the tiles were to encounter delays.

In addition to this, while the tiles were to cover large areas, they could not cope with the reentry temperatures of a shuttle's nose and wing leading edges, which would range from 2,500 to 3,500 °F. For these limited regions, still another alternative was under development: carbon composites. Carbon had an excellent ability to withstand high temperatures; vanes of graphite, dipping into the rocket exhaust of the V-2, had steered that missile as far back as 1942. Being brittle, graphite was unsuitable for use in thermal protection. The new carbon composites, however, were resilient, and reusable.

These composites drew on a recent invention: carbon fiber, which was not fragile, but possessed some strength. Such fibers could be woven into cloth, then impregnated with a specialized resin. A contractor would pile layers of this resinous cloth within a mold, forming a layup. Heated to high temperatures in the absence of oxygen, the resin would pyrolize, emitting gases and turning into carbon as well. The resulting article, treated with a coating to resist oxidation, showed promise at temperatures up to 4,000 degrees.[9]

A strong technical background was also emerging in the third of Mueller's critical areas, which he had described as "a checkout and control system which provides autonomous operation by the crew and which will allow low cost of maintenance and repair." Mueller had outlined a basis for such a system in a May 1969 briefing to shuttle contractors. He had called for an array of sensors and onboard computers that could diagnose the health of a shuttle's engines and other subsystems, returning such messages as "I am well" or "I am going to be sick."[10]

Computerized checkout offered an important path toward low-cost space flight. Cost meant people, and it was taking a ground crew of 20,000 NASA and contractor employees to prepare and launch a Saturn V with its Apollo

---

9. Proceedings, NASA Space Shuttle Symposium, October 16-19, 1969, pp. 581-591; NASA TM X-52876, vol. III, pp. 185-200; AIAA Paper 73-31, pp. 14-15, 35-36; Jenkins, *Space Shuttle*, pp. 129-131.
10. See chapter 3 pp. 133–134, for a more complete description.

moonship. If computers were to eliminate some of these jobs, here was a good reason for this to happen. Computerized checkout also promised more effective maintenance, a topic on which people in the airline industry had a number of pointed comments.[11]

In 1968, with computerized checkout still off in the distance, two maintenance managers at American Airlines noted "the difficulty of quickly and accurately locating a fault in our complex airplanes. As a consequence, much of our current troubleshooting efforts are ineffective." Many aircraft components received an allotted time in service prior to removal for test or overhaul; yet over 80 percent of these "time-controlled" units did not run to their approved time limits. Yet, it was not desirable to reduce the time between overhauls. Experience had shown that when items were removed for test or for major maintenance, they tended to fail more frequently after being reinstalled.

Troubleshooting also was hit-and-miss. We all have had the experience of taking a car to a garage for repair, having a mechanic replace a part, paying the bill—and finding that the problem remains unsolved. Such experiences were also common in the airline industry. The American Airlines managers wrote that

> *over a recent six-month period, 44 percent of the components replaced during maintenance of the air conditioning system did not eliminate the pilot's complaint. Fifty-two percent of the replacements in the autopilot system did not eliminate the pilot's complaint.*

The nation's airlines thus had a particularly strong interest in computerized checkout. While NASA was quite prepared to develop its own system for the shuttle, the airlines and their contractors could offer valuable experience, while subjecting such systems to the demands of daily use in large fleets over long periods. Pan American World Airways was emerging as an industry leader in this area; in 1970, it was providing onboard fault detection and analysis for cockpit instruments and items of flight equipment. These included the radio altimeter, radio receivers used for navigation and low-visibility landings, transponders that returned a radar signal to make a plane show up brightly on a radar screen, and electrical generating systems.

---

11. Mueller, *Briefing*, May 5, 1969; Heppenheimer, *Countdown*, p. 254.

Pan Am was also extending the use of such airborne monitoring systems to detect faults in engines. Sensors took data on engines during flight; an onboard computer used this data to determine solutions to equations that calculated engine performance. It also compared the solutions to stored values to establish trends in performance. If a trend was unfavorable—if an engine was beginning to deteriorate—a printer on the flight deck would prepare a message and warn the crew. In 1970, a prototype had already been flight-tested aboard a Boeing 707 and was slated for similar testing on a Boeing 747.

During that same year, those airlines became part of the teams that conducted the principal space shuttle design studies. North American Rockwell worked with American Airlines, leading a team that also included General Dynamics, Honeywell, and IBM. McDonnell Douglas linked up with Pan Am, while also bringing in TRW and Martin Marietta. Hence, in seeking airline-style operations for a shuttle, these teams had the counsel and experience of the airlines themselves.[12]

Of course, NASA was going to have to spend money to achieve low-cost space flight, and development of the Shuttle would not be cheap. This was worrisome, for in pushing the frontiers of technology during the 1960s, the agency had often encountered cost overruns. An in-house review, which Paine received in April 1969, showed that NASA's principal automated spacecraft programs had increased in price by more than threefold, on average, since their initiation. The costly programs in piloted flight had performed similarly. Gemini had gone from an initial estimate of $529 million, late in 1961, to a final expenditure of $1.283 billion. Apollo, with a program cost estimated at $12.0 billion in mid-1963, ballooned to $21.35 billion by the time of the first Moon landing in July 1969. That program indeed had fulfilled President Kennedy's promise by reaching the Moon during the decade of the 1960s, but only because it had drowned its problems in money.[13]

What had caused these overruns? Here, too, cost meant people. Major overruns resulted when large technical staffs drew salaries to little effect, as when projects encountered technical stumbling blocks, forcing major redesigns. Such difficulties brought delays and pushed up costs by wasting much of the earlier work. Other delays stemmed from unanticipated failures,

---

12. *Astronautics & Aeronautics*, July 1968, pp. 42-51; NASA TM X-52876, vol. V, pp. 1-32; *National Journal*, April 24, 1971, p. 875.
13. NASA SP-4102, p. 155; NASA SP-4012, vol. III, p. 61.

such as the Apollo fire in early 1967; this alone accounted for much of Apollo's overrun. The Shuttle was all too likely to encounter such issues, for it offered technical challenges aplenty. Budget officials therefore were well aware that the cost estimates of the day represented estimates made at the start of a program and were subject to potentially large increases several years down the road. Even so, low-cost space flight indeed appeared feasible.

In summary, people believed in this feasibility because leaders such as Mueller had identified the most promising routes to low cost: engines with high performance and long life, reusable heat shields, and onboard checkout. Experience in engine development, supplemented by a wealth of design alternatives for critical technical problems, promised assurance of a good SSME. Similar alternatives existed for thermal protection, again promising multiple routes to low cost. Major airlines, working in partnership with shuttle contractors, were already taking the lead in developing onboard checkout.

Nevertheless, a question remained: Even if NASA could build its Shuttle, was it in the national interest for the agency to do this?

## *The Shuttle Faces Questions*

In carrying through the increasingly detailed studies that were to precede a major program commitment, NASA had adopted a phased approach, which Paine described in a letter to Congressman Teague:

> *The first phase (Phase A) consists primarily of an in-house analysis and preliminary study effort to determine whether the proposed technical approach is feasible. Phase B consists of detailed studies and definition, comparative analyses, and preliminary design directed toward facilitating the choice of a single approach from among the alternate approaches selected through the first phase. Phase C involves detailed systems design with mockups and test articles to assure the hardware is within the state-of-the-art and that the technical milestone schedules and resource estimates for the next phase are realistic. The final phase (Phase D) covers final hardware design development and project operations.*[14]

---

14. Letter, Paine to Teague, May 28, 1970.

## Economics and the Shuttle

Like the progression of a personal friendship through dating, engagement and onward to marriage, this phased sequence carried increasing levels of commitment at each step. The most noticeable sign of this commitment was the budget. The Shuttle studies of 1969 had held the level of Phase A; they had initially been funded at $300,000 for each of four contractors. The studies of 1970 would constitute Phase B, and were considerably more costly. The SSME alone would receive three such studies, at Rocketdyne, Aerojet General, and Pratt & Whitney, funded at $6 million each. The Shuttle itself, as a two-stage fully-reusable design, would be the subject of two Phase B investigations, conducted by teams led by North American Rockwell and by McDonnell Douglas. Their funding was initially set at $8 million each, and subsequently raised to $10.8 million.[15]

This increasing commitment was sure to bring increasing scrutiny from the Budget Bureau, whose analysts were prepared to seek justification of the Shuttle by applying a standard economic approach. This approach relied on constant or uninflated dollars, thus making it possible to ignore the effects of inflation. Its point of departure lay in the indisputable fact that during the 1970s, the Shuttle program would require substantial outlays of funds to pay for its development. In exchange for this, the program could hope to reap valuable savings by lowering the cost of space flight, during the 1980s. One then could ask if would not be better and more cost-effective to use the Titan III family instead. As an alternative to the Shuttle, the Titan III was already in hand, and could readily receive technical improvements that would allow it to carry heavier payloads.

On a straight dollar-for-dollar basis, the answer to this question clearly was "no." The Titan III was an expendable launch vehicle, thrown away after each flight. Hence, even a modest level of space activity would give advantage to the Shuttle, for the continuing costs of Titan III production would quickly exceed the one-time-only cost of shuttle development. With the Shuttle being reusable, its cost per flight, once operational, would be minimal by comparison.

The BoB, however, was not about to assess the merits of the Shuttle in this straightforward way. Instead, it insisted on the use of discounted dollars, reflecting the time value of money. To economists, this concept reflected accepted professional practice and was not a subject for argument. It stemmed

---

15. "NASA Space Shuttle Studies" (summary of contracts), April 16, 1971.

from a simple principle: the dollar of next year is worth less than the dollar one holds today, even without inflation.

We apply this principle in our own personal investments, when we purchase a certificate of deposit (CD). This CD ties up money for years, and we will not buy the CD if it will merely keep that money safe for that duration, returning it with no interest earned. Similarly, we will not buy it if it only returns interest at a ridiculously low rate such as two percent per year. We insist on a reasonable rate such as six percent, or four percent after allowance for inflation. That four percent represents the true rate of return, in constant dollars.

On this basis, again in constant dollars, a ten-year CD with value at maturity of $1,000 will cost $675 in money we hold today. This is as much as to say that the sum of $1,000, payable in 10 years, has a value at present of $675. This also illustrates that not only is tomorrow's constant dollar worth less than today's, but that money markets act to determine how much less, and to set the price of securities accordingly.

In working with discounted dollars, the BoB applied a discount rate, analogous to the interest rate on that CD. The Bureau was prepared to set this rate by invoking a concept analogous to investment risk. In securities trading, it is commonplace to demand higher return on investments that carry greater risk. Thus the corporate bonds of AT&T, which are very safe and highly rated, may return no more than six percent, while bonds of riskier companies may pay over eight percent.

For the BoB, the analogous concept was national priority. Many federal programs could be viewed as investments, laying out money in the short-term in hope of realizing a social or economic return in the future. Programs holding high priority—interstate highways, construction of schools and colleges—could receive a low discount rate, analogous to the low interest rate of bonds rated AAA. By contrast, programs of low priority resembled speculative investments, and demanded a high discount rate. Because of the priorities of the Nixon administration, the Bureau gave the space program a particularly low priority, and imposed discount rates as high as 10 percent. This was as much as to say that the Space Shuttle, viewed as an investment, was no better than an issue of junk bonds.

The BoB's analysts were prepared to compare the Shuttle and Titan III in a variety of ways. The comparison would depend closely on the assumed

## Economics and the Shuttle

level of activity, or number of flights per year. For a given level, these analysts could determine the discount rate at which the cost of the Shuttle, in discounted dollars, would be low enough to save money. For a given discount rate, such as 10 percent, the BoB could also show whether the Shuttle indeed would be the less costly way to proceed—or whether the Titan III would hold the advantage.

Such analyses, using discounted dollars, in no way amounted to a simple comparison of shuttle development cost to Titan III production cost. If the Shuttle would ever pay for itself, it would do so when operational during the 1980s, using the discounted dollars of the 1980s. Because the discount rate was high, those dollars would have little present value at the immediate moment, in 1970. Hence, the Shuttle would have to promise a discounted cash flow that would be enormous indeed. Its discounted cash savings, achievable during the 1980s and hence worth very little in 1970, would nevertheless have to exceed the cost of development, which NASA would incur during the 1970s.

Therefore, in dealing with the BoB, NASA was in the position of a corporation whose officials hoped to finance a major development program by issuing bonds. With the program being speculative, the bonds would feature high risk and would carry high interest. Investors then might readily fear that the company would go broke paying interest before it could realize the return from a successful program. To guard against this, company executives would have to give those investors excellent reasons to believe that the benefits from this return, far off in the future, would be large enough to be worth the wait.

If NASA had held higher priority, qualifying for a lower BoB discount rate, it would have been in the position of a solidly-managed corporation with a gilt-edged credit rating. Such a corporation, paying low interest on its bonds, might readily carry its indebtedness while awaiting the benefits of its new projects. This interest rate would correspond precisely to the BoB's discount rate, for with those benefits being discounted less heavily, they would have greater present value and would more convincingly justify the short-term project expense.[16]

Thus, in August 1969, the BoB had carried through a comparison of the Shuttle and Titan III using discounted dollars. This analysis presented low, medium, and high scenarios for NASA activity, respectively at 15, 20, and 25 flights per year. It also presented low, medium, and high scenarios for Air Force

---

16. AIAA Paper 71-806; *Standard & Poor's Bond Guide* (any month), p. 3; letter, Mayo to Paine, January 20, 1970.

## Shuttle vs. Titan III:
Outlays, 1970–1985 (billions of dollars)[17]

| Requirements | | Gross Outlay | Cash Outlays Discounted to 1970 Present Value | | Shuttle Discount Rate |
|---|---|---|---|---|---|
| | | | 5% rate | 10% rate | Rate of Return |
| 1. NASA High, DoD High (Averages 55 flights per year) | Shuttle | $ 9.0 | $ 6.8 | $ 5.2 | 9% |
| | Titan III | 15.0 | 8.0 | 5.0 | |
| | Benefits | 6.0 | 1.2 | - 0.2 | |
| 2. NASA High, DoD Medium (Averages 45 flights per year) | Shuttle | 7.9 | 5.9 | 4.5 | 8% |
| | Titan III | 13.0 | 7.1 | 4.1 | |
| | Benefits | 5.0 | 1.2 | - 0.4 | |
| 3. NASA Medium, DoD Medium (Averages 40 flights per year) | Shuttle | 8.6 | 6.5 | 5.2 | 5% |
| | Titan III | 10.6 | 6.5 | 4.1 | |
| | Benefits | 2.0 | 0.0 | - 0.9 | |
| 4. NASA Medium, DoD Low (Averages 36 flights per year) | Shuttle | 8.0 | 6.1 | 4.8 | 4% |
| | Titan III | 9.5 | 5.8 | 3.7 | |
| | Benefits | 1.5 | - 0.3 | - 1.1 | |
| 5. NASA Low, DoD Low (Averages 28 flights per year) | Shuttle | 7.2 | 5.6 | 4.4 | 1.5% |
| | Titan III | 7.7 | 4.8 | 3.1 | |
| | Benefits | 0.5 | - 0.8 | - 1.3 | |

and Defense Department activity, at 15, 20, and 30 flights per year. The 15-per-year rate was close to the current DoD level; the high rate was twice this level.

The analysis showed that at the lowest level of activity, averaging 28 flights per year, the Shuttle would barely compete with the Titan III even on a straight dollar-for-dollar basis, without discounting. The Shuttle would save only half a billion discounted dollars, and would break even with the Titan III at a discount rate of only 1.5 percent. This was as if a Las Vegas hotel and casino, a speculative venture if ever one existed, were to try for a loan with interest at 1.5 percent. If the Shuttle had to do this, it would not fly.

---

17. Budget Bureau, "NASA Issues Paper," August 1969: attachment, "Space Transportation System," August 22, 1969.

## Economics and the Shuttle

Its prospects, however, improved markedly at the highest activity level of 55 flights per year. Now the Shuttle would break even at a discount rate of nine percent, encouragingly close to the BoB requirement that the Shuttle justify itself at a 10 percent discount rate. True, this projection raised the question of whether it was anything more than blue sky and hype, for it would call for doubling the recent Air Force activity level. Much of that activity had involved the launch of large numbers of Corona reconnaissance satellites, which were about to give way to the far more capable Big Bird spacecraft—with Big Bird flying in much-reduced numbers. Nevertheless, under these BoB ground rules, it was clear that the best way to justify a shuttle program was to project the largest possible number of operational flights.[18]

The Bureau's analyses carried a thoroughness that could put a tax audit to shame. Even so, its analysts would give NASA full opportunity to argue in favor of the Shuttle, and particularly of the two-stage fully-reusable configuration that now was the agency's desire. In doing this, the BoB would repeat the experience of 1969. Its director, Robert Mayo, had given Paine free rein to develop the post-Apollo plan of his dreams and even to see it endorsed by the Space Task Group, largely without change. Then Mayo had lowered the boom, cutting NASA's budget and putting that plan out of reach. His colleagues now were ready to give NASA similar leeway during the studies of 1970, amid a general awareness that their budget axe was close at hand.

Robert Lindley, an engineering director at NASA Headquarters in Washington, held the initial responsibility for studies of shuttle economics. Though he was well aware that the Shuttle would have to make its living by providing low-cost space transportation, he appreciated that even this might not be enough. President Nixon's budget for FY 1971, which went to Congress early in 1970, provided $125 million for procurement of expendable launch vehicles. This was 3.7 percent of the total request of $3.333 billion and offered a useful estimate of the amount NASA might have saved that year if it already had a shuttle. This was not an aberration. The nuclear physicist Ralph Lapp, a Manhattan Project veteran and a leading critic of the Shuttle, would shortly note that during the eight years of 1964-1971, procurement of expendables had

---

18. *Ibid.*

cost an average of some $130 million per year, or 2.9 percent of NASA's cumulative budgets.[19]

NASA certainly was not going to justify the Shuttle on such a basis, particularly since these minimal savings would fall much further in present value through use of the 10 percent discount rate. It was not clear how to invent additional savings, and some officials seemed constrained to conjure them out of little more than thin air. Dale Wyatt, an assistant administrator, put his hope in the fact that just then, in early 1970, the Shuttle still held close links to the space station. He assumed that the nation and not just NASA would need this station. He further assumed that it would demand logistic support at the rate of a resupply mission every two weeks. If those missions were to use conventional expendables, including an enlarged Gemini capsule for the crews, they would cost $1.625 billion per year. If, however, they were to use the Shuttle, their cost would drop to $480 million. Thus, out of these calculations, Wyatt came up with savings of over a billion dollars per year, more than enough to justify shuttle development.[20]

This, however, represented a retreat toward viewing the Shuttle once again as a vehicle for use in space station logistics. During 1969, the Shuttle had gained considerable headway through a different approach, which had presented it as the linchpin of a program of automated rather than of piloted spacecraft. Lindley saw that he could provide a more convincing justification by extending this approach. He asserted that the Shuttle could achieve additional cost savings not only by reducing launch costs, but by cutting the cost of the payloads themselves.

Lindley proposed that the availability of such inexpensive payloads could stimulate new uses for space, encouraging satellite contractors to build more such spacecraft. The Shuttle could thus promote the growth of its own traffic, for it would carry not only the planned payloads of 1970 but many others besides. The Shuttle then might repeat the experience of commercial aviation, which had achieved vast growth by cutting the prices of its passenger tickets.

How could the Shuttle achieve such "payload effects?" It would do this by completely revamping standard practice in satellite design and development. The spacecraft of 1970 faced stringent limits on weight and volume,

---

19. Letter, Paine to Shultz, September 1, 1970; U.S. Senate Committee on Aeronautical and Space Sciences, *Hearings*, FY 1973, pp. 1079-1086.
20. Wyatt, memo, "Cost Effectiveness of the Shuttle," February 12, 1970.

## Synchronous Equatorial Orbiter (SEO)
A Medium-Size, Semi-Complex Payload
Operating in Synchronous Equatorial Orbit

Conventional Launched Configuration (Minimized Weight and Volume)

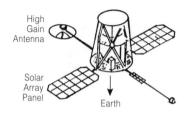

Weight: 495 kilograms
Volume (Undeployed): ~ 3 cubic meters
Cost: ~ $36 million

Space Shuttle Launched Configuration Using Low Cost Module Construction

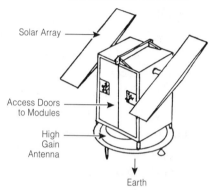

Weight: 1420 kilograms
Volume (Undeployed): ~ 18 cubic meters
Cost: ~ $21 million

*The concept of payload effects. The large volume and payload capacity of the shuttle's cargo bay made it plausible that spacecraft might cut cost by relaxing constraints on weight and size. (NASA)*

imposed by the restricted capacities of that era's launch vehicles. Because there was no way to recover a failed satellite for study, much effort went into extensive ground tests that could assure reliability. Quality assurance demanded extensive documentation, to assure that engineers could use limited data from telemetry to trace and recreate the cause of an in-flight failure. To cope with such a failure, the project staff had to remain on call, drawing salaries all the while. A large technical staff would also be necessary to assure success, conducting extensive pre-launch checkouts and then working with the spacecraft after it reached orbit.

The Shuttle offered a completely different outlook. Already its capacious payload bay was promising to ease restrictions on weight and volume. To Lindley, this meant that the electronics of future spacecraft might be packaged in modules mounted in racks, having standard connections for power and data. Like the cockpit instruments of Pan Am with their provisions for onboard fault detection, these modules would indicate their health to the Shuttle flight crew.

Satellite checkout would occur after reaching orbit, not on the ground. Astronauts would locate problems using the satellite's fault detection system, removing faulty modules and replacing them with spares. A satellite also would incorporate other systems: solar panels, power conditioning, attitude control, and data and telemetry. These could also receive on-orbit checkout. In addition to this, because they would provide standard functions, they could be built to standardized designs. They would take shape as additional modules, listed in a catalog.

Existing practice called for new spacecraft to have new subsystems in all these areas, designed from scratch and meticulously tested. The use of standard subsystems, however, would turn satellite design into an exercise in choosing and assembling these off-the-shelf components. They would usually demand more weight, volume, and power than custom versions, but would offer great cost savings through their standardization. Other savings would accrue through the Shuttle's low cost per flight. When a spacecraft began to fail after years of service, a shuttle mission might refurbish and restore it for a fraction of the cost of a replacement.[21]

Lindley's work received an attentive audience at the BoB, where Mayo wrote a letter to Paine in mid-March that called for NASA to prepare a detailed economic analysis of the Shuttle. Mayo accepted that payload effects represented a promising route toward justifying the Shuttle, and called on Paine to conduct a study that would define their cost savings. He also urged NASA to compare the merits of four alternative programs:

1. *Full scale development of a fully reusable space shuttle.*
2. *Develop a hybrid system with a reusable spacecraft and expendable booster.*
3. *Develop a fully expendable low-cost launch system.*
4. *Continue to rely on the current family of launch vehicles or improved versions of these vehicles.*

---

21. Low, Personal Notes No. 16, March 28, 1970; Report LMSC-990594 (Lockheed); *Astronautics & Aeronautics*, June 1972, pp. 50-58.

*Economics and the Shuttle*

Mayo wanted estimates of the expected potential demand for payloads in orbit, with the understanding that payload effects could increase this demand. He also wanted estimates of the cost of development of his four alternatives. Finally, he requested calculation of the discount rate for each alternative, equivalent to a rate of return. He described this as "the discount rate which equates the annual benefits to the annual program costs through 1990." He added in his cover letter that "we request general use of a 10 percent discount rate"; it was up to Paine to show that the Shuttle could achieve this.[22]

NASA was to conduct the analyses in-house while working with a BoB staff member, Earl Rhode. Though Lindley was the man in charge, it soon became clear that he was getting in over his head. Joseph McGolrick, a manager of advanced programs, later recalled what happened:

> *Lindley had this group of people from all over Headquarters, and he was drawing from people their estimates of "How much could be saved?" He was an extremely charming and extremely shrewd man who was getting out of this group of people a set of numbers for what the economics of the shuttle might be downstream. I mean, people would object about "This is not knowable, or if it is knowable, we don't have the information yet; we would have to do a study." But he really charmed them and said, "Hey, you know, let's just get an estimate."*
>
> *It was obvious to me what he was doing was focusing, steering this group of Headquarters people into a totally subjective, qualitative kind of justification of the shuttle, without any real basis at all. And he went through about four or five iterations of this thing, finding out where the critical problems were, and finding solutions to these little problems. [The] problems, from their point of view, in justifying the shuttle.*[23]

Lindley knew that he needed more than arm-waving. He required an assessment of payload effects by an aerospace corporation with actual experience in building spacecraft. He wanted "mission models," projections of the specific spacecraft, and payloads that the shuttle might carry, and he needed such mission models for the Air Force as well as NASA. The BoB also encouraged him strongly to have the economic analysis—including the vital

---

22. Letter, Mayo to Paine, March 18, 1970.
23. John Mauer interview, Joseph McGolrick, October 24, 1984, pp. 22-24.

determination of discount rates—conducted by professional economists with experience in this area.

He proceeded to set up a series of studies. For mission modeling and for payload and launch vehicle cost estimates, he turned to the Aerospace Corp., which had strong ties to the Air Force and was widely known as a center of expertise. Lockheed, builder of the Corona spacecraft, took charge of work on payload effects. For the overall economic evaluation, which these other contracts would support, Lindley followed recommendations from the BoB and approached the firm of Mathematica, Inc., in Princeton, New Jersey.

Mathematica was the lengthened shadow of its founder, the economist Oskar Morgenstern. Expelled from a professorship at the University of Vienna following the Nazi occupation in 1938, Morgenstern had taken a post at Princeton's Institute for Advanced Study, where he proceeded to work with John von Neumann, one of the world's leading mathematicians. Together they developed the theory of games, which provided mathematical analysis of situations where competitors act independently and with conflicting interests, while influencing one another's actions. Their book of 1944, *Theory of Games and Economic Behavior*, became a landmark. In turn, its mathematical methods proved applicable not only in business and economics but in military planning and nuclear arms negotiations. Morgenstern set up his firm of Mathematica to pursue such applications.[24]

In addition to analysis that might justify NASA's Shuttle, BoB officials also wanted further studies of alternate shuttle configurations. Though NASA might be ready to push ahead at full speed with a detailed study of two-stage fully-reusable designs, as early as February 1970, agency officials assured industry representatives that NASA would pursue other concepts as well. These might offer lower development cost, or reduce the outlays in the near term.

In mid-May, NASA awarded the main Phase B contracts to North American Rockwell and McDonnell Douglas. The chairman of Grumman, a losing bidder, responded with a vigorous protest. In phone calls and meetings with NASA officials, he stated that he opposed NASA's preferred shuttle concept, that the Request for Proposal had been faulty, and that NASA's decision was tantamount to declaring that Grumman would have to get out of the

---

24. *National Journal*, March 13, 1971, p. 540; August 12, 1972, p. 1292; Report LMSC-A990594 (Lockheed), p. 1-1; Blaug, *Economists*, pp. 172-174; *New York Times*, February 13, 1977, section 3, pp. 1, 9.

## Economics and the Shuttle

business of piloted space flight. He also complained that Grumman lacked strong support from the senators of New York, its home state, and that the company's top executives lacked rapport with their NASA counterparts. Paine's colleagues responded in kind, noting that selection of contractors was not a popularity contest and adding, frankly, that in its technical aspects, Grumman's proposal had been the worst of the four received.

At the same time, NASA was ready to supplement the main Phase B contracts with additional Phase A studies of alternatives. Grumman walked away with the largest of these, funded at $4 million. This company had a strong background in piloted space flight, having built the lunar module that had carried Apollo astronauts to the Moon's surface. Grumman did lack experience, however, with large rocket stages. Its management redressed this deficiency by teaming with Boeing, which had designed and built the first stage of the Saturn V. Boeing's own Phase B bid had failed, but this team was potentially as strong as that of North American Rockwell or McDonnell Douglas.

The Grumman/Boeing alternatives included the use of expendable propellant tanks, in the fashion of Lockheed's Star Clipper. They also included several approaches to phased development, whereby an initial version of the Shuttle would fly with interim systems. Rather than use the SSME for the main engines of both stages, a two-stage fully reusable shuttle might use a different engine, Rocketdyne's J-2S. This was a simplified version of the standard J-2, with its thrust increased to 265,000 pounds. At 14,030 feet per second, its exhaust velocity would not match that of the SSME. Still, it was 2.6 percent higher than that of the standard J-2, representing a modest but useful increase.

A more far-reaching approach to phased development called for the initial use of an expendable booster. This could be a Saturn V first stage; it also might be a large new stage using solid propellant. This approach would allow NASA to delay the development of a reusable shuttle first stage while allowing the stage that counted—the orbiter—to enter initial service.[25]

Two other companies also received Phase A contracts, each worth $1 million. Lockheed was to study new versions of Star Clipper, including a variant that might fly as a second stage atop the reusable booster of McDonnell

---

25. Low: Personal Notes No. 22, May 16, 1970; No. 23, May 19, 1970; Rocketdyne, *Expendable Launch Vehicle Engines*; *Aviation Week*, June 22, 1970, p. 257.

# THE SPACE SHUTTLE DECISION

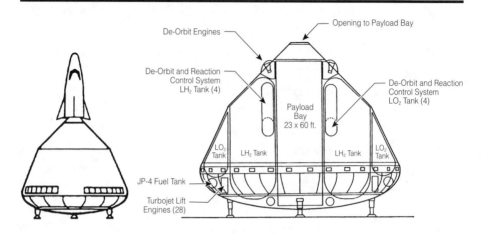

*Shuttle concept of Chrysler Corporation: left, as a booster stage with a conventional shuttle orbiter; right, cutaway view, which allegedly could carry payload to orbit. (Art by Dennis Jenkins)*

Douglas. The second such contractor, Chrysler Corp., had operated since the early 1950s as Wernher von Braun's manufacturing arm and had built most of the first stages of the Saturn I-B. Its alternative shuttle concept was strange indeed, with a reusable first stage powered by Rocketdyne's aerospike engines and shaped like an enormous Apollo capsule. Though it was definitely a wild card in NASA's deck, it showed that even at this late date, Paine was still willing to look at concepts that did not reflect the views of Max Hunter or Max Faget.[26]

The following is a summary of the studies that were under contract by mid-1970.[27]

### Phase B—Fully Reusable Space Shuttle
   North American Rockwell: $8 million (later, $10.8 million)
   McDonnell Douglas: $8 million (later, $10.8 million)
### Phase B—Space Shuttle Main Engine
   Pratt & Whitney: $6 million
   Rocketdyne: $6 million
   Aerojet General: $6 million

---

26. *Aviation Week*, June 22, 1970, p. 96; Jenkins, *Space Shuttle*, pp. 96-100.
27. "NASA Space Shuttle Studies," summary, April 16, 1971.

*Phase A—Alternate Space Shuttle*
    *Grumman/Boeing: $4 million*
    *Lockheed: $1 million*
    *Chrysler: $1 million*
**Economic Studies and Analysis**
    *Aerospace Corp.: Payloads and launch costs, $1,625,000*
    *Lockheed: Payload effects, $399,000*
    *Mathematica: Cost-benefit analysis, $400,000*

The Phase B contracts, initially totalling $34 million, reflected NASA's hope that the detailed study of vehicle and engine designs could lead relatively quickly to award of contracts for Phase C and D, covering mainstream design and development. NASA would fulfill this wish for the SSME by granting its development contract to Rocketdyne in July 1971. The agency hoped to choose a single main contractor for the Shuttle itself soon afterward.

Under the spur of questions from the BoB, however, NASA now would give the Shuttle an unusually close level of scrutiny. Its economic analysis would go beyond standard cost-benefit analysis, with its emphasis on discounted cash flows, by introducing the new topic of payload effects. This topic, with its promise of sweeping changes in methods of satellite development, promised to broaden anew the significance of the Shuttle. Studies of alternative designs would go beyond the earlier issue of reducing development cost while accepting higher operational cost. These studies would now include phased development, with the prospect of treating the shuttle effort as three separate projects—booster, orbiter, SSME—that might go forward in a sequence rather than simultaneously. The resulting program stretchout then might allow NASA to proceed with the Shuttle while fitting its year-to-year costs within a tight budget ceiling.

## Change at NASA and the Bureau of the Budget

Tom Paine, who had been reporting to James Webb as Deputy Administrator during much of 1968, took over as Acting Administrator following Webb's resignation that October. During 1969, as Paine became full Administrator, he served without a deputy. In September, he moved to remedy this situation by recommending George Low for that post. Low, who had been managing the

# THE SPACE SHUTTLE DECISION

George M. Low, NASA Deputy Administrator, 1969–1976. (NASA)

Apollo spacecraft program at NASA's Manned Spacecraft Center, started work as Paine's deputy in December. When Paine left NASA in September 1970, Low became Acting Administrator in his turn.[28]

The BoB saw considerably more far-reaching changes that grew out of a revamping within the Executive Branch. Nixon had a strong interest in management and policy development; in April 1969, he set up an advisory council to recommend changes within the White House that could enhance its effectiveness in these areas. During the following year, his personal experience stimulated his desire for change. As Vice President under Eisenhower, and now as President, he had worked with the National Security Council (NSC), which had dealt in an orderly fashion with contrasting recommendations from the Pentagon, the State Department, and the intelligence agencies. Nixon felt the lack of any similar institution to coordinate policy in domestic affairs.

In March 1970, he announced his decision. He would set up a Domestic Council within the White House, as a Cabinet-level counterpart of the NSC. Its membership would include the Vice President as well as nine Cabinet sec-

---

28. Press release, White House, November 13, 1969; Low, Personal Notes No. 1, January 1, 1970.

## Economics and the Shuttle

retaries. Nixon chose his assistant John Ehrlichman to direct it, thus giving him power on a par with that of Henry Kissinger, the National Security Advisor and head of the NSC.

Nixon also proposed to reorganize the BoB, to strengthen an emphasis on management while enhancing its activity in program evaluation and coordination. The BoB would also receive a new name: the Office of Management and Budget (OMB). In Nixon's words, "The Domestic Council will be primarily concerned with what we do; the Office of Management and Budget will be primarily concerned with how we do it, and how well we do it."

The BoB had long since established itself as an elite group within Washington. Its responsibilities covered the whole of the federal budget, with its myriad of programs and agencies. Yet the BoB staff rarely numbered more than 550, with some 350 being professionals, many with two or more college degrees. They stayed away from the media; they were not a good source of leaks. Veteran staffers, proud of the BoB's small size and central responsibilities, viewed themselves as unique. They said that if an army of Martians marched on the Capitol, while everyone would flee to the hills, the BoB staff would stay behind and prepare for an orderly transition in government.

Robert Mayo, head of the BoB, did not stay on. He had worn out his welcome by interceding in the shaping of the Pentagon budget, which Nixon had sought to develop through talks with only Henry Kissinger, the Joint Chiefs, and the Secretary of Defense. John Ehrlichman would recall that "Nixon felt that he understood enough of the general budget process that he didn't need Mayo. Nixon just froze him out. And he also just plain didn't like Mayo."

The new OMB had plenty of clout. Mayo's successor, George Shultz, had been Nixon's Secretary of Labor. After taking over as head of the OMB, he received an office within the White House itself. He started work on July 1, 1970, the day the OMB formally came into existence, and quickly emerged as one of Nixon's closest intimates.

"I think he has the most important position in government," AFL-CIO president George Meany said in late August. "He is, in my book, the executive vice president of the corporation. In other words, he is the guy who runs the corporation from day to day. He is without question over all the Cabinet members. They are just department heads under him." In January 1971, *National Journal* reported that "Shultz sees and communicates with Mr. Nixon on official business more than any other senior White House aide.

## THE SPACE SHUTTLE DECISION

Several times a day he is summoned to the President's oval office; the two talk by phone frequently. He regularly receives memoranda from the President with the notation penned in the margin, 'What do you think, George?'"[29]

For NASA, during the summer and fall of 1970, the immediate matter at hand was the budget for FY 1972. On August 7, Shultz sent a letter to Paine that set a target of $3.215 billion in outlays, representing a further cut from the already much-reduced level of FY 1971. Low responded on September 30, replying: yes, we could meet this mark with additional cancellations, including termination of NERVA and scrapping plans for Apollo 17. He would much prefer, however, an outlay of $3.411 billion. "We strongly advise against the actions that would be required to achieve the target level," he concluded.[30]

Subsequent exchanges with the OMB did not go well. In late November, Low hosted a meeting at his home to discuss what to do next. Participants included Dale Myers, head of the Office of Manned Space Flight, and Charles Donlan, head of the Space Shuttle program. As Low noted in a personal memo, a few days later:

*We held the meeting because of our collective concern that the shuttle program, as now constituted (two-stage fully reusable vehicle), would probably cost more than we could afford on an annual basis in the middle of the '70s. A phased program, wherein we would first procure only the orbiter and launch it on a modified S-IC stage[31] and only subsequently build a booster, would make more sense from the point of view of annual funding. It might also make more sense technically because we would face only one major problem at a time. At the same time, we could also adopt a Block I/Block II approach, wherein many of the "nice to have features" would be reserved for Block II and would not be incorporated into Block I.[32]*

On December 7, late in the evening, Low received a phone call from Donald Rice, an OMB assistant director. Rice said that NASA would receive outlays of $3.206 billion, less than Shultz's mark of four months earlier, with

---

29. Berman, *Office*, pp. ix-x, chapter 5; *National Journal*, March 21, 1970, pp. 620-626; January 23, 1971, pp. 151-165; June 12, 1971, pp. 1235-1244; December 13, 1975, pp. 1690-1691; John Logsdon interview, John Ehrlichman, Santa Fe, New Mexico, May 6, 1983, pp. 17-18.
30. Letters: Shultz to Paine, August 7, 1970; Low to Shultz, September 30, 1970.
31. The first stage of the Saturn V.
32. Low, Personal Notes No. 36, November 28, 1970.

## Economics and the Shuttle

new obligational authority of $3.283 billion. The latter represented the request for appropriations that Nixon would send to Congress. Apollo 17 remained in the budget, while NERVA would survive with support from key senators. The Shuttle, however, took a heavy hit.

Low still hoped to receive approval to proceed with shuttle development during FY 1972. Four days later, he sent Rice a letter proposing language for a formal statement on shuttle policy:

> *The FY 1972 budget provides for proceeding with the development of a space shuttle system.*
>
> *Detailed design and development of the shuttle engine—the longest lead time component—will be begun in FY 1972.*
>
> *Airframe design and development will proceed on an orderly step-by-step basis leading to detailed design or initiation of development during FY 1972 depending on progress in studies now underway.*

Rice sent this letter down through channels to Daniel Taft, an OMB staffer who worked with this issue. Taft replied with his own letter to Rice, noting that Low was proposing that "the Administration has approved proceeding with the space shuttle system." Taft continued:

> *We recommend that the Administration preserve flexibility by:*
> *A. Making no commitment to proceeding with the development of the entire shuttle system.*
> *B. Making no commitment to an FY 1972 decision on initiation of development of the airframe.*

Taft added a draft of a letter with which Rice replied to Low. While Rice agreed to proceed with development of the SSME, he made no such commitment to the Shuttle itself. This had obvious potential for an embarrassing situation wherein people would regard a commitment to the engine as a backdoor commitment to the shuttle. The SSME therefore received the provisional name of Advanced Space Engine, with the understanding that it might power a new low-cost expendable launch vehicle instead.[33]

---

33. Low, Personal Notes No. 37, December 20, 1970; letters: Low to Rice, December 11, 1970; Taft to Rice, December 15, 1970; Rice to Low, December 17, 1970; Niskanen to Rice, December 26, 1970; Taft to Rice, January 8, 1971.

## THE SPACE SHUTTLE DECISION

Rice's phone call of December 7 did not mark the end to the budget negotiations. In Nixon's formal budget, late in January, NASA received another cut. It now was slated for $3.271 billion in new obligational authority. The only hopeful note was that at least this matched the $3.269 billion appropriated for FY 1971, with no allowance for inflation. NASA, however, was to set aside over $100 million of this to spend in future years; its outlays were to total only $3.152 billion.

In initial discussions, NASA had sought $220 million for the Shuttle. A preliminary review at OMB had cut this to $195 million. In a subsequent review, however, the OMB swung its budget axe much harder, chopping the final number to $105 million, or $100 million in budget authority. This included $35 million for studies of the orbiter and booster along with $44 million for engine development. The space station was still alive within the FY 1972 budget and would receive $15 million, to cover continuing studies. While the shuttle was healthier than the station, it now would be on hold for another year.[34]

The Administration's intention to keep NASA on a tight leash was reemphasized as the agency received a new Administrator, James Fletcher. Fletcher had been president of the University of Utah. While his predecessor, James Webb, had spent much of his career as a high-level Washington apparatchik, and while Tom Paine had been a research manager at General Electric, James Fletcher brought hands-on experience in aerospace development. During the 1950s, he had headed a guidance-system group at the firm of TRW. That company had provided a technical staff that managed the development of the Air Force's major missiles: Atlas, Titan, Thor, and Minuteman.

The aerospace industry was well aware of the leadership that had come out of the Navaho missile program at NAA: NASA's Dale Myers, Paul Castenholz of the SSME, and Sam Hoffman who had been president of Rocketdyne. TRW had produced leadership that was even more stellar, for this company's Air Force work had placed it at the center of the nation's most important military efforts.

George Mueller had been a TRW manager before coming to NASA. Other TRW alumni included Richard DeLauer, who became an under-secretary of defense; Louis Dunn, who had headed the Jet Propulsion Laboratory; Ruben

---

34. Low, statement, January 28, 1971; letter, Shultz to Low, February 19, 1971; *Aviation Week*, January 25, 1971, p. 13; February 1, 1971, pp. 18-19.

James C. Fletcher,
NASA Administrator,
1971–1977. (NASA)

Mettler, who stayed at TRW and became its chairman and chief executive; George Solomon, who went on to serve as TRW's executive vice president; and Albert Wheelon, who pushed the development of the Big Bird reconnaissance satellite within the CIA, and later became chairman of Hughes Aircraft.[35]

Having served amid such company, Fletcher's pedigree was sterling. To Nixon's colleagues, however, he was merely a man who might help NASA toe the line. In February 1970, Peter Flanigan of the White House asked his staffer Clay Whitehead to prepare a memo on the subject of NASA, which Flanigan sent over to Ehrlichman:

*This Administration has never really faced up to where we are going in Space. NASA, with some help from the Vice President, made a try in 1969 to get the*

---

35. Heppenheimer, *Countdown*, pp. 80, 149, 259; Ramo, *Business*, p. 102.

*President committed to an "ever-onward-and-upward" post-Apollo program with continuing budget growth into the $6-10 billion range. We were successful in holding that off at least temporarily, but we have not developed any theme or consistency in policy. As a result, NASA is both drifting and lobbying for better things—without being forced to focus realistically on what it ought to be doing....*

*NASA is—or should be—making a transition from rapid razzle-dazzle growth and glamor to organizational maturity and more stable operations for the long-term. Such a transition requires wise and agile management at the top if it is to be achieved successfully. NASA has not had that. (Tom Paine may have had the ability, but he lacked the inclination - preferring to aim for continued growth.) They have a tremendous overhead structure, far too large for any reasonable size space program, that will have to be reduced....*

*We need a new Administrator who will turn down NASA's empire-building fervor and turn his attention to (1) sensible straightening away of internal management and (2) working with OMB and White House to show us what broad but concrete alternatives the President has that meet all his various objectives. In short, we need someone who will work with us rather than against us, and will seek progress toward the President's stated goals, and will shape the program to reflect credit on the President rather than embarrassment....*

*We really ought to decide if we mean to muddle through on space policy for the rest of the President's term in office or want to get serious about it.*[36]

## The Fall of the Two-Stage Fully-Reusable Shuttle

Mayo's letter of March 18, 1970, which directed Paine to compare the merits of four alternatives, set a date of May 1 for an interim report. With a week to spare, Lindley concluded that economic criteria showed that the fully-reusable shuttle was best. It would cost the most to develop, but even when using discounted dollars, its low cost per flight would yield the largest savings. The report ranked the other choices in order: a new low-cost expendable booster, a partially reusable shuttle, and continued use of current expendables. In sum, NASA's preferred design offered the greatest advantage, but anything would be better than the expendables that represented current practice.

---

36. Memo, Whitehead to Flanigan, February 6, 1971. Reprinted in NASA SP-4407, vol. II, pp. 50-52.

*Economics and the Shuttle*

This interim report impressed officials at the OMB; it also drew favorable attention at Mathematica. Early in July, representatives from that firm met with counterparts from NASA and OMB. Pending further study, initial results from Mathematica's analysis proved to agree with those of Lindley's interim report. In mid-August, NASA sent a second report to OMB, in which the ranking of the alternatives again was unchanged. This new report asserted that payload effects would lead to larger savings than reduction in launch costs. In a bold ploy to remove the bothersome distraction of alternate designs, the report presented the fully-reusable Shuttle as the ultimate goal. Therefore, the "hybrid (partially reusable shuttle) has been dropped from contention."

The next nine months, from August 1970 to May 1971, were a time of preparation. NASA carried through the negotiations with OMB that defined the budget for FY 1972, while proceeding with arrangements for award of the SSME development contract. The studies of 1970 went forward: Phase B for the fully-reusable Shuttle and main engine, Phase A for alternative designs. The economic analyses also went ahead. In turn, the OMB strengthened its hand by bringing in an economist, John Sullivan, to review the analysis of Mathematica. Then during May and June, a flurry of contractor reports presented the most extensive work to date on the Shuttle design and its economics. If NASA was to win over the skeptics at OMB, these reports would be the key.[37]

Designers expected to launch the Shuttle using the Apollo facilities at Cape Canaveral. Those facilities were alive with activity during 1971. Visitors came away with a strong conclusion: This is not the delicate precision of a laboratory; this is heavy industry. The centerpiece was the Vehicle Assembly Building, one of the largest structures on the planet, with nearly eight acres under roof and an enclosed volume of close to five million cubic yards. It had four bays, each with a door as tall as a 45-story office building, and each able to accommodate a complete Saturn V—or a shuttle.

When launching a shuttle, work crews would begin with the booster, longer than a Boeing 747 and considerably fatter in the fuselage. They would swing it to an upright position and mount it to a Launch Umbilical

---

37. Letter, Mayo to Paine, March 18, 1970; NASA interim report, "Alternate Systems for Reducing the Cost of Payload in Orbit," April 24, 1970; memo, Rhode to Young (OMB), July 23, 1970; NASA report, "Economic Analysis: Alternate Systems for Reducing the Cost of Payloads in Orbit," August 15, 1970; NASA, "Documentation of the Space Shuttle Decision Process," February 4, 1972.

275

## THE SPACE SHUTTLE DECISION

*Saturn V, on its launch platform with tower, aboard an enormous tracked vehicle as it exits the VAB. NASA expected to use these facilities and equipment in launching the shuttle. (NASA)*

Tower, a massive steel platform carrying a red-painted tower with arms that would connect to the Shuttle. Next, an orbiter would go onto the booster's back. A crawler, a diesel-powered vehicle weighing 3,000 tons, then would move beneath the platform and lift it to ride atop a flat surface the size of a baseball infield.

## Economics and the Shuttle

Eight tractor treads would begin to clank, each 40 feet long and 10 feet high. The entire array—crawler, platform, tower, booster, orbiter—would make its way through the immense door. The crawler would proceed down a roadway paved with crushed rock and resembling an interstate highway, with two sets of lanes divided by a median. That entire roadway, however, would serve the crawler alone, with those divided sections accommodating the crawler's widely-separated treads. The crawler would head for a launch pad, three miles away, and in this fashion the Shuttle would set out for orbit at the speed of those treads—one mile per hour.

Fueled and cleared for launch, the Shuttle would thunder into the air with the thrust of 12 SSMEs in the base of the booster. The orbiter would separate and fly onward, propelled by its own SSMEs. The booster, empty of fuel, would come down through the atmosphere and return to a runway at the Cape, with power from up to 12 jet engines. After completing its mission in space, the orbiter would reenter and land on the same runway. Preparations for the next flight, covering both the booster and orbiter, would take as little as two weeks.[38]

North American Rockwell and McDonnell Douglas both presented estimates for the development costs of their shuttles, and for operational costs. Analysts at the Aerospace Corp. prepared separate estimates, which fed into the economic studies of Mathematica. These projected $9.92 billion for shuttle development along with a cost per flight of $4.6 million. The latter number represented "marginal cost," which would pay for one additional shuttle flight once the system was up and flying routinely. It was less than the cost of a Delta expendable vehicle, $5.4 million. While a Delta could carry some 5,000 pounds to orbit, the Shuttle would carry 65,000. This capacity would show a substantial increase over the 40,000 pounds of a Saturn I-B, with this Saturn showing a cost per flight of some $55 million. Hence, if all went well, the Shuttle indeed would cut the cost of space flight by an order of magnitude.[39]

The Aerospace Corp. received wish lists from NASA, the Air Force, and from the commercial Communications Satellite Corp., describing the payloads these agencies hoped to launch during the years 1978-1990. It is

---

38. Heppenheimer, *Countdown*, p. 224; Jenkins, *Space Shuttle*, pp. 86-94; *Aviation Week*, June 7, 1971, pp. 55-61; AIAA Papers 71-804, 71-805; Reports MDC E0308 (McDonnell Douglas); SD 71-114-1 (North American Rockwell).
39. Report ATR-72 (7231)-1 (Aerospace Corp.); AIAA Paper 71-806; Jenkins, *Space Shuttle*, p. 125.

difficult to characterize those payload lists as reflecting more than wishes, for while these agencies had some sense of what they hoped to do during the 1970s, the 1980s were too far off for serious attention. Nevertheless, although NASA and the Pentagon were subject to year-to-year funding approvals, their planners blithely proceeded to describe what they hoped to be doing some 20 years in the future.

The ensemble of these lists defined a baseline mission model, with each individual payload described in at least an introductory fashion. Each payload had an estimated weight which Aerospace Corp. analysts used as a basis for work that would compare the Shuttle to alternatives calling for expendable launch vehicles. Based on its weight, each payload could be assigned to a particular rocket having a known launch cost, such as a Delta or Titan III. These analysts also determined a cost for each payload by using "cost estimating relationships," which drew on historical data. These relationships gave good approximations of the actual costs of existing spacecraft and satellites, which served as benchmarks.

In treating shuttle payloads, payload effects were at the forefront. Lockheed selected four spacecraft, three being satellites that together spanned a broad range of sizes and the fourth being a Mars orbiter. Company engineers examined these spacecraft in considerable detail, at the level of subsystems. They then prepared a design guide for low-cost shuttle payloads, which would emphasize minimal weight and volume constraints, modular electronics and subsystems, and on-orbit checkout and refurbishment. The Aerospace Corp. then used this Lockheed work to derive new payload cost and weight estimates, redefining the entire baseline mission model in the light of payload effects. This baseline mission model thus called for 736 shuttle flights during 1978-1990, or some 57 flights per year.[40]

The Aerospace analysis also presented year-by-year costs for both payloads and launch vehicles, treating three alternatives: a fully-reusable shuttle, a new low-cost expendable booster, and the continued use of existing expendables. These covered three of the four alternatives that Mayo had presented to Paine in his letter of March 18, 1970. The fourth alternative, a hybrid or partially-reusable shuttle, was not treated, a point that would not escape the

---

40. Reports ATR-72 (7231)-1 (Aerospace Corp.); LMSC-A990556, -A990558 and -A990594 (Lockheed); *Astronautics & Aeronautics*, June 1972, pp. 50-58; AIAA Paper 73-73.

## Economics and the Shuttle

### Costs of Three Space Vehicles
(in millions of dollars):

|  | Current Expendable | New Expendable | Fully Reusable Shuttle |
|---|---|---|---|
| **Expected Launch Vehicle Costs** | | | |
| Research and development | $ 960 | $ 1,185 | $ 9,920 |
| Facilities and fleet | 584 | 727 | 2,884 |
| Total incurring costs | 1,544 | 1,912 | 12,804 |
| Launch-by-launch recurring costs | 13,115 | 12,981 | 5,510 |
| Total launch costs | $14,659 | $14,893 | $18,314 |
| **Expected Payload Costs** | | | |
| Research and development | 12,382 | 11,179 | 10,070 |
| Production and recurring costs | 31,254 | 28,896 | 15,786 |
| Total payload costs | 43,636 | 40,075 | 25,856 |
| **Expected total space program costs** (Aerospace Corporation) | **$58,295** | **$54,968** | **$44,170** |

attention of the OMB. Aerospace Corp. presented its findings in undiscounted dollars, leaving the discounted cash-flow analyses to Mathematica. Even so, these findings were valuable, and revealing (above).

Current expendables would also require funds for research and development, to enhance their capabilities. Their launch facilities would receive their own enhancements. Nevertheless, even with an extraordinarily generous mission model and with undiscounted dollars, the Shuttle would fail to pay its way—if one considered launch costs only. The large size of the mission model would amplify the substantial difference between recurring launch costs of the Shuttle versus those of expendables. Still the high cost of shuttle development and procurement, some $12.8 billion, would swamp these savings derived from lower launch costs, to give the current expendables a strong advantage.

Payload effects promised to change this picture dramatically. These effects promised modest but welcome reductions in the cost of payload research and development. They also promised savings of some 50 percent in payload recurring costs, which represented the largest single item in all three space programs. As a consequence, when one extended the comparisons to include payload costs and not merely launch costs, the Shuttle-based space program gave a projected saving of as much as $14 billion when compared to a program based on current expendables.

# THE SPACE SHUTTLE DECISION

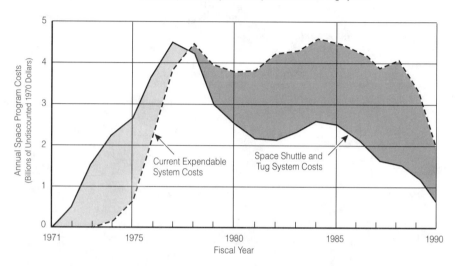

*Economic justification for the Shuttle. The dotted area at left, 1971–1977, represents the cost of Shuttle development. The much larger shaded area at right, 1978–1990, represents cost savings thought to be achievable using the Shuttle. (Mathematica Corporation)*

A year-by-year comparison showed that the Space Shuttle would operate at a strong financial disadvantage between 1971 and 1977, for it would incur its cost of development during those years. As early as 1979, however, the Shuttle-based program would show a billion-dollar saving in annual outlays, due to payload effects as well as lower launch costs. That advantage would top $2 billion per year during most of the following decade.

In discounted dollars, the heavy outlays of the 1970s would be equivalent to money borrowed at an interest rate of 10 percent. The savings of the 1980s then would amount to repayments of these borrowed sums, with interest plus principal. Using discounted cash flow, however, it would not be necessary for the Shuttle to show an advantage of $14 billion. It would suffice if the Shuttle could break even. Its $14 billion in undiscounted savings then would amount to the total of interest paid on the initial borrowings. Put another way, that $14 billion would provide a cushion, ensuring that the Shuttle would save enough discounted and therefore less-valuable dollars during the 1980s to repay in full the more-valuable dollars that would pay for its development during the 1970s.

At Mathematica, analysts led by the economist Klaus Heiss considered a broad range of mission models. They did not treat only the baseline model with its 736 shuttle flights, but examined other models that called for from 500 to 900 shuttle flights, again between 1978 and 1990. In addition to this, they carried through two types of economic analysis. "Equal capability" analysis, using discounted dollars, assumed that a particular mission model represented the whole of the nation's space activity, with shuttle and expendables launching the same payloads and therefore having the same capability. In this view, the economic benefits of the Shuttle would consist entirely of cost savings, due to both launch cost reductions and to payload effects, in flying this standard set of missions. The table, "Cost of Three Space Programs" (p. 279), gives an example of an equal-capability analysis, in undiscounted dollars.

This type of analysis could be described as highly conservative, for it ignored the likelihood that with the Shuttle chopping the cost of space activity, government and commercial agencies would seek to do more in space. The second type of economic analysis, "equal budget," addressed this issue. It took the view that total space spending would stay the same, even with the Shuttle bringing lower costs. Shuttle-derived cost savings then would not represent cash returned to the United States Treasury, but instead would pay for additional spacecraft and their Shuttle flights. In turn, those additional missions would have economic value, and Mathematica could estimate what this value would be.

Equal-budget analysis captured the cost savings of the equal-capability case. It asserted, however, that the Shuttle's economic benefits would also include the value of those additional payloads, with the Shuttle spurring the growth of its own traffic. This type of analysis broadened the Shuttle's economic rationale—and increased the cost that the program could incur for development and procurement, while still breaking even through discounted cash flow.

Mathematica's equal-capability studies showed that with the Shuttle incurring $12.8 billion in total nonrecurring costs, it would pay for itself with 506 flights between 1978 and 1990, or 39 flights per year. The baseline mission model with its 736 flights, in turn, would support a shuttle program with nonrecurring costs as high as $18 billion, in undiscounted dollars. Results from equal-budget analysis were even more hopeful, showing that this baseline mis-

# THE SPACE SHUTTLE DECISION

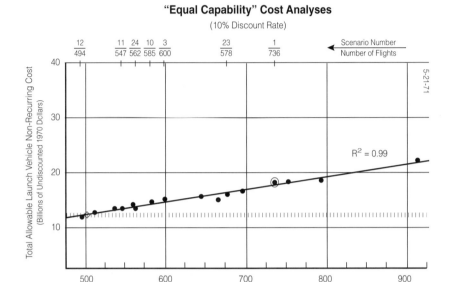

*Equal-capability cost analysis. By projecting a large number of Shuttle flights, economists could justify a large expense for Space Shuttle development. (Mathematica Corporation)*

sion model could justify nonrecurring costs approaching $24 billion, nearly twice NASA's planned level.[41]

The work of Mathematica was brilliant. If its sole purpose had been to allow one of Klaus Heiss's graduate students to win a PhD, it would have succeeded magnificently. At the OMB, however, key people hardly believed a word of it.

These people were prepared both to criticize the Mathematica report severely, and to conduct their own independent assessments as well. This drew on strengthening capabilities within the OMB for program management

---

41. Morgenstern and Heiss, *Analysis*, May 31, 1971; *Astronautics & Aeronautics*, October 1971, pp. 50-62; AIAA Paper 71-806.

*Economics and the Shuttle*

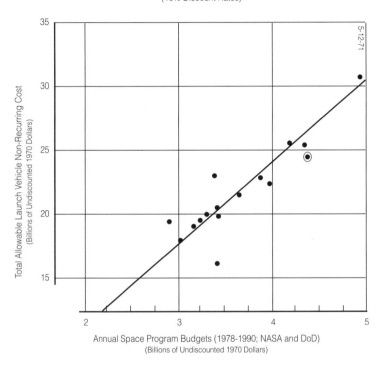

Equal-budget cost analysis. This approach appeared capable of justifying even larger Shuttle development costs. (Mathematica Corporation)

and evaluation. Those capabilities had long been present within the BoB; NASA's James Webb, who had headed the BoB during the Truman years, often said that "its name conceals its function." Under Lyndon Johnson, the BoB included an Office of Program Evaluation. Newly strengthened, its counterpart within the OMB became the Evaluation Division, with the economist William Niskanen as its director.

Niskanen, from the University of Chicago, counted himself as a disciple of the economist Milton Friedman, a leading advocate of the free market and a strong critic of government programs. Niskanen himself went on to build a reputation as a supporter of tax cuts, heading the libertarian Cato Institute. He did not love the Space Shuttle in 1971, and his criticisms were blunt. He dismissed out of hand the Aerospace Corp.'s mission models: "My impression is that the mission models that NASA is projecting for the 1980s are unrealistic. They start at a number that strains credibility and go up from there."

The Mathematica report had tried to make such models appear plausible by noting that "the 1964-1969 U.S. traffic equivalent is represented by an annual traffic of 51 Space Shuttle flights." This was close to the 57 flights per year of the baseline mission model. Because a rising tide lifts all boats, NASA's flight rates during the 1960s had been buoyed powerfully by the agency's generous budgets. The OMB had no intention of granting such largesse during the 1970s. In addition to this, the Air Force had flown large numbers of Corona reconnaissance satellites, modest in size. These were about to give way to the much larger Big Bird, which would fly far less frequently.

Niskanen also struck at the heart of NASA's rationale for the Shuttle, as he rejected the idea that payload effects would lead to large cost savings: "A large part of the presumed savings come from relaxed design, repair, and refurbishment of satellites. I was struck, however, with the fact that payload design is so far down the road—in miniaturization, sophistication, and reliability—that you wouldn't get manufacturers or users to go for much relaxation."[42]

The payload-effects concept amounted to asserting that the Shuttle indeed would meet its cost goals, including cost per flight as low as $4.6 million, and hence would spark a revolution in spacecraft design. The first statement was a speculation; the second then amounted to a speculation that rested on a speculation. Moreover, while payload effects drew strong enthusiasm from a coterie of supporters, this concept flew in the face of the hard-won lessons through which engineers indeed had learned to build reliable spacecraft.

Much of this experience had accumulated within Lockheed itself, which had struggled through a dozen failures in the Corona program, during 1959 and 1960, before finally achieving success. "It was a most heartbreaking business," Richard Bissell, the program manager within the CIA, would later recall. "In the case of a [reconnaissance] satellite you fire the damn thing off and you've got some telemetry, and you never get it back. You've got no hard-

---

42. John Mauer interview, Willis Shapley, October 26, 1984, p. 29; *National Journal*, August 12, 1972, p. 1296; *Who's Who in Economics* (1986), p. 641; Morgenstern and Heiss, *Analysis*, May 31, 1971, p. 0-37.

ware. You never see it again, and you have to infer from telemetry what went wrong. Then you make a fix, and if it fails again you know you've inferred wrong. In the case of Corona it went on and on."[43]

NASA's Jet Propulsion Laboratory had a similar experience. This lab had developed after World War II as a center for the Army's battlefield missiles, which did not demand particular care in development. Failure was acceptable in test flights of such missiles because other rounds were readily available for future launches. During the 1950s, a new director, William Pickering, redirected the lab as it became a center of expertise in electronics. By 1960, the Jet Propulsion Laboratory was ready to proceed with Ranger, a series of lunar spacecraft that would carry television cameras to photograph the Moon's surface at close range.

The program suffered three highly embarrassing failures during 1962, all involving onboard circuitry within the spacecraft. Pickering reshuffled his managers and ordered a standdown that lasted over a year. It didn't work; the next Ranger failed as well. Now the lab was really up against it and heads rolled anew, for everyone understood that another such failure could bring cancellation of the program. Fortunately, the next one indeed succeeded, as did two later ones. This showed that the Jet Propulsion Laboratory indeed had learned to build successful spacecraft.

The historian Clayton Koppes notes

> *the precision engineering and quality assurance necessary for spacecraft which had to operate nearly perfectly every time. God was in the details—in spotless cleanliness, in thorough testing, and in ruthless follow-up to make sure each failure report was corrected rather than accepted on faith.*

This new culture carried over to subsequent projects. When the Jet Propulsion Laboratory built the Mariner 4 spacecraft that flew to Mars in 1965, it became necessary to remove an onboard instrument that had been prone to electrical arcing. Jet Propulsion Laboratory technicians then installed a dummy instrument of the same weight, polished to give the same reflectivity, and engineered to use the same electric current.[44]

---

43. Ruffner, ed., *Corona*, pp. 16-24; Mosley, *Dulles*, p. 432.
44. Heppenheimer, *Countdown*, pp. 291-295; Koppes, *JPL*, p. 164; *Time*, July 23, 1965, p. 37. See also NASA SP-4210, index references.

## THE SPACE SHUTTLE DECISION

Such experiences flew in the face of the payload effects concept, which amounted to asserting that spacecraft of the future would resemble stereo systems assembled from components. Yet there also was excellent reason to believe that even if the users of satellites were in a position to do so, they would not want to pursue payload effects.

On-orbit refurbishment of spacecraft represented an important aspect of payload effects. NASA's Joseph McGolrick noted that "the users that were contacted indicated no interest in doing that. Usually, what you were talking about was a satellite that was at the end of its life or was partway through its life, and they really didn't want it back. It was, effectively, garbage." Refurbishment on the ground was even less promising: "you're bringing back junk and relaunching, and you've got an extra launch in there to be paid for."

On-orbit checkout of payloads was another important concept. It drew fire from NASA's Philip Culbertson, Director of Advanced Manned Missions:

> *We asked the communications satellite people if they expected to check their payloads out in low earth orbit. And the answer came back that they would not anticipate doing an extensive test of the satellites, if for no other reason than that would require deploying solar arrays and then retracting them and putting them back together again. They felt that the benefit from that was outweighed by the additional risks that they would go through in going through that additional deployment and retraction.*[45]

Niskanen was one influential skeptic within the OMB; another was Donald Rice. He had served as director of cost analysis within the Office of the Secretary of Defense, at a time when Defense Secretary Robert McNamara viewed cost-benefit analysis as a key to successful procurement and military management. In 1972, Rice left OMB to become president of the Rand Corp. Willis Shapley, a senior NASA official, would later describe Rice's approach to the Space Shuttle: "He was very much enamored of, and very capable in, Rand-style systems analysis. And so he treated this as well he might: as a classic case for a Rand-style study of what system to select."

---

45. John Mauer interviews: Joseph McGolrick, October 24, 1984, pp. 34-36; Philip Culbertson, October 29, 1984, p. 15.

## Economics and the Shuttle

At the OMB, between 1970 and 1972, Rice served as an assistant director. The Economics, Science, and Technology Programs Division, which covered NASA, was part of his purview, and he developed his own complaints about the Mathematica study. He was not pleased that NASA had refused to allow that study to treat alternate shuttle designs. He also would have liked to see "a good careful scrubdown of the operating costs. That number [$4.6 million per flight] that NASA was carrying around was absurd."[46]

Cost overruns represented another sticking point. If the Shuttle did no worse than Apollo, which increased in cost by over 75 percent between 1963 and 1969, then its expense would leap to well over $20 billion. Mathematica itself admitted that it was difficult to see how any plausible level of space activity would justify such outlays.[47]

The whole of that company's analysis thus was open to challenge: Shuttle development costs, launch costs, mission models, payload effects. The one point that was beyond challenge was OMB's insistence on a 10 percent discount rate. This was bad news for NASA officials, who had cherished the hope that if only they could present a good justification of the Shuttle on economic grounds, OMB would allow this program to proceed.

Significantly, however, the OMB did not reject the basic premise of NASA and Mathematica: that technical means existed to build a shuttle capable of low-cost and routine access to space. Those means—including rocket engines with long life, onboard systems for automated checkout, and reusable thermal protection—all appeared within reach. Rather, the OMB argued that the Shuttle would cost too much when compared to the benefits it promised, and that those benefits had been overstated. This argument invited a response wherein NASA would seek shuttle designs having lower development costs, which might win the OMB's favor. Such a strategy would carry NASA through the months that lay ahead.

In May 1971, in a briefing to OMB, NASA officials stated that, as early as August, they intended to issue a Request for Proposal covering mainstream development of a fully-reusable shuttle. This Request for Proposal then would follow closely the completion of the Phase B design studies, at North American Rockwell and at McDonnell Douglas. Those studies, complemented by work

---

46. John Mauer interview, Willis Shapley, October 26, 1984, pp. 29-30; *National Journal*, January 23, 1971, p. 161; John Logsdon interview, Donald Rice, November 13, 1975, p. 5.
47. *National Journal*, August 12, 1972, p. 1293.

at the Aerospace Corp., were projecting peak funding requirements of as much as $2.3 billion in the mid-1970s. NASA hoped for a budget to match this.[48]

Though Apollo was nearing its end and Skylab would soon pass the peak of its funding, NASA had wide-ranging responsibilities in space flight and could not give itself over predominately to the Shuttle. The OMB was about to present NASA with a preview of its FY 1973 budget allowances. On May 14, George Low met with Daniel Taft, an OMB staffer who dealt with NASA. Low stated that he viewed annual funding levels of $4.5 to $5 billion as reasonable to expect. This would represent a marked rise from recent levels, with $3.27 billion appropriated in FY 1971 and requested in Nixon's budget for FY 1972. Low, however, was an optimist.

Three days later, his hopes received a rude shake. An in-house OMB analysis repeated the findings of the economic review of August 1969: The fully-reusable shuttle would not be cost-effective when compared with the Titan III family. NASA had done its best, using mission models and payload effects to get the answer it wanted in the Mathematica study. The OMB, however, was well aware that economic analysis was certainly not a disinterested exercise seeking demonstrable merit. It was highly political, and the OMB's new in-house study reflected this.

On that same day, May 17, Rice sent a letter to Fletcher that rejected all hope for Low's increased budgets. Rice would not budge from his proposed five-year NASA plan with a peak at $3.2 billion per year. Fletcher would later recall him wanting a budget that would "trail off to nothing, really. He didn't ever say 'nothing,' but at various times he said, 'well, why don't you just make it constant in terms of absolute dollars and let inflation take it down.' Don Rice wouldn't give up. He wanted to do away with the manned space program; and quite honestly, I think he probably wanted to close down Marshall Space Flight Center."[49]

NASA had now taken three heavy hits in only a year and a half. The first blow, in the fall of 1969, had come when the BoB had cut the agency's budget request by over a billion dollars. This had forced Tom Paine to abandon his hopes for Mars and to fall back on the Shuttle/station. While this combined

---

48. NASA, "Documentation of the Space Shuttle Decision Process," February 4, 1972; Morgenstern and Heiss, *Analysis*, p. 0-15; AIAA Paper 71-804, p. 19; Report MDC 0308 (McDonnell Douglas), p. 28.
49. NASA, "Documentation of the Space Shuttle Decision Process," February 4, 1972; letter, Rice to Fletcher, May 17, 1971; John Logsdon interview, James Fletcher, September 21, 1977, pp. 8, 15.

## Economics and the Shuttle

program survived near-death experiences in the House and Senate during 1970, as the budget received new cuts during that summer, Paine chose to emphasize the Shuttle while deferring the space station for the indefinite future. Now, with Rice bringing more bad news, Fletcher saw that even the Shuttle was about to fall, at least as a two-stage fully-reusable design. The Shuttle concept that could fit the budget was nowhere in sight.

On May 22, George Low summarized his agency's dilemma:

> *The question, therefore, is, "is there a phasing of the shuttle or, alternatively, a cheaper shuttle that will not reach the very high expenditures in the middle of the decade?" In spite of the fact that I have been pushing this point for about six months now, we have not yet been able to come up with an answer. It may well be that there is no viable answer. One then has the choice of foregoing the shuttle altogether for the 1970s and starting it in the 1980s. In that case and with the argument that manned space flight must go on, one would go back to something like a "big G" approach[50] and a cheap space station. Of course, I'm not sure whether that alternate approach would be any more acceptable in this period of time.[51]*

---

50. "Big G" was a proposal for an enlarged Gemini spacecraft carrying up to nine people.
51. Low, Personal Notes No. 47, May 22, 1971.

# CHAPTER SEVEN

# Aerospace Recession

The poor prospects for the Space Shuttle, midway through 1971, emerged within a broad and sweeping downturn within the aerospace industry as a whole. This industry has long been highly cyclical. For instance, orders for military aircraft have soared during wars and Air Force buildups, only to fall off sharply in times of peace. The nation's aerospace efforts have also included major activities in civil aviation and in space flight, which at times have tended to counteract downturns on the military side.

During the early 1970s, however, all three of these industry components went into downturn simultaneously. The waning of the Vietnam War brought a sharp falloff in military procurement, which dropped from $23.3 billion in 1968 to $18.4 billion only three years later. The waning of Apollo led to a similar falloff in NASA contractor employment, which plunged from 394,000 in 1966 to 144,000 in 1971. The nation as a whole went into a recession during 1970, which caused new orders for airliners to dry up as well. This brought extensive layoffs at Boeing, along with severe distress within its home city of Seattle.[1]

Total aerospace employment reached a peak of over 1.4 million in 1967. It then slid downhill very rapidly, dropping to 900,000 in mid-1971. Employment of production workers fell by nearly 50 percent, from nearly 800,000 to just over 400,000. It was nearly as bad for scientists and engineers, as their ranks dwindled from 235,000 in 1968 to 145,000 four years later.[2]

---

1. *Astronautics & Aeronautics*, February 1972, p. 27; NASA budget data, February 1970.
2. *Astronautics & Aeronautics*, February 1972, pp. 27, 28.

## THE SPACE SHUTTLE DECISION

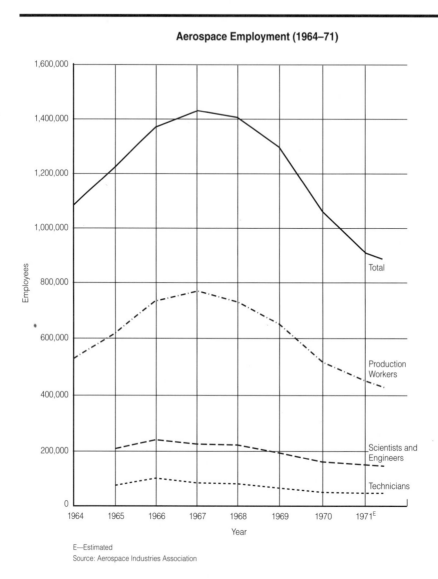

*Aerospace recession, which followed the boom of the 1960s. (Aerospace Industries Association)*

Much of this was unavoidable. The nation certainly was not about to keep the NASA and military programs at 1967 levels merely to maintain full employment within the industry. In addition to this, the layoffs at Boeing stemmed from a cyclic downturn in civil aviation. Indeed, the aerospace

*Aerospace Recession*

recession cut deepest on the commercial side. Three programs held particular significance: the Boeing 747, the supersonic transport (SST), and the Lockheed L-1011. The SST and L-1011 also brought unprecedented forms of federal involvement in commercial planebuilding.[3]

This involvement proved highly controversial, and led to a series of close congressional votes during 1971. These votes made it clear that neither Congress nor the Nixon Administration would sit back and allow major aerospace corporations to wither on the vine. Rather, despite heated controversy, Washington would step in to offer support. With this, the prospects shifted for aerospace, and for the Shuttle. With the industry taking its lumps, it lost something of its reputation as a recipient of undeserved largesse. This made it politically feasible to support the Shuttle, not with interim funding from one year to the next, but as a long-term national effort.

## *The Boeing 747*

The background to the 747, and the source of most of its troubles, lay in its engines. These were of a new type, known as the high-bypass turbofan. In contrast to earlier jet engines, which had the long and slender shape of a cigar, they introduced an enormous and gaping mouth, with a very large rotating fan in the front. This arrangement produced high thrust with relatively low noise and excellent fuel economy. The term "high-bypass" reflected the fact that most of the air blown by the fan would bypass the engine core, allowing the fan to act as a high-speed propeller.[4]

During the mid-1960s, the Air Force held a burgeoning interest in such engines, which were to power transport aircraft of unprecedented size. These would support the policy of Defense Secretary Robert McNamara, whereby the U.S. was to build up its airlift and sealift capacity to be ready to carry troops and equipment wherever America might choose to intervene. In August 1965, the Air Force picked a high-bypass design from General Electric, the TF-39, and marked it for development. Mounting a fan with diameter of eight feet, this engine was to produce 40,000 pounds of thrust.[5]

---

3. Discussion of these three programs, within the present chapter, generally follows Heppenheimer, *Turbulent*, chapters 8 and 9.
4. NASA SP-468, pp. 225-227; Newhouse, *Sporty*, pp. 111-112; *Eight Decades*, p. 152; Bathie, *Gas Turbines*, pp. 167-170.
5. Newhouse, *Sporty*, p. 113; *Eight Decades*, p. 131; Rice, *C-5A*, p. 3.

*THE SPACE SHUTTLE DECISION*

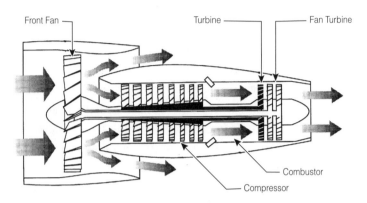

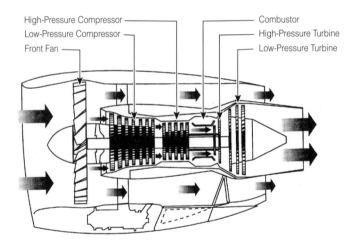

High-bypass turbofan engines, key to the widebody airliners that entered service after 1970. (Art by Don Dixon and Chris Butler)

A month later, in September, the Air Force awarded a contract to Lockheed for the C-5A, the transport that this engine would power. *Newsweek* later referred to this aircraft as Moby Jet. Placed within a football stadium, it would stretch from the goal line to the opponents' 18-yard line. Its wings

## Aerospace Recession

would overhang both teams' benches. Its cargo would accommodate heavily-armored tanks. Fully loaded, its weight of 769,000 pounds would double that of the largest commercial jetliners.[6]

Douglas and Boeing had competed with Lockheed for this award, coming forth with similar designs. At Boeing, the immediate question was how to turn such a concept into something that looked like an airliner. For several reasons, the C-5A as such would not do. It was too big; it could carry far more passengers than the market would provide. Its planned cruising speed, 506 mph, would also be too slow. It would be costly to operate, and its design was to emphasize the military requirements necessary for operation from short and unpaved landing strips rather than from hard-surface airports. Nevertheless, Boeing's work on the C-5A offered a basis for an airliner that took shape as the 747.[7]

Right at the start, this project held the strong interest of Juan Trippe, chairman of Pan American World Airways. Pan Am was the nation's leading overseas carrier. While it held no domestic routes, in its chosen realm of international travel, it carried more passengers than all other U.S.-flag airlines combined. Within the aviation industry, Trippe was a power in his own right. He had single-handedly launched the jet age, in October 1955, by placing a $269 million order for Boeing 707 and Douglas DC-8 jetliners. Now, a decade later, he had the strength to launch the 747.[8]

His vice president of engineering, John Borger, began talking with Boeing officials on the very day that Lockheed won the C-5A. Now, by prearrangement, Boeing's advanced-design policy shifted focus to the 747. The group's manager, Joseph Sutter, knew he would head up the new project as soon as the Air Force gave the C-5A to Lockheed. Market projections also favored the 747—and called for it to be huge in size.

As the Boeing vice president John Steiner describes it, aircraft are designed to fit the market four years after they enter service. For the 747, that was to be 1973 or 1974. Traffic had been shooting up for several years at annual rates of increase of 10 to 12 percent; lately those rates had gone up further. Pan Am, carrying nearly six million passengers in 1965, would top the ten million mark only four years later. Yet, if one projected no more than that,

---

6. Rice, *C-5A*, p. 1; *Pedigree*, p. 57; NASA SP-468, p. 497.
7. Rice, *C-5A*, p. 2; Daley, *Saga*, p. 435; *Aviation Week*, April 26, 1965, pp. 35-38.
8. Lehman Brothers, *Prospectus* (Pan Am); Daley, *Saga*, pp. 411-412.

295

## THE SPACE SHUTTLE DECISION

growth rates would stay below 12 percent, then airliners delivered in 1970 should accommodate 350 to 375 people.[9]

An early issue called for selection of an engine. The big General Electric TF-39 was far from being the obvious choice. GE's management, believing that this engine would find a civilian market, expected to pitch it to the airlines after the Air Force had paid for its basic development. It quickly became apparent, however, that for the 747, the TF-39 was too noisy. While this problem was far from insuperable, it would demand a major rework, a significant civilian effort that would run parallel to the military's. GE took the view that the Air Force would receive priority. That did not suit Boeing, and the chance for a deal fell through.

GE's main competitor, Pratt & Whitney, had a different spirit. That firm had built a high-bypass turbofan of its own for the Air Force engine competition. While it had lost to GE's TF-39, Pratt's design won new life as an engine for the 747. The initial concept called for 41,000 pounds of thrust, matching the performance of the TF-39. Pratt called its engine the JT-9D. In turn, the selection of Pratt was propitious, for this contractor had a virtual monopoly (with a market share of some 90 percent) on jetliner engines that were already in production. By contrast, GE's strength lay in military engines.[10]

At Boeing, Joe Sutter's engineers proceeded to prepare design concepts for Trippe's latest world-beater. Initial thinking, during 1965, held the view that the 747 might emerge as something resembling a big 707 with a double-deck cabin. Trippe, who had a strong interest in air freight, insisted, however, that the 747 was to permit easy conversion for use in cargo hauling. His requirement called for the plane to accommodate two side-by-side rows of containers of the type that were traveling by ship, rail, and truck. Their standard dimensions included width and height of eight feet. To fit them into a fuselage of circular cross section, then, would require a diameter of 21 feet. Here was the origin of the wide-body cabin, with its double aisles and ten-abreast seating. It would give a feeling of spaciousness that travelers would greatly appreciate.

From this basic decision came others. The 747 might sustain hard landings in which those containers would rip free of their moorings and hurtle forward with crushing force. Hence it would be a good idea to put the pilot

---

9. Newhouse, *Sporty*, p. 113; Steiner, *Jet Aviation*, p. 25.
10. Newhouse, *Sporty*, pp. 117-120; *Aviation Week*, April 25, 1966, p. 41.

and crew out of the way, with the flight deck high above the cargo deck. That would also offer the opportunity to install a big upward-swinging nose door for easy loading of freight, as on the C-5A.

For aerodynamic reasons, however, this flight deck could not simply sit atop the front fuselage like a camel's hump. It would have to be faired smoothly with the rest of the fuselage, sweeping gently to the back. This meant there would be a good deal of extra space to the rear of the cockpit. Sutter thought this would be a good place to put air-conditioning ducts. Trippe had other ideas.

Some 20 years earlier, Trippe had supported Boeing in another venture that had built a four-engine airliner called the Stratocruiser. It had featured a downstairs cocktail lounge that longtime travelers remembered with pleasure. He now took the view that a similar lounge in the 747, reached by a spiral stairway, would be just the thing. In subsequent versions of the 747, it would grow into a true passenger deck. But even in the earliest models, this lounge would offer a popular center of cheer.[11]

Meanwhile, other Boeing officials were addressing the question of where to build their leviathan. The company's existing production facilities were busy building the 707, 727, and 737; hence Boeing would need new facilities for the 747. The search for a new plant site led to Everett, a lumber town 30 miles north of Seattle. The new factory, quite simply, would feature the largest enclosed space in the world, within a building spanning 40 acres. At Cape Canaveral, the Vehicle Assembly Building had held the previous record, with 130 million cubic feet under roof; the new Boeing plant would be 50 percent larger. The completed production center would span more than a square mile and would have a concrete apron with room for 20 of the big jets.[12]

Then in April 1966, Trippe formally ordered 23 passenger and two freight versions of the 747, for a total of $531 million. A decade earlier, his $269 million jetliner order had set a record for dollar value; this 1966 order now set a new record. Over the next three months, five other airlines signed on for 28 more 747s. With these orders, the die was cast.[13]

---

11. Newhouse, *Sporty*, pp. 115, 116, 163; Steiner, *Jet Aviation*, p. 26; Bender and Altschul, *Instrument*, p. 504; Irving, *Wide-Body*, p. 204; *Pedigree*, p. 44; Kuter, *Gamble*, pp. 7-9, 19, 24; *Aviation Week*, November 20, 1967, pp. 60-61.
12. Serling, *Legend*, pp. 287-290; *Pedigree*, p. 62; Steiner, *Jet Aviation*, p. 26.
13. Newhouse, *Sporty*, pp. 113-114, 120-121; *Aviation Week*, April 18, 1966, pp. 38-40.

## THE SPACE SHUTTLE DECISION

As the orders rolled in, however, managers at Boeing and Pan Am followed a course that soon had the weight of the 747 running badly out of hand. Initial design decisions had been relatively straightforward, with the planned weight being 655,000 pounds as of April 1966. Pratt & Whitney was prepared to accommodate such a design using its proposed engine, the JT-9D, and expected to increase its power according to a careful plan.

When it entered service in 1969, the engine was to produce 41,000 pounds of thrust. This thrust would increase to 44,000 in new versions planned for 1972. Pratt's designers would do this by pushing up the turbines' operating temperatures; in essence, this engine would deliver more power by running hotter. There are, however, a number of other ways to boost an engine's rated power. The view within Boeing, strongly encouraged by Pan Am, was that Pratt could deliver a 44,000 pound engine a lot sooner and enable the 747 to grow larger still.

Once this point of view took hold, Boeing's managers began acting like kids in a candy store. As early as April 1966, as Trippe was placing his order, Boeing was already anticipating that the plane's weight would run to 680,000 pounds. There were plenty of opportunities to go further. For a while, people talked of putting a swimming pool in the upstairs lounge. Though that notion fell by the wayside, the cocktail lounge by itself added more than two tons to the empty weight. More tons went in when Boeing lengthened the fuselage to accommodate extra seats. The additional passengers meant larger and heavier galleys for the food service, which in turn called for weightier structural bracing. The British airline BOAC declared that noise rules of the London Airport Authority would demand quieter engines. The 747's engine pods took on an additional half-ton of sound-absorbing linings.

Pratt & Whitney now had to play catch-up. Its basic engine would now be quite inadequate; it had to offer more thrust, and quickly. In October 1966, Pratt achieved a small rise in the turbine temperature, pushing the thrust to 42,000 pounds. This was pushing limits as well and would be all it could offer for a while.

In June 1967, Bruce Connelly, Boeing's vice president of sales, sent a letter to Pan Am's chief technical managers. He stated that the 747's weight was on its way to 710,000 pounds. Even then, the 747 would be lighter than the C-5A. It was to fly considerably faster, however, which is why its engines needed more power. To Pan Am, the 747's weight meant a cut in the passenger capacity that would slice the profit on each flight by as much as $20,000. Alternately, the plane would fall short in range on a number of key overseas routes. Either way, this design would be unacceptable.

Boeing nevertheless hoped that Pan Am would accept such limitations on the grounds that better engines soon would restore the 747 to its full promise. Yet, in the words of Laurence Kuter who headed Trippe's technical staff,

> *There was no doubt that Pan Am was convinced that it was Boeing, not Pan Am, that became pregnant when the 747 was conceived. Pan Am expected Boeing to make good on all commitments to the time of delivery and all elements of guaranteed airplane performance that were specified in the half billion dollar contract.*[14]

Fortunately, Pratt had some power in reserve. By strengthening the engine's compressor and turbine, it could arrange for the engine to run at higher rotational speeds, processing more airflow and yielding more thrust. This would boost takeoff power to 43,500 pounds. Late in 1967, Pratt offered more. By providing water injection, that firm would boost the takeoff thrust to 45,000 pounds. Pratt promised to deliver such engines late in 1969.

Water injection was a specialty of the house at Pratt, dating to the piston-driven aircraft motors of World War II. Small quantities of water injected into an engine's airflow would evaporate within the engine, cooling the air and making it denser. This denser air then could burn more fuel, for extra power. This same principle had carried over to jet engines. Pratt had used water injection on the engines of the Boeing 707. Its additional thrust helped assure safe takeoffs. One senior Pan Am captain declared that he would rather lose an engine on takeoff than lose his water supply.

In 1967, however, Pratt, too, was overextending itself. It was promising a hotter, heavier engine of greater complexity: the plumbing and controls needed for water injection would not be simple. Yet this firm was holding to the same delivery schedule of a year and a half earlier, when the design of the JT-9D had been so much less demanding. It was these engines that would determine whether Boeing could build complete airplanes

---

14. Kuter, *Gamble*, pp. 23-32; Newhouse, *Sporty*, pp. 162-165; Irving, *Wide-Body*, p. 277; *Aviation Week*, April 18, 1966, pp. 38-40, 42-43; November 20, 1967, pp. 79-85.

rather than gliders, and whether Pan Am and other airlines could put the 747 into service.[15]

In pursuing this program, Boeing faced difficulties that went beyond the sheer size of the aircraft and the need for its vast new Everett facility. The 747 set new marks in complexity. For instance, it was so large that not one of its control surfaces, such as ailerons or rudder, could be deflected through the use of a pilot's muscles. The demands of safety then required four independent hydraulic systems. Earlier jetliners, such as the 727, had gotten along with only two. The demands on suppliers also were correspondingly greater than on the earlier programs. In turn, the task of assembling wings and tail surfaces was that much more complex.[16]

At the outset, Boeing's senior management had been well aware that the 747 would soak up money for several years before it could begin to generate revenue by delivering complete aircraft. The up-front expenses would include building and equipping the Everett factory, paying wages and benefits for its workforce, and assembling the first flyable 747 aircraft. Yet even then the company would not be ready to deliver them to customers such as Pan Am. Those initial production aircraft would first undergo extensive flight tests that would win a federal Certificate of Airworthiness for the 747. During these tests, Boeing would have to continue paying salaries as well as interest on borrowed money. This process of certification represented a legal requirement that the 747 would have to meet before it could see use in scheduled service. Only after completion of this process would Boeing be free to deliver those airliners and receive payment.

Hence, during 1966, the company laid financial groundwork by assembling a billion-dollar kitty. It raised $420 million through sale of notes, convertible debentures, and stock. Boeing's bankers helped as well, with a $400 million line of credit. The firm owned a subsidiary that was building gas turbines; the president, William Allen, ordered it sold. Airlines, placing orders for the 747, also contributed. They had usually paid no more than one-fourth of the purchase price prior to delivery. For the 747, however, they would pay half. Pan Am, for one, would pay as much as $275 million in advance. Then in 1967, company underwriters converted recently-issued

---

15. Solberg, *Conquest*, pp. 396-397; Irving, *Wide-Body*, p. 306; Kuter, *Gamble*, pp. 62, 72-73; *Bee-Hive* (Pratt & Whitney), January 1947, p. 3; *Aviation Week*, November 20, 1967, pp. 79-85.
16. Newhouse, *Sporty*, p. 102; Eddy et al., *Disaster*, pp. 30, 98.

debentures into new stock, thus placing Boeing in a position to sell still more securities.[17]

As work began at Everett, however, the 747's assemblers proved not necessarily to be the highly skilled production workers upon whom Boeing had long relied. The mid-1960s had brought a boom and had taken available aircraft assemblers for existing programs, leaving relatively few for the 747. During 1967, amid the buildup for this newest effort, Boeing hired 37,000 employees and let 25,000 go. The company was resting its prospects on its most inexperienced workers.

Then the engine problems hit home. Coming to the fore following the rollout of the first 747 in September 1968, these problems dogged the program as it proceeded through initial production and flight test. No one ever expected that the rollout would lead in mere weeks to commercial service, for Boeing had planned from the outset to use the entire year of 1969 in testing five such aircraft. Still, in the words of John Newhouse of the *New Yorker*:

> *William Allen, now the honorary chairman, says that what he remembers best about the engines is that "they didn't work." Boeing used eighty-seven engines in testing the 747; sixty of them were destroyed in the process. At one time, Boeing had four 747s to be tested, and couldn't get more than one of them off the ground at a time, because so few of the engines were working. By 1969, finished 747s were rolling off the line, but there were no engines for them. Instead, Boeing was obliged to hang cement blocks on the wings so as to balance the airplanes and prevent them from tipping over.*[18]

The flight tests disclosed a new engine problem known as "ovalization," which cropped up only after hundreds of hours in the air. It resulted from wear in the compressor assemblies that distorted the circular cross sections of elements of the compressor into an oval shape, with loss of power and considerable increase in fuel consumption. This resulted from the engines' high thrust, which reacted against their supports and bent the engine casings.

---

17. Newhouse, *Sporty*, pp. 120-121; *Business Week*, December 24, 1966, p. 44; *Aviation Week*, November 20, 1967, p. 59.
18. Newhouse, *Sporty*, p. 166; Serling, *Legend*, p. 335; *Astronautics & Aeronautics*, June 1969, pp. 26-29.

Though cure emerged in the form of a steel yoke that would stiffen the case, it took time to apply.[19]

Meanwhile, new orders were drying up. During 1967, 1968, and 1969, the total value of airliners on order from Boeing, of all types, fell from $3.2 billion to $1.1 billion. This did not reflect a falloff in passenger demand, for airline traffic was zooming. The carriers, however, had anticipated this demand and had provided for it with their earlier purchases. Then, in 1970, as a nationwide recession blew in, passenger traffic went flat. It would not rise again until 1972. Airlines responded by cutting new orders close to zero. Boeing's John Steiner notes that "at the bottom, we did not sell a single commercial airplane to a U.S. trunk carrier for a period of seventeen months."

The 747 took its lumps as well. Airline executives, sensing an opportunity, moved to sweeten their terms of purchase. Instead of paying 50 percent of the purchase price prior to delivery, they dropped the amount to 30 percent. It did not help; in the year and a half after September 1970, Boeing sold only two 747s in the world, and went nearly three years without a single sale to a domestic carrier. Total orders were barely 200, too few to cover the program's costs.

Even when the Everett facility rolled out production 747s, they were not always in condition for service. In March 1970, two dozen of these craft were parked outside the factory waiting for their engines. Together with other 747s in final preparation, Everett had a total of $800 million worth of aircraft on hand. Boeing could not receive the airlines' checks, for payments due on delivery, until these planes were actually ready for commercial use.

These cash-flow problems brought dreadful consequences for the company's debt. Following conservative accounting practices, Boeing had maintained the trust of its bankers. This helped as the firm's debt, owed to a syndicate of banks, topped the billion-dollar mark. In 1970, however, William Allen and Hal Haynes, his chief financial officer, tried for a further increase in their credit line and met defeat.

To win further leeway, Boeing had few choices. Its executives could not seek a merger, for the firm was heavily burdened with debt; who would want to buy it? Nor could the company raise capital by issuing new stock; its shares on Wall Street were in a slump. Because it was indebted beyond the value of

---

19. Newhouse, *Sporty*, p. 164; AIAA Paper 2987 (1991), p. 8.

its net worth, there was no equity on which to base an offering of new bonds or debentures.

Bankruptcy loomed. "We have never revealed how close we got to the edge," wrote Steiner. In speaking of the 747, William Allen noted that "the magnitude of the risk and the capital required were sufficiently great that, at best, we knew that it would strain the Boeing Company. It was really too large a project for us." Though he had hoped to keep his debt below the billion-dollar mark, the actual amount topped $2 billion. Much of the difference lay in nearly-complete but undelivered aircraft that sat outside the Everett plant, waiting for their engines. At the worst, Boeing's syndicated debt, owed to its banks, reached $1.2 billion. This set a record, not only within the aviation industry, but for all corporate borrowing.[20]

The company could do little more than to fall back on its own resources, instituting sweeping reorganizations aimed at boosting efficiency. Massive layoffs paced these changes. The Commercial Airplane Group was by far the largest part of Boeing, and its employment peaked at 83,700 during 1968. Layoffs proceeded at a modest pace during 1969 but stepped up abruptly during 1970. The number of employees fell below 30,000 by year's end, dropping toward a nadir of 20,750 late in 1971. This was part of an industrywide trend, for from December 1970 to June 1972, employment in the commercial airplane industry fell by nearly one-third.

During one week alone, some five thousand of Boeing's people received pink slips. Firings reached to the top of major organizations; even vice presidents got the axe. People took to saying that an optimist was someone who brought a lunch to work; a pessimist kept his auto engine running while he went inside.

In the Seattle area, the consequences were devastating. Each unemployed Boeing worker cost the job of at least one other person, due to the loss of the worker's purchases and spending. The resulting multiplier effect sent unemployment to 14 percent, the highest in the nation, according to the Department of Labor. About the same number of people were on welfare or receiving food stamps. Enrollment in a free-lunch program for schoolchildren soared more than fiftyfold.

---

20. Steiner, *Problems*, pp. 1, 15; Serling, *Legend*, p. 333; Newhouse, *Sporty*, pp. 168-169; *Forbes*, July 1, 1970, pp. 33-34; *Business Week*, March 28, 1970, pp. 124-128; April 1, 1972, pp. 42-44; author interview, John Steiner, Bellevue, Washington, April 13, 1991.

*THE SPACE SHUTTLE DECISION*

A brand-new car went on sale at half price—and drew no takers. A former Boeing employee had to back out of a deal to buy a house with a federal low-interest loan, for nothing down. Apartment managers offered a month's free rent along with a free stereo. Nevertheless, vacancy rates reached 40 percent in some suburbs and topped 16 percent within the area, up from one percent during the boom of 1967. Night after night, near the main airport, fewer than half the available motel rooms were full. The operator of one motel, the Sky Harbor, declared that he would "rent any room for any price right now."

Auto sales dropped by as much as 50 percent, and more than a dozen dealerships went under. Seattle's sister city, Kobe in Japan, sent food parcels and relief funds. As people fled the area in droves, the demand for U-Haul trailers grew so large that local agencies ran out of equipment to lease. Two real-estate men put up a billboard near the airport, showing a light bulb hanging on a wire and captioned:

> *Will the last person*
> *leaving SEATTLE—*
> *Turn out the lights*

As lights dimmed across the city, another Boeing project, the Supersonic Transport (SST), was flying toward its own day of decision.[21]

## *The Supersonic Transport (SST)*

The SST took shape as a response to a joint Anglo-French venture, the Concorde. Like the 747, the push for supersonic commercial flight demanded heavy dollops of advanced technology. While the 747 developed into an exercise in corporate management and finance, the Concorde and SST programs were marked by politics. The politics featured international agreements, competing centers of influence in Washington, congressional hearings, and the rise of environmentalism as a major popular movement.

The Concorde grew out of a strong base of experience, in both Great Britain and France, in commercial aviation as well as supersonic flight.

---

21. Steiner, *Problems*, p. 2; Serling, *Legend*, pp. 334-337; *Astronautics & Aeronautics*, February 1972, pp. 27, 32; *Aviation Week*, June 29, 1970, pp. 14-17; July 6, 1970, pp. 44-46; *Time*, January 4, 1971, pp. 28-29; April 5, 1971, pp. 76-82; *Newsweek*, August 17, 1970, pp. 56-57; August 28, 1972, pp. 72-73.

## Aerospace Recession

Britain's Sir Frank Whittle had invented the jet engine; the Yankees had for a time been little more than apt pupils of the British, with General Electric building British-designed engines under license. Sir Geoffrey de Havilland, a leading planebuilder, then had parlayed this engine technology into the Comet, the world's first commercial jetliner. Though it aimed at the transatlantic market, it proved uneconomical and failed to compete with the 707 and DC-8. It did, however, demonstrate a clear penchant for pioneering.

The French followed with the Caravelle, a small short-range jetliner built by Sud Aviation in Toulouse. Significantly, its engines also were British: Avon turbojets from Rolls-Royce, with 12,600 pounds of thrust. In this fashion, the Caravelle set a precedent for future Anglo-French cooperation. It sold well in Europe, and won sales in America as well. United Airlines bought 20 of them, putting the first ones in service in mid-1961. For France, this was a breakthrough; never before had a French manufacturer sold aircraft to a U.S. airline.[22]

Another French planebuilder, Marcel Dassault, spent the 1950s leading his country into supersonic flight. The company he headed, Avions Dassault, built the Mystere IV-B, the first European plane to break the sound barrier in level flight. It accomplished this feat in February 1954, only nine months after an American fighter, the F-100, did the same. Then, in October 1958, another Dassault aircraft, a Mirage III-A, became the first European aircraft to fly at Mach 2.

The British were also making sonic booms. The firm of Fairey built an experimental jet, the Delta FD-2. In March 1956, it set a world speed record at Mach 1.71, or 1,132 mph. Another company, Bristol Siddeley, developed a highly capable engine called the Olympus; an upgraded version would power the Concorde. In addition, the Royal Aircraft Establishment at Farnborough was a world-class center of aeronautical research.[23]

Two planebuilding firms, Sud Aviation and British Aircraft, carried through the design studies that led to the Concorde. For the engine, Bristol Siddeley cooperated with SNECMA, a French firm that had built engines for the Mirage fighters of Dassault. As design concepts took shape, leaders in both countries cherished the hope that they might leap past the era of subsonic

---

22. *Eight Decades*, pp. 30, 42-55; Davies, *History*, pp. 451-455, 487-489; Wilson, *Fiasco*, p. 16.
23. Gunston, *Fighters*, pp. 38, 46, 171; Wilson, *Fiasco*, p. 17; Burnet, *Concorde*, pp. 19, 151; Costello and Hughes, *Concorde*, p. 43.

jets, in which America had taken a strong lead, and take the initiative in a new realm of supersonic flight.

France, led by the strongly nationalistic Charles de Gaulle, had reasons of its own to proceed. De Gaulle had vowed to challenge what he called "America's colonization of the skies," and won strong support from his nation. There was widespread resentment of American corporations that were dominating a host of European markets, including commercial aviation. This resentment was quite similar to what Americans themselves would feel, two decades later, as Japan took over increasing shares of the automobile and electronics industries.

The joint commitment to Concorde took the form of an intergovernmental agreement in November 1962, with the force of a treaty. Each nation agreed to carry half the cost. In turn, the four participating companies—Sud, British Aircraft, Bristol Siddeley, SNECMA—would all work as contractors to their respective governments.[24]

This challenge was too serious for President Kennedy to ignore. America's planebuilders had nothing like Concorde in the offing. Moreover, there was never any prospect that an American SST would go forward as a purely commercial venture, with corporations raising the needed funds through bank loans and sales of securities. The costs of an SST would be too great, as were the technical uncertainties. In addition to this, airline executives, busily purchasing the current generation of jets, were far from thrilled at the thought of being stampeded into a supersonic era. Within the Kennedy Administration, however, the SST found a persuasive champion in Najeeb Halaby, the head of the Federal Aviation Agency.[25]

Halaby started in early 1961 by winning a congressional appropriation of $11 million with which he launched feasibility studies. Late in 1962, with the study results in hand and the Concorde under way, he urged JFK to initiate a major SST program in response. Though Kennedy was not quite ready just then, he responded by commissioning an interagency review headed by Vice President Lyndon Johnson, a strong SST supporter. While this review proceeded, Juan Trippe proceeded to stir the pot.

---

24. Dwiggins, *SST*, pp. 197-198, 201-202; Newhouse, *Sporty*, pp. 193-194; Wilson, *Fiasco*, pp. 24-32; Knight, *Concorde*, pp. 21-31; Costello and Hughes, *Concorde*, pp. 39-52; Owen, *Concorde*, pp. 44-58, 262; *Aviation Week*, September 17, 1962, pp. 34-36; December 3, 1962, p. 41.
25. The FAA changed its name to Federal Aviation Administration in 1967, upon formation of the Department of Transportation. Kent, *Safe*; see index references.

During the spring of 1963, Trippe let it be known that he intended to place a "protective order" for six Concordes. He, however, would much prefer to purchase American SSTs, should they become available. In June, he announced that he was taking options on the European airliner, putting down money to reserve positions on the production schedule, though he was not actually committing to make the purchases. By then, Kennedy had the favorable results of the interagency review. On the day after Trippe's announcement, he also made a favorable statement of his own. Addressing the graduating cadets of the Air Force Academy, he declared:

*It is my judgement that this Government should immediately commence a new program in partnership with private industry to develop at the earliest practical date the prototype of a commercially successful supersonic transport, superior to that being built in any other country in the world.*[26]

In his formal message to Congress, sent in mid-June, he emphasized that the government would put up no more than $750 million, while the manufacturers would carry at least 25 percent of the development costs.

Halaby got the program off to a running start in August, as the FAA issued a formal Request for Proposal to interested companies. As they prepared their proposals, however, executives of major planebuilders also came forward with complaints. They objected strongly to the cost-sharing arrangements, under which they were to put up 25 percent of the program expense. This was their way of declaring that the SST looked like a fine way to lose money. Nevertheless, they would do their duty as patriots if Uncle Sam would carry more of the financial load. Boeing's William Allen was particularly blunt: "Government must be prepared to render greater financial assistance than presently proposed."

Kennedy responded by commissioning an outside review of the issue, putting it in the hands of Eugene Black, former president of the World Bank, and Stanley Osborne, chairman of Olin Mathieson. He asked them not only to review the cost-sharing issue but also to cast a broad net by talking as well to government officials. Their report reached the White House a week before Christmas, with Lyndon B. Johnson now holding the presidency following the death of Kennedy.

---

26. Dwiggins, *SST*, pp. 1-9, 118-126; Horwitch, *Wings*, pp. 53-54; *Aviation Week*, June 10, 1963, p. 40.

## THE SPACE SHUTTLE DECISION

The report's conclusions were devastating to Halaby. It rejected his view that the SST should go forward as a race with Concorde. Instead, the effort was to focus initially on building a test aircraft to serve for research. The report went so far as to recommend that the program should be taken out of Halaby's hands altogether, for the FAA had no staff ready to manage such a task. On the cost-sharing issue, it recommended that the government should pick up 90 as opposed to 75 percent.

These conclusions generally suited the preferences of another player: Robert McNamara. He had faced down the Air Force in dealing with a technically similar program, the North American B-70, that sought to build supersonic bombers with the size and speed of an SST. Though Air Force generals had called for its rapid development and production, McNamara endorsed an Eisenhower Administration decision to build only three prototype craft, XB-70s, for use in flight test. McNamara also had developed an interest in the SST itself, and had served as a member of Vice President Johnson's interagency review panel.

The Black-Osborne report set in motion a Washington debate that eased Halaby toward the margins of SST management and made McNamara a central figure. In April 1964, Lyndon B. Johnson picked him to head a presidential advisory committee on the SST. Though the program remained within the FAA, high-level decisions went into the hands of this advisory panel. As defense secretary, McNamara had insisted that new military programs were to receive extensive study and analysis before their managers could cut metal for prototypes. He now approached the SST from the same perspective, arguing that the FAA should commit to building a prototype only after suitably refined designs were in hand and only after serious economic analyses showed a reasonable prospect for success.[27]

It would take nearly three years, till the end of 1966, before SST studies would reach this level of depth. An initial issue for research involved public response to sonic booms. These are different from ordinary loud noises, as from a jackhammer. A sonic boom arises from an airplane's shock wave, which spreads behind the aircraft like the bow wave of a ship. The shock produces a moving wall of compressed air that trails along the ground, sweeping

---

27. Pace, *XB-70*, pp. 15-19; Horwitch, *Wings*, pp. 64-73; Dwiggins, *SST*, pp. 12, 15-16, 108-111, 128-133, 138-143, 147, 149-152; *Fortune*, February 1964, pp. 118-122, 168-172.

out a swath up to 50 miles wide and the full length of the supersonic flight-path. Within this swath, every person feels the boom when the shock passes. The pressure rise is not large, rarely more than a thousandth of atmospheric pressure. It is, however, both sharp and sudden; hence it can startle people and crack plaster. The strength of a sonic boom is measured as an overpressure; designers expected that an SST would produce values of around two pounds per square foot during cruise. By contrast, loud noises have their intensity measured in decibels, a completely different unit. Hence the FAA wanted to know how boomy an SST could be and still produce no more annoyance than conventional subsonic jets.

An initial exercise, Operation Bongo, took place around Oklahoma City during 1964. It was a joint FAA-Air Force experiment that sought to determine whether people could learn to accept sonic booms as just another type of noise, akin to that of railroad trains or trucks on a highway. For six months the Air Force sent supersonic F-104 fighters over the city, day after day and at specified times. Observers found reason to believe that there might indeed not be much of a problem, for a number of people put the booms to their advantage.

A secretary used the recurring booms as an alarm clock. She got out of bed at the window-rattling crack of the seven a.m. boom, then took a shower. She shut off the water when she heard the next boom, for this meant it was 7:20, time to start her day. Other people also treated the eight daily booms as if they were blasts from a factory whistle. One group of construction workers used the 11 a.m. boom as their signal for a coffee break. Animals as well went undisturbed. In El Reno, a nearby town, a farmer saw a tom turkey chasing a hen. Though a boom rattled the barn, the tom never broke stride.

In several respects, these tests were biased toward minimizing citizen complaints. Oklahoma City was strongly aviation-minded, with a major FAA center and an Air Force base. The booms came by day, never at night, and people knew when to expect them. They also knew that the test would run for only a few months. The booms themselves were weaker than those of an SST and carried less energy, though they did increase in strength over the months.

Nevertheless, the results were enough to give pause, as some 4900 people filed claims for damages. Though most involved little more than cracked plaster, one man did receive a payment of $10,000. Two high-rise office towers sustained a total of 147 cracked windows. During the first three months of the tests, polls indicated that 90 percent of the people felt they could live with the booms. After six months, this number was down to 73 percent. This meant that some one-fourth of these citizens believed they could not live with them and would regard them as unacceptable.

This was bad news at the FAA in Washington. The news soon grew worse, as a second series of tests, at Edwards Air Force Base, introduced the use of larger supersonic aircraft. These included the XB-70, the only plane in the world with the size and speed of an SST. The workhorse of the new studies, the B-58 bomber, was only slightly smaller. Already it had shown its uses in sonic-boom tests, flying from Los Angeles to New York in two hours. Unfortunately, it had shattered windows as well as speed records, showering offices and living rooms alike with broken glass. Police switchboards from coast to coast had lit up with calls as frightened people reported they had heard a terrible explosion.[28]

The tests at Edwards took place during 1966, and Karl Kryter, a sonic-boom specialist at Stanford Research Institute, summarized the findings in the journal *Science*: When both European and American SSTs are fully operational, late in the 1970s,

> *it is expected that about 65 million people in the United States could be exposed to an average of about ten sonic booms per day.... A boom will initially be equivalent in acceptability to the noise from a present-day four-engined turbofan jet at an altitude of about 200 feet during approach to landing, or at 500 feet with takeoff power, or the noise from a truck at maximum highway speed at a distance of about 30 feet.*

The historian Mel Horwitch would note that when these results reached an SST coordinating committee, "an almost instant consensus developed that the American SST could never fly overland."

This did not rule out going ahead with the program. Boeing and the FAA estimated that even if the SST was restricted to overwater flights, it could still sell 500 airplanes. That would suffice to ensure commercial success. With no restrictions, Boeing's managers believed they could sell as many as twelve hundred. Even so, *Business Week* noted that "at $40 million per SST,

---

28. *Fortune*, February 1967, p. 117; Shurcliff, *S/S/T*, pp. 3-5, 21-38; Dwiggins, *SST*, pp. 57-62, 69-73, 77-78, 80.

a ban would mean a sales penalty of $28 billion—greater than Boeing's total sales for the last fifteen years."

Similar warnings came from Senator William Proxmire, an eventual opponent of the Space Shuttle who was already taking the lead as a strong opponent of the SST: "The SST will start by flying the ocean routes. Soon the economic pressures of flying these high-cost planes on limited routes will force admission of the planes to a few scattered land routes. And ultimately they will be flying everywhere."[29]

Also during 1966, design studies and analyses reached a level that allowed the FAA to select contractors through a design competition. Boeing won, with a proposal that called for engines from General Electric. This contract award came through on the last day of that year; a four-year program now lay ahead, aimed at building two prototype aircraft. This selection of contractors was crucial. The program now was in a new phase, no longer one of endless study but rather of mainstream airliner development.

This shift in status brought a quick response from SST critics, as the beginnings of organized opposition took form. The man who did the organizing was William Shurcliff, a physics professor at Harvard. Early in 1967, he set up the Citizens League Against the Sonic Boom. His son and sister were founding members; its office was in his home. He did not set out to arrange protest demonstrations. Instead, he proceeded to run a clearinghouse for critics, taking out newspaper ads, writing letters, raising questions, and generally working to argue that the emperor had no clothes. His organization was never large, its peak membership running to only a few thousand. The rudder of a ship is also quite small. Like that rudder, Shurcliff would prove to be highly influential in steering the SST to its fate.

Shurcliff's activities unfolded within a burgeoning environmental movement that was about to rise to a height of influence. This movement drew strength from a surge in public outrage against air and water pollution. As early as 1965, the Opinion Research Corp., a polling organization, found that up to one-third of the American people viewed such issues as serious. Here was a level of concern that no political leader could ignore. By 1970, nearly three-fourths of the public shared this attitude, representing a power capable of sweeping everything before it.

---

29. Horwitch, *Wings*, p. 148; *Science*, January 24, 1969, p. 359; *Business Week*, October 28, 1967, pp. 64-68.

## THE SPACE SHUTTLE DECISION

Matching this rise was a dramatic increase in the prominence and clout of leading environmental organizations. In 1967, the Sierra Club, then with only 55,000 members, was already one of the largest and most active of these groups. Though its emphasis was on protecting wilderness areas, its focus at the time was on a regional issue, fighting the construction of Marble Canyon Dam on the Colorado River. To win political support, it had to bend to the needs of such powerful senators as Henry Jackson, chairman of the Senate Interior Committee and a strong SST supporter. By 1971, its membership was at 200,000 and rising, and its leaders were taking pivotal roles in the fight against the SST.

The rapid growth in environmental concern during the late 1960s recalls the widening power of the civil rights movement. A turning point for that movement had come in Birmingham, Alabama in May 1963, when the nation watched as that city's commissioner turned police dogs and fire hoses against protesting black citizens. For the nation's environmentalists, a similar moment came early in 1969 in Santa Barbara, California.

The Santa Barbara Channel is rich in offshore oil; a line of drilling platforms stands six miles out to sea. Early that February, an oil-well blowout sent vast flows of crude into the water, where it quickly drifted onto the beaches. The Santa Barbara beaches, as highly prized as those of Malibu, now turned from shining white to gummy black. The very waves of the ocean lay unformed as they drowned beneath the thick suffocating scum. Its stink blew into the canyons, a mile and more inland. It took live steam to remove this ugly mess from the hulls of boats, and the toll of birds and sea life was immense. The historian William Manchester would write that "pelicans drove straight into the oil and then sank, unable to raise their matted wings, and the beaches were studded with dead sandpipers, cormorants, gulls, grebes, and loons, their eyes horribly swollen and their viscera burned by petroleum."[30]

Shurcliff had been proceeding with his anti-SST activism. In July 1969, he received valuable support as David Brower, who had been executive director of the Sierra Club, founded Friends of the Earth. It took a strong stand against the SST. The following March, a wealthy Baltimore man, Kenneth Greif, took the lead in organizing a nationwide coalition of SST opponents.

---

30. Wattenberg, *America*, pp. 226-227; Manchester, *Glory*, pp. 1173-1174; Horwitch, *Wings*, pp. 221-224, 233-239, 310; *Fortune*, February 1967, pp. 113-116, 227-228.

The Sierra Club now signed on. So did the National Wildlife Federation, the Wilderness Society, and the Consumer Federation of America. In this coalition, opponents now had an instrument suited for work in the political arena.

A nucleus of anti-SST sentiment already lay at hand within the Senate, where William Proxmire regarded its economics as most curious. The plan called for the FAA to put up $1.3 billion to carry the program through the construction and test of two prototypes. The SST then would go into production, and Boeing would pay the government a royalty on each plane sold. The federal outlay thus was "not a subsidy, it's a loan," said William Magruder, a Lockheed man who had taken over as SST program manager. "By the time the 300th airplane is sold, all of the Government's investment will be returned to the U.S. Treasury, and when we sell five hundred airplanes, there will be a billion dollars in profit to the Government."

Proxmire responded by arguing that Uncle Sam was not a venture capitalist. If this "loan" was so profitable, then Boeing should tap into its banks instead, as it had done in financing the 747. Referring to Nixon's SST budget request for fiscal 1971, Proxmire added,

> *We are being asked to spend $290 million this year for transportation for one half of one percent of the people—the jet setters—to fly overseas, and we are spending $204 million this year for urban mass transportation for millions of people to get to work. Does that make any sense?*

His colleague Gaylord Nelson, another Senate opponent, described the SST as

> *a high-cost, high-fare plane being built to serve a small constituency that may be willing to pay a substantial extra fee to save three hours' travel time to Europe. These people are flying on expense accounts or fat pocketbooks. If there is sufficient demand to support such a plane, it should stand on its own and be built without subsidy.*[31]

The immediate focus of attention was a congressional hearing held in May 1970, with Proxmire as chairman. He chose the witnesses with care. Among

---

31. Horwitch, *Wings*, pp. 276-278; *Newsweek*, December 14, 1970, p. 83; *U.S. News & World Report*, March 15, 1971, pp. 68-69.

them was Richard Garwin, a senior physicist at IBM who had participated in a White House review of the program. Calling for an immediate end to its federal support, Garwin asserted that "the SST will produce as much noise as the simultaneous takeoff of fifty jumbo jets." He drew concurrence from Russell Train, a member of Nixon's Council on Environmental Quality, who described such noise as the SST's "most significant unresolved environmental problem."

Train also opened a new attack by introducing the issue of whether a fleet of SSTs might damage the ozone layer in the upper atmosphere. The air at its cruising altitude, some 65,000 feet, is very dry and low in humidity. It also is rich in ozone, which forms a layer that protects the Earth from the Sun's dangerous ultraviolet rays. The atmospheric scientist Conway Leovy, writing in the *Journal of Geophysical Research*, had set forth a "wet photolysis" theory whereby water vapor in the stratosphere could speed the destruction of ozone.

Train stated in his testimony that the SST would discharge "large quantities of water vapor, carbon dioxide, nitrogen oxides and particulate matter." He added that "500 American SSTs and Concordes flying in this region of the atmosphere could, over a period of years, increase the water content by as much as 50 to 100 percent." This water vapor, formed copiously from the burning of jet fuel, could destroy some of the ozone, putting the world at greater risk from the ultraviolet. Proxmire welcomed Train's statement as a "blockbuster."

The turn of the tide quickly became evident. During the previous autumn, SST funding had passed by large margins in both the House and Senate. On May 27, however, voting on the 1971 budget, the House passed the bill by only 13 votes, 176 to 163. Opponents took new heart, for they understood that with the margin of victory having narrowed so dramatically, the SST might quickly fall during the next round of congressional action.

During the summer of 1970, critics sprouted anew. In July, the Airport Operators Council, representing all major airports, stated that the SST should receive funding only if it could meet stringent noise standards. In August, a group at MIT, conducting the Study of Critical Environmental Problems, gave further support to concerns about the upper atmosphere. It stated that a fleet of SSTs could produce effects similar to those of the 1963 eruption of the volcano Mt. Agung, which had increased stratospheric temperatures by as much as 12 degrees. In September, the prestigious Federation of American Scientists came out against the SST. So did the mayor of New York, John Lindsay, who was widely viewed as the Republicans' answer to the Kennedys.

*Aerospace Recession*

Also in September, Kenneth Greif's coalition orchestrated a devastating attack on the SST's economic prospects. Over a dozen prominent economists signed individual statements stating their criticisms. The group included Paul Samuelson, Milton Friedman, Kenneth Arrow, John Kenneth Galbraith, Wassily Leontief, Walter Heller, and Arthur Okun, who had chaired the White House's Council of Economic Advisors. The group thus spanned the political spectrum from Friedman on the right to Galbraith on the left. Only one leading economist, Henry Wallich, came out in favor of the SST.

Senate leaders put off their vote until after the November election, a move that SST supporters hoped would allow some senators to vote with less fear of public pressure. Instead, the delay gave opponents more time to organize. Leading supporters included the senators from Washington state, Warren Magnuson and Henry Jackson. On November 30, sensing defeat, they introduced a last-minute bill to ban overland flights that would produce sonic booms. It was too late; such bills had been in the congressional hopper since 1963, and the fact that this one passed unanimously was not important. After all, it would have to pass the House as well, where it quickly died. Early in December, the Senate voted to kill funding for the SST, 52 to 41.

This was not the end of the matter. The House, after all, had passed the bill in May, albeit narrowly. Now a conference committee recommended a compromise: to continue the SST program, but with reduced funding. The issue was not settled; it now would take the form of whether Congress would accept or reject this new arrangement. The vote would not take place for three months.

Again, though, time worked for the opponents. In January 1971, the citizens' group Common Cause, which was growing in influence, announced its opposition. So did Charles Lindbergh, the man who had flown to Paris in 1927. Still active after all those years, he had long held a seat on Pan Am's board of directors, and had become an ardent environmentalist.[32]

Another round of hearings would precede the votes, and again the opponents had new ammunition. James McDonald of the University of Arizona, a member of a National Academy of Sciences panel on climate modification, asserted that 500 SSTs could deplete enough ozone to produce 10,000 cases

---

32. Kent, *Safe*, pp. 302-306; Horwitch, *Wings*, pp. 282-289, 303-311; Dwiggins, *SST*, pp. 68-69, 81; Bender and Altschul, *Instrument*, p. 501; *Newsweek*, December 14, 1970, p. 83; *Aviation Week*, December 14, 1970, p. 18; *Science*, July 24, 1970, pp. 352-355; *Time*, December 14, 1970, pp. 13-14; *Journal of Geophysical Research*, January 15, 1969, pp. 417-426.

# THE SPACE SHUTTLE DECISION

*Artist's conception of American SST in the late 1960s. (NASA HQ RA69-15944)*

per year of skin cancer within the U.S. This would result from the increased power of the solar ultraviolet. His statement caused a sensation.

McDonald had based his conclusions on the threat to ozone from water vapor. Ironically, this wet-photolysis theory was overturned within months, as new research in atmospheric science showed that the effects of water vapor on ozone were all but nil. Another scientist, Harold Johnston of the University of California at Berkeley, rode to the rescue by asserting that nitrogen oxides would also damage the ozone layer. SST engines would produce such oxides in large quantities. Johnston calculated that 500 SSTs would destroy up to half the ozone in the air over the United States.[33]

---

33. Horwitch, *Wings*, pp. 319, 327; *Astronautics & Aeronautics*, December 1972, pp. 56-64; *Science*, August 6, 1971, pp. 517-522; *Journal of Planetary and Space Science*, April 1971, pp. 413-415.

Not all the arguments were on Proxmire's side. During 1970, the pro-SST forces had consisted largely of the usual corporate interests. By early 1971, however, these forces were stiffening their strength. A key argument involved jobs: With the Concorde as an SST in being, an American riposte was essential. That argument had failed to win more than divided support among union leaders, but now George Meany, head of organized labor's powerful AFL-CIO federation, came out in favor of the SST. Nixon Administration officials also weighed in with endorsements. Even William Ruckelshaus, director of the new Environmental Protection Agency, argued in favor of building at least the two prototypes.

Acoustics expert Leo Beranek, chief scientist of the firm of Bolt, Beranek, and Newman, concluded that production SSTs could be quiet enough to meet FAA noise restrictions. There also was countering testimony on the atmosphere, as William Kellogg, associate director of the National Center for Atmospheric Research, stated that effects due to SSTs would be imperceptible amid those due to natural causes.

Yet, by 1971, the issue was well past being one of whether design refinements might address specific objections or whether new research might lay scientists' concerns to rest. The public was simply against the SST, by over 85 percent in opinion polls. In 1971, barely half of all Americans had ever flown in any kind of airplane; supersonic flight to Europe was as far beyond most expectations as a visit to Shangri-La. Many people thus viewed the SST as useless, as well as being harmful to the environment. The *Los Angeles Times* cartoonist, Paul Conrad, caught this spirit neatly by showing an SST's four engines as garbage cans spewing refuse that included a dead cat.

Even so, the final vote was close. As recently as December 1970, the House had maintained its narrow margin of support. Now, however, Congressman Sidney Yates, a key SST opponent, took the floor and said, "I demand tellers with clerks." This set in motion a new procedure, in use only since the beginning of the year, whereby the votes would be recorded. Unable to vote in secrecy, as it had done before, the House turned thumbs down on the SST, 215 to 204. The Senate repeated its earlier no vote, and it was all over.[34]

---

34. Horwitch, *Wings*, pp. 314-327; *Time*: March 22, 1971, p. 15; March 29, 1971, pp. 13-14; April 5, 1971, pp. 11-12, 77; *Newsweek*: March 8, 1971, pp. 81-82; March 29, 1971, pp. 23-24; April 5, 1971, pp. 19-21.

THE SPACE SHUTTLE DECISION

These votes eliminated further federal funding for the SST. They did not ban the construction of SSTs using private-sector funding; Boeing was perfectly free to proceed with the program, if it could win the necessary support through banks or sale of securities. The company, however, was already mortgaged to the hilt; its financial leeway was close to zero. When the SST died on Capitol Hill, it died for good.

This congressional action had important consequences. It marked an end to the policy of having the FAA take on a new role by underwriting the development of new jetliners. The funding of such projects now returned to the private sector. The FAA returned to its permanent responsibilities, which included air traffic control and certification of airliners and their equipment.

The demise of the SST also brought an end to a half-century of continuing advance in the performance of commercial airliners. The industry would continue to come forth with new designs, but these would be conventional in form. The nation's airlines would find their future below the speed of sound and at altitudes well under the ozone layer.

In the struggle over the SST, the environmental movement came of age and took its place as a major and powerful political force. In defeating the SST, the nation's environmentalists showed that they had the clout to block such a program even when it held support from the AFL-CIO, the Administration, and the aerospace industry with its well-funded lobby.

With its votes against the SST, the House and Senate showed that they would cancel an important aerospace program even in the face of an industry-wide recession, and with the national economy as a whole in a slump. This raised the question of whether Congress as a whole would continue to oppose the interests of this industry. This question would not take long in receiving an answer, for in the immediate wake of the SST controversy, Congress faced a debate over another project: the Lockheed L-1011 airliner.

## *The Lockheed L-1011*

The new high-bypass turbofan engines, in launching the Boeing 747, also launched a parallel effort that proved less ambitious but better-suited to the workaday needs of the nation's domestic carriers. At American Airlines the vice president of engineering, Frank Kolk, was responsible for determining what type of equipment his airline would need and for working with the manufacturers to get it. When Juan Trippe ordered his 747s, in April 1966, Kolk saw that this aircraft was far too large for his market. He quickly took the initiative in recommending the development of another new airliner, one that would offer wide-body comfort along with the economy of the new turbofans. His plane, however, would be intermediate in size between the earlier jets and the 747.

Kolk's initial concept was well suited to American's route structure, which featured large numbers of flights between New York and Chicago. Indeed, it was a little too well suited; it lacked the size and performance that other airlines demanded. Kolk held discussions with his counterparts at Eastern, TWA, United, and Delta, and together they agreed that the new airliner was to have three engines and a larger passenger capacity. These four carriers along with American would be the initial customers, and Kolk and his colleagues proceeded to develop a common set of requirements.

Two planebuilders, Lockheed and McDonnell Douglas, proceeded to craft designs. This, however, was no federal competition for a contract, wherein one would win and the other would lose; this was an exercise in free-market competition, in which both firms had the opportunity to vie for success. The designs that emerged, the DC-10 and L-1011, were highly similar in size, performance, and general appearance, reflecting their compliance with Kolk's specifications.

During 1966, Lockheed was matched with Boeing in a federal competition that was the mirror image of the one in 1965. That earlier bidding war had involved the C-5A; when Boeing lost, its management immediately moved to pursue the 747. In 1966, the focus of attention was the SST, with these same firms competing for the FAA contract, and this time it was Boeing's turn to win. Lockheed's president, Daniel Haughton, learned the news on the last day of the year. Like his counterparts at Boeing, he immediately ordered that the people who were working on the SST shift gears and turn their attention to Kolk's airliner.

In aerospace design, small details can have large consequences, and this would be true of the L-1011. This airliner was to install one of its engines at the rear end of the fuselage, receiving its air through a curving duct that ran beneath the vertical fin. At the outset, Lockheed's engineers knew that they needed a short engine to fit this installation. Neither General Electric nor Pratt & Whitney had what they wanted, but a third player was at hand: Britain's Rolls-Royce. That company had a design on paper for a new engine, the RB-

*Lockheed L-1011, showing its rear engine installation. (Lockheed)*

211, along with a very aggressive head of its Aero Engine Department, David Huddie. Above all, he wanted to place his company's engines within America's new generation of wide-body jetliners. Rolls had never cracked the domestic market in America, the world's most lucrative, but Huddie saw his opportunity in the L-1011. He succeeded, and in return he later received knighthood from the Queen.[35]

By 1971, however, Lockheed's Dan Haughton was finding that he had hatched some chickens that now were coming home to roost. This had happened during 1965, when he had presided over his company's bid for the C-5A. The company had needed the work quite badly; if it had lost the contract, it would have had to shut down a division in Georgia, a major operating arm. To guard against this, Haughton had "bought in," submitting an unrealistically low bid of $1.95 billion. Even the Air Force had estimated that $2.2 billion would be more like it.

Then, amid escalation of both inflation and the Vietnam War, the C-5A program encountered major strains and delays. Costs went through the roof.

---

35. Newhouse, *Sporty*, pp. 122-123, 141-155; *Astronautics & Aeronautics*, October 1968, pp. 64-69; *Fortune*, May 1968, pp. 61-62; June 1, 1968, pp. 80-85, 151-154; March 1969, pp. 123-128, 136-140.

## Aerospace Recession

By 1971, the Pentagon had budgeted $1.3 billion to cover Lockheed's share of the overrun. Though most of this would be charged to the taxpayers, Lockheed would take its lumps as well. Early in 1971, Haughton, now chairman, agreed to accept an additional loss of $200 million. That wiped out a modest profit; it even cut into the company's net worth. This news would not be welcome at the annual meeting, but business was business, and this transaction meant that Lockheed could begin to put the messiness of the C-5A behind it.[36]

Haughton executed the agreement, headed for the airport, and flew to London to talk about the L-1011 with people from Rolls-Royce. As he later put it, "For about fourteen hours I felt good." But Rolls had been buying in as well, and for the same reason: it needed the business. Its 1968 contract with Lockheed had committed Rolls to develop its turbofan, the RB-211, for a fixed price of $156 million and Lockheed to pay $840,000 for each engine. Rolls was also pushing onto new technical ground. This became apparent as the development of the RB-211 proceeded.

Rolls had been pioneering in the use of carbon fiber, a strong and very lightweight material. In selling the RB-211, a key point had been the firm's intention to build its fan of Hyfil, a proprietary carbon-reinforced epoxy. Hyfil resembles plastics used in today's tennis rackets, and its use in the three engines of an L-1011 stood to save 900 pounds of weight. Such fans must stand up to collisions with seagulls in flight. Hyfil's merits would rest on its ability to pass the chicken test. This involved a cannon that would fire four-pound chicken carcasses at an engine operating at full speed on a test stand. The blades broke under the impact, which meant that these blades would have to use the conventional material, titanium. Titanium was heavier than Hyfil, and this change marked a sharp setback for the RB-211 program.

It was one of a number of problems that drove up the program's cost. As this cost escalated, Rolls reported a loss of $115 million for the first half of 1970. Its chairman, Sir Denning Pearson, turned to the recently-elected Tory government of Prime Minister Edward Heath. The Tories responded by offering a subsidy of $100 million. Pearson, however, had failed to control his costs and hence he would have to go; the firm would have a new chairman, Lord Cole. His board members would include a representative of the govern-

---

36. Rice, *C-5A*, pp. 8-16, 18, 25, 27, 195; *Time*, May 31, 1971, p. 78; *Fortune*, June 1971, p. 69.

ment, Ian Morrow, who specialized in healing sick companies. Morrow soon arranged for an independent accounting firm, Cooper Brothers, to audit Rolls' books.

There was ample opportunity for questions, for Pearson had been using accounting practices that made bankers wince. Since 1961, he had avoided debiting the expenses of jet-engine development in the years they were incurred. Rather, he held them over and debited them in subsequent years, as these engines reached their customers. This practice amounted to prorating the development cost against income from sales. In this fashion, Rolls had reported a string of profits prior to 1970. Now it was difficult to know the firm's total liabilities.

The Cooper audit even had difficulty in estimating the cost of completing the development of the RB-211. The 1968 contract had specified $156 million. Early in 1971, it was at least $408 million. In turn, Lockheed had contracted to pay $840,000 for each engine, a price that supposedly would allow Rolls to make a profit. However, the bare-bones cost of production, even without profit, would now be $1.1 million. In addition to this, Rolls would deliver the engines late. As a consequence, it faced penalties for late delivery of an additional $120 million.

All this meant that Rolls was well past the point where an extra $100 million from the government, or even $200 million, could make a difference. Late in January 1971, Lord Cole learned that he lacked the funds to proceed with the RB-211. His board of directors promptly voted to place the entire company in receivership. In a word, Rolls was bankrupt.

This would be very bad news for Haughton. Britain's bankruptcy laws are far more stringent than those in the United States. American law works to protect a company against its creditors, shielding the firm against debts and legal claims while seeking a reorganization that can open a path to profitability. In Britain, however, creditors come first. A company is not permitted to operate if it has no prospect of success. Rather, it must sell off its assets and go out of business.

Though the Rolls-Royce board reached this decision on January 26, it did not announce it publicly. A week later Haughton, newly arrived at the Hilton Hotel, received a phone call from Lord Cole of Rolls: Could they meet privately at the Grosvenor House? Cole proceeded to tell him the news, which was both unexpected and crushing. When other executives arrived, for a previously scheduled luncheon, they found Haughton looking "as if he had got a bullet between the eyes."

The bullet was aimed more at Lockheed than at its chairman, for those engine intakes on the L-1011 now were all too likely to suck the company into its own bankruptcy. There simply was no easy alternative to the Rolls engines. To turn to Pratt & Whitney for its JT-9D turbofan or to General Electric for its own commercial engine, the CF-6, would cost a year in time and $100 million in development costs. That was because neither of these engines would slip in neatly as a replacement. There would be need for extensive redesign of engine housings and installations, starting with wind-tunnel tests, proceeding through reconsideration of weight distributions, and ending with extensive new tests necessary to win FAA certification. Lockheed would receive a triple blow: a massive overrun, a set of prices charged to airlines that would bring further losses on each sale, and penalties payable to the airlines for late delivery.

In addition to this, Lockheed already was deep in hock, having drawn $350 million from a $400-million credit line held by a syndicate of its banks. It could not seek help from the Defense Department; the settlement of the C-5A had also settled other outstanding issues. The company's stock was depressed. Worse, the L-1011 itself was stirring little interest. Though it had pulled in as many as 168 orders back in 1968, the total since then had grown by only ten more. Lockheed had not booked a single order for it in over a year. Yet to abandon the L-1011 was unthinkable. Its overhang of bank debt could drive Lockheed into insolvency as well.

Rolls' receiver, Rupert Nicholson of Peat, Marwick, and Mitchell, took control of that company on February 4. On the same day, the bankruptcy was announced in the House of Commons. As one official told the magazine *Fortune*, "The news was like hearing that Westminster Abbey had become a brothel." Prime Minister Heath might have bailed everyone out by nationalizing the whole of Rolls, but he had excellent reason not to do so. His legal advisers held that by doing so, the government could become liable for Rolls's debts, the magnitude of which was unknown even to the auditors from Cooper Brothers. Instead, Heath would take over only the portions of the company that were building military equipment. The receiver could sell off the division that was building the famous motorcars, which was profitable and would readily find a buyer. As for the RB-211, Heath would leave it to twist slowly in the wind.

## THE SPACE SHUTTLE DECISION

This approach drew vigorous objection in Parliament. Jeremy Thorpe, leader of the Liberal Party, stated that the L-1011 would then be "the largest glider in the world." Worse, a default on Rolls' contract with Lockheed would "throw into doubt our credibility, our commercial competence and our good faith in all spheres of advanced science." Labour M.P.'s raised the issue of jobs, for some 24,000 people were working on the RB-211 at Rolls and at its subcontractors and suppliers.

Faced with such arguments, Heath unbent slightly, agreeing to have his defense minister take a closer look at the engine's prospects. This minister, Lord Carrington, appointed three investigators that he called his "ferrets," whose report a few weeks later struck a more hopeful note. The RB-211 was meeting its performance goals in runs on the test stand. This was important; it meant the engine after all could be a technical success. Moreover, its development could go to completion for an extra $288 million.

Even so, the odds were formidable against saving the RB-211, and hence Lockheed. Twenty-four banks were directly involved as Haughton's creditors. All were highly averse to risk. Nevertheless, they would have to live with it and accept more; they might even have to throw good money after bad. Nine customers also had ordered the L-1011. Each had its own financial problems and could solve them in part by enforcing contract provisions requiring Lockheed to pay out money as a penalty for late delivery.

Though his hand was weak, Haughton was not without cards of his own to play. The banks, after all, wanted him to succeed; a Lockheed bankruptcy would leave them with bad debts, whereas with forbearance they might yet continue to hold profitable loans. The customers also had reason to stick with the L-1011, for they had already laid out substantial down payments. They also had purchased this airliner on highly favorable terms. This had resulted from Lockheed's competition with the McDonnell Douglas DC-10, wherein Lockheed had won orders by lowering its price and sweetening the terms of sale.

Even under the best of circumstances, the problems with the program would bring delays of several months in delivering the L-1011. However, most major airlines had lost money in 1970. They were in no hurry to receive the new airliners in accordance with the contracted schedule. To the contrary, delays in delivery would also put off the dates when they would have to pay the balance of the purchase price. The chairman of TWA went so far as to suggest that "a delay of a year would have as many advantages as disadvantages, maybe more."

Hence, the report to Lord Carrington meant that the outlines of a deal could begin to emerge. In essence, it would call on everyone to go back to Square One and renegotiate their contracts, paying little heed to the legal commitments of the previous years. Heath would need assurance that Lockheed would indeed stay in business and would not abandon the L-1011. Haughton would need more money from his bankers to give him a base from which to offer such guarantees. He also would have to pay more for his engines, while waiving penalties for late deliveries. For their part, the airlines would have to accept higher prices and later deliveries for their airplanes, again without receiving penalty payments.

Haughton now was the man who had to make it come together. He had a prodigious capacity for work, on which he now drew. Often he had flown in from the East Coast in his Lockheed JetStar, sleeping en route on a couch, checking in at home for a quick shower, then reaching his desk at three or four in the morning to begin his day's work. He also had extensive experience as a salesman. In this business this certainly did not make him a Willy Loman in the play by Arthur Miller, riding on a smile and a shoeshine. It meant, rather, that although he was Lockheed's chairman, he had a strong personal involvement in its sales. If an airline executive raised a question, Haughton himself might turn up the next day in that person's office to answer it.

In dealing with his banks and airlines, Haughton had to do a lot of hand-holding. Two financiers, one a vice president from Bank of America and the other a vice president from Bankers Trust, accompanied him on his travels, as representatives of the entire banking syndicate. Still, each airline and every bank would have to agree that such a deal would represent the best possible outcome for its investors and stockholders. Each of them would naturally prefer to hold back and try for better terms. All would have to agree at the same time, however, or the chance for a deal would fall through. As Nixon's treasury secretary, John Connally, put it, "Dan, your trouble is you're chasing one possum at a time up a tree. What you've got to do is get all those possums up the tree at the same time."

The most elusive of those possums would be the U.S. government. Early that spring, Haughton became aware that he could build a fragile arch that might support Lockheed, Rolls, and the L-1011. Its keystone, however, would be a new line of bank credit totaling $250 million. Lockheed lacked the assets

## THE SPACE SHUTTLE DECISION

to pledge as collateral, and its creditors would certainly demand security. That might be available, however, through a federal loan guarantee, a pledge that the Treasury would reimburse the banks if Lockheed should fold. On May 6, Connally met with Nixon at the White House and announced that the Administration would send the necessary legislation to Congress.

There it would face a minefield of opposition. Congressman Wright Patman, chairman of the House Banking Committee, had blocked federal support for the bankrupt Penn Central Railroad only a year earlier. He was highly skeptical of the proposed Lockheed loan guarantee. Senator William Proxmire, slayer of the SST and a harsh critic of Lockheed, was ready to filibuster against the bill. Though Lockheed was an important defense contractor, the L-1011 was entirely a commercial venture. If the firm went bankrupt, the Pentagon would find a way to rescue its military projects, most likely by having other aerospace firms buy up the pertinent company divisions. Moreover, the L-1011 was to use British engines, a point that did not escape the attention of lawmakers with ties to General Electric and Pratt & Whitney. An alternative, the DC-10, was already on the verge of entering service.

Weighing against these arguments was a single word: jobs. Haughton, testifying before Patman's committee, stated that as many as 60,000 people would be out of work if the L-1011 were to fail. The Democratic Party, which controlled both House and Senate, was still the party of Senator Hubert Humphrey, the presidential nominee of 1968 and a strong labor man. Having shot down the SST as recently as March, Congress could not lightly affront the unions a second time, particularly since the country was still in a recession. Moreover, 1972 would be an election year.

The outcome was thin indeed. On July 30, the House approved the bill, 192 to 189. The measure then moved to the Senate, which was to recess for a month on Friday, August 6. Haughton, however, had warned that by September, Lockheed would be out of cash. The Senate leadership responded by bringing the bill to a vote the previous Monday.

California's Senator Alan Cranston, a principal backer, had been doing the nose-counting and calculated that it would lose by the margin of a single vote. He tried to win over Lee Metcalf of Montana, whose no vote seemed soft, and as the calling of the roll reached its conclusion, Metcalf saw that his vote was likely to be decisive. He told Cranston, "I'm not going to be the one to put those thousands of people out of work." He voted yes, and the loan guarantee passed by a margin of 49 to 48.[37]

With this, the main stone of Haughton's arch fitted into place. The threat of a Lockheed bankruptcy receded, while Rolls now could emerge from its own receivership. With its RB-211, it would become a leader in the business of building engines for wide-body airliners. In turn, Lockheed now was free to proceed with its L-1011.

## *Aftermaths*

The L-1011 did not succeed in the market. Though the program went through development and production, Lockheed went on to construct only 252 of these airliners, rolling out the last in 1983. The program did not earn back its development costs; in fact, this firm sold few if any at a profit, for this company faced strong competition first from the DC-10 and later from the Boeing 767 and Airbus A-300. Hence to win further sales, Lockheed had to offer prices that were very low. The program had received over $1.7 billion at the time of the near-collapse of Rolls; the final losses, at the time of program cancellation, came to $2.5 billion. With this, Lockheed retired from the ranks of the commercial plane-builders and proceeded to make its living almost entirely as a military contractor.[38]

By contrast, Boeing came back strongly following its own brush with bankruptcy. Though the company's sweeping layoffs were painful, they were part of a set of management reforms that brought sharp reductions in the time necessary to build a 707, 727, or 737. In 1966, this had averaged 17 months, from customer order to delivery. By 1972, it was down to 11 months. "You may ask why the hell we didn't do that earlier," said Jack Steiner. "We never had to. We could have done better. Any time you're threatened with extinction you develop abilities you didn't know existed."

In turn, the company saved itself by offering new versions of its narrow-bodied 727 and 737. To compete with the wide-bodies, they needed new

---

37. Newhouse, *Sporty*, pp. 48, 153, 173-183; Eddy et al., *Disaster*, pp. 100-104, 120-121; *Fortune*, August 1, 1969, p. 77; June 1971, pp. 66-71, 156-160; *Business Week*, February 13, 1971, pp. 64-68; March 13, 1971, pp. 42-43; January 29, 1972, pp. 72-74; *Time*, February 15, 1971, pp. 68-69; February 22, 1971, pp. 84-86; May 31, 1971, pp. 78-79; August 9, 1971, p. 57; August 16, 1971, pp. 70-72; *Newsweek*, August 9, 1971, pp. 51-53; August 16, 1971, pp. 65-66.
38. Newhouse, *Sporty*, p. 4; *Fortune*, June 1971, p. 68.

features: longer range, quiet engines, low operating cost, plenty of seats for the purchase price. Boeing introduced such improvements, which amounted to offering more airplane for the purchaser's dollars, and won new income through increased sales.

At the nadir, in 1971, Boeing indeed had been close to ruin. Production of the 707, 727, and 737 was forecast to fall to three per month during 1972 (not three of each model but three of the entire group). The SST was dead, and sales of the 747 were flat. As sales of the improved 727 and 737 took hold, prospects brightened. By late 1974, production of the three narrow-bodies was up to 15 a month. Debt went down rapidly; in 1973 alone, Boeing paid off nearly half a billion dollars. Better yet, orders for the 747 finally picked up. In 1978, the company was back on its feet and was strong enough to launch not one but two new programs: the 757 and 767.[39]

The demise of the SST might have opened a major opportunity for the Concorde. Early in 1973, however, Pan Am declined to exercise its option to purchase these airliners, noting "significantly less range, less payload and higher operating costs that are provided by the current and prospective widebodied jets." TWA, Pan Am's principal rival, followed suit by declining to exercise its own option, with its chairman noting Concorde's "dismal economics."

Significantly, these U.S. carriers made these decisions nearly a year before the energy crisis sent the price of fuel soaring. The airlines of the 1960s had grown rapidly in an era of cheap fuel; the price of jet fuel was only 11 cents a gallon in 1973, and builders of the SST expected the price to stay at this level for the next two decades. Needless to say, it did not; to the contrary, the second and more severe oil crisis, in 1979, pushed this price above a dollar per gallon. Though this hurt all of commercial aviation, it particularly hurt Concorde whose supersonic speed demanded high fuel consumption. In the end, only 14 of these aircraft entered service, divided equally between British Airways and Air France. Taxpayers' subsidies built those planes, and to paraphrase Sir Winston Churchill, rarely have so many given so much for so little.[40]

---

39. Steiner, *Problems*; Steiner, *Jet Aviation*, pp. 31-34; *Pedigree*, pp. 67-68; *Business Week*, April 1, 1972, pp. 42-46.
40. Knight, *Concorde*, p. 100; Newhouse, *Sporty*, pp. 12, 227; Owen, *Concorde*, p. 235; Costello and Hughes, *Concorde*, p. 11; *Astronautics & Aeronautics*, April 1970, p. 50.

## Aerospace Recession

These developments unfolded in the wake of the House and Senate votes of 1971. As exercises in raw vote-counting, the narrow margin of the Lockheed loan guarantee—192 to 189 in the House, 49 to 48 in the Senate—recalls the near-death of NASA's shuttle/station a year earlier, which survived by 53-53 in the House and 32-28 in the Senate. Even the players were the same, with Senators Walter Mondale and William Proxmire playing active roles in both controversies. Proxmire had also taken the lead in the fight against the SST.[41]

The two votes, however, had very different meanings. The shuttle/station was a standard federal project, of the type that NASA had been set up to pursue. Though critics challenged the wisdom and desirability of such an enterprise, no one sought to repeal the Space Act of 1958, which had created NASA and gave it the charter to pursue such initiatives. In turn, this challenge proved to be addressable through such means as having the shuttle stand on its own, supporting it with an Air Force endorsement, and allowing the station to fade in significance. By mid-1971, the Shuttle was well past its time of danger in Congress, as its funding authorization passed the House on a voice vote and the Senate by a vote of 64 to 22.[42]

In both the SST and Lockheed debates, however, the issues were more far-reaching. Though the drama of environmentalism captured the headlines, the SST debate also introduced a disturbing economic question: Was the federal government to provide funding for this project as a risky venture that could not win financial support in the private sector? Similarly, the Lockheed loan guarantee amounted to proposing that Washington should underwrite a line of credit that this firm could not back with collateral, and that therefore was also too risky for banks.

Neither of these ventures were simple exercises in corporate welfare. Federal support for the SST was to be repaid through royalties from sales. Recall the remarks of the project manager, William Magruder, that the SST

---

41. These senators were liberal Democrats. More importantly, they were from Wisconsin and Minnesota, which lacked important aerospace corporations. Other liberal Democrats took different views, reflecting the interests of their states. Senator Alan Cranston (D-Calif.) took the lead in fighting for the Lockheed loan guarantee, because that company was a major employer in his home state. Senator Henry Jackson (D-Wash.) was known as "the senator from Boeing" because of his strong support for that company. *Newsweek*, October 8, 1962, pp. 25-28.
42. *Aviation Week*, July 5, 1971, p. 19.

would return a profit to the government by selling 500 of these airliners. Similarly, the funds advanced to Lockheed came from banks, not from the Treasury, with the loan guarantee merely providing security in lieu of collateral. Both ventures, however, were controversial because they opened the door for the government to assume risks that had traditionally rested within the domain of corporate finance, with its banks and securities markets.

In the face of these well-founded objections, the House and Senate nevertheless voted to support Lockheed, even though their members were well aware that they might be setting an unwelcome precedent for further such interventions. In doing this, Congress showed that it would fight the aerospace recession by passing a measure—the Lockheed loan guarantee—that went well beyond the usual demand for pork-barrel spending to provide jobs during hard times. This meant that to support the aerospace industry in its time of difficulty, Washington would go the extra mile and would enact legislation that ordinarily it would not consider. Against this background of industry woes, the Shuttle, which had stirred such controversy during 1970, appeared in 1971 as a straightforward initiative that could win backing on its merits. In this spirit, though continuing to face opposition within the Office of Management and Budget, the shuttle would gain support where it counted most: from Nixon himself.

CHAPTER  EIGHT

# A Shuttle to Fit the Budget

In May 1971, the Office of Management and Budget (OMB) proposed to limit NASA's spending to a peak of $3.2 billion during the next five years. George Low, working with his NASA associate Willis Shapley, responded by proposing a constant budget, with shuttle spending rising no higher than a billion dollars per year. Low would later describe this approach as "the only one which could convince the OMB that we could do the shuttle and at the same time have a balanced space program." Unfortunately, this billion-dollar limit left NASA with funds to build an orbiter but not a booster. Could the agency find a way to build both?

## *The Orbiter: Convergence to a Good Solution*

As early as 1969, during the initial Phase A studies, Lockheed had taken the initiative in proposing a two-stage fully-reusable design with both stages built of aluminum and using silica tiles for thermal protection. While the final design for the shuttle orbiter would in fact use this approach, one must not think that Lockheed was prescient. Though that company indeed was in the forefront in developing such tiles, they were items for laboratory research. A design that specified their use had no more intrinsic credibility than one that proposed to use the miracle metals Unobtanium and Wishalloy. Nevertheless, the work at Lockheed suited an emerging preference within the Air Force.

Late in 1969, Air Force officials stated that they wanted to build the orbiter using a conventional aluminum airframe, along with whatever form of

331

## THE SPACE SHUTTLE DECISION

thermal protection would be appropriate. In contrast to strong reliance on titanium in hot structures, this preference for aluminum stemmed from an Air Force finding that the aerospace industry faced a shortage of the specialized machine tools needed to fabricate large structural parts from titanium alloy.

Within NASA and its contractors, design studies weighed the relative merits of aluminum and titanium as primary structural materials. The aluminum airframe promised to be lighter in weight, reflecting the fact that aluminum is lighter than titanium. It also would be less costly to build, reflecting the industry's long experience with aluminum. By contrast, titanium structures promised to cost up to three times as much as their aluminum counterparts, and would carry greater risk in development.

Titanium, however, could overcome these disadvantages with its ability to withstand temperatures of 650 °F, compared with 300 degrees for aluminum. This brought a considerable reduction in the weight of the thermal protection, for two reasons. The temperature resistance of titanium would make it possible to build the top areas of the wing and fuselage of this metal alone, without additional thermal protection, for they would be shielded against the extreme temperatures of re-entry by the bottom of the vehicle. In addition to this, a titanium structure could function as a heat sink, absorbing some heat and thereby reducing the thickness and the effectiveness of thermal protection where it would be needed.

Overall, the advantages of titanium promised a complete orbiter, including thermal protection, that would weigh some fifteen percent less than a counterpart built of aluminum. With the titanium orbiter requiring less thermal protection, it also would cost less to refurbish between missions. Though the higher cost and risk of titanium would militate in favor of aluminum once NASA faced the OMB's cost ceiling, the merits of titanium encouraged its use during NASA's design work of 1970 and 1971.[1]

The Phase B studies represented the main line of effort, with John Yardley directing the work at McDonnell Douglas and Bastian Hello managing the activity at North American Rockwell. Both contractors proposed fully-reusable orbiters, carrying all propellant tankage within the fuselage. They looked like large delta-winged fighter planes; indeed, the McDonnell concept

---

1. *Aviation Week*, January 18, 1971, pp. 36-39; Jenkins, *Space Shuttle*, pp. 67, 95; letter, Low to Professor John Logsdon, January 23, 1979.

## A Shuttle to Fit the Budget

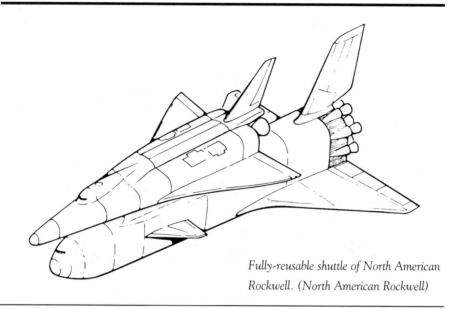

Fully-reusable shuttle of North American Rockwell. (North American Rockwell)

somewhat resembled that company's F-4 Phantom aircraft, which were seeing extensive use in Vietnam.[2]

Both orbiter concepts called for building the propellant tankage of aluminum. Though the tanks would be empty of propellant during reentry, they would require insulation to keep their contents—liquid hydrogen and liquid oxygen—from evaporating. That same insulation then would protect the aluminum from heat that would soak through the thermal protection. Similarly, the crew compartment was to be of aluminum, with the crew members riding amid an insulated and air-conditioned coolness that would protect the adjacent structure from overheating as well.

In the structural frames and outer skin of the wings and fuselage, the contractors proposed to use titanium freely. They differed, however, in their approaches to thermal protection. McDonnell Douglas continued to favor hot structures, with insulation to protect the underlying framework and temperature-resistant metal panels facing the heat of reentry. Metallurgists had developed specialized alloys of nickel and cobalt for the turbine blades of jet engines; these metals resisted oxidation when hot, making them suitable for

---

2. *Aviation Week*, April 5, 1971, pp. 36-38; June 7, 1971, pp. 55-61.

# THE SPACE SHUTTLE DECISION

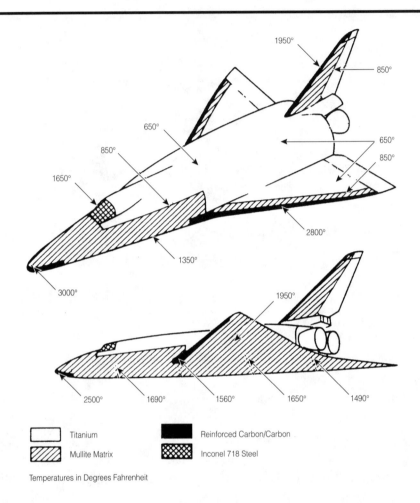

*Thermal protection on the North American orbiter. (North American Rockwell)*

the skin of this orbiter. For instance, most of its underside was to be covered with shingles of Hastelloy-X, a chrome-nickel alloy. The wing leading edges would use struts and beams of columbium; shingles of coated columbium protected areas that were too hot for Hastelloy. Though only a small coterie of engineers had experience with such materials, the McDonnell Douglas designers were not reticent about pushing the state of the art.[3]

---

3. AIAA Paper 71-804; Report MDC E0308 (McDonnell Douglas).

## A Shuttle to Fit the Budget

**High Cross Range System**

| | Orbiter | Booster |
|---|---|---|
| Dry Weight (lb.) | 226,400 | 511,000 |
| Landing Weight (lb.) | 264,700 | 533,000 |
| Gross Weight (lb.) | 836,000 | 3,764,000 |
| GLOW | | 4,600,000 |

*Fully reusable shuttle of McDonnell Douglas. (NASA)*

North American also had a strong interest in titanium hot structures, expecting to use them as well on the upper wing surfaces and the upper fuselage. Everywhere else possible, this company's design called for applying thermally-protective tiles directly to a skin of titanium. Such tiles would cover the entire underside of the wings and fuselage, along with much of the fuselage forward of the wings. The main exceptions included the nose and leading edges, protected with carbon composite, and the vertical fin, designed as a hot structure with a skin of Inconel 718, a heat-resistant nickel steel.[4]

These orbiter concepts represented the fruit of several years of experience in the design and study of two-stage fully-reusable shuttles. They also were quite representative of what NASA now would *not* be able to build, under the fiscal limits of the OMB. The agency, however, had an ace in the hole, for in parallel with these Phase B efforts, Lockheed and Grumman had been pursuing studies of alternatives.

---

4. AIAA Paper 71-805; Reports SV 71-28, SD 71-114-1 (North American Rockwell).

# THE SPACE SHUTTLE DECISION

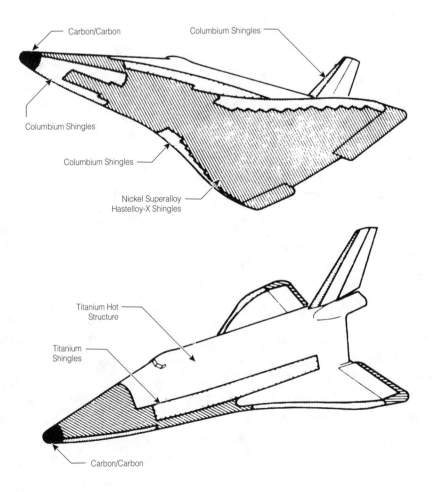

*Thermal protection on the McDonnell Douglas orbiter. (McDonnell Douglas)*

For both companies, the point of departure lay in partially-reusable configurations that would carry their propellant in expendable tanks. This offered a route to lower development cost because the orbiter could shrink in size by carrying its propellant externally. The tanks could take form as simple aluminum shells, while the orbiter would have much less volume to enclose within its hot structures, and much less surface area to protect thermally.

## A Shuttle to Fit the Budget

Such approaches dated to the original Star-Clipper concept of 1965. They had lost favor because the throwaway tankage would cost money, in dollars that literally would be thrown away. Partial reusability was attractive to the OMB because it would cut the all-important development cost, the year-to-year line item in the federal budget. NASA's fully-reusable approach promised huge up-front costs in return for the goal of large savings through reduced cost per flight. The OMB insisted on smaller short-term outlays in exchange for the prospect of a somewhat greater cost of launch, 15 years down the road. The OMB held most of the cards.

Among the alternative shuttle designs, Lockheed was continuing to examine variants of Star-Clipper, with a lifting body for its core vehicle and external tanks flanking this core. The new versions included a two-stage fully-reusable design that placed the propellant tanks within the core and used a McDonnell Douglas shuttle booster as the first stage. NASA by now had little interest in lifting bodies, for they promised difficulty in development along with high landing speeds. With the main line of design activity now defined by delta-winged versions of Max Faget's two-stage rocket airplanes, NASA's continuing support of Lockheed amounted to telling the OMB: Since you want us to look at alternatives, here is what we have.[5]

By contrast, Grumman began its examination of alternative designs with no preconceived views as to how to proceed. When the work began, in mid-1970, this company started with 29 configurations in three categories: two-stage fully reusable, reusable orbiter with expendable booster, and reusable orbiter with expendable propellant tankage. Like everyone else, the Grumman manager, Lawrence Mead, concluded that full two-stage reusability would be best in the long run. His report, however, noted pointedly that the other approaches offered promising means of reducing the peak funding.

Meanwhile, during the fall of 1970, the Grumman group supplemented its NASA funds with company money, to broaden further the range of alternatives. The most promising modified the basic two-stage fully-reusable approach by removing the orbiter's liquid hydrogen fuel from within its fuselage, and storing it in a pair of expendable tanks. Grumman managers presented this concept to officials at the Manned Spacecraft Center in

---

5. Report LMSC-A989142 (Lockheed); Jenkins, *Space Shuttle*, pp. 98-103.

*THE SPACE SHUTTLE DECISION*

November 1970. Within weeks, these managers were instructed to concentrate their efforts on further and more detailed study of this expendable-tank orbiter, and to compare it with a fully-reusable variant having internal tankage. This approach gained further favor in March 1971, as NASA instructed McDonnell Douglas and North American Rockwell to develop variants of their fully-reusable configurations that would also place the orbiter's liquid hydrogen within expendable tankage.[6]

Why was this approach so promising? Liquid hydrogen is bulky, having only one-fourteenth the density of water. Thus, although it makes up only about one-seventh of a shuttle's propellant load by weight, with six-sevenths being liquid oxygen, liquid hydrogen accounts for nearly three-fourths of the volume. Being low in density and hence light in weight, this fuel could be carried in external tanks of similar light weight. Being bulky, its removal would bring a welcome reduction in the vehicle size and surface area.

In addition to this, Grumman's approach brought a useful shrinkage in the size and weight of the complete two-stage shuttle, including the booster. In designing two-stage vehicles, standard procedures exist for choosing the best staging velocity, so as to achieve the lowest weight of the two stages together. At a higher staging velocity, the first stage becomes excessively large and heavy; at a lower velocity, the orbiter requires more size and weight. Ground rules set at Marshall Space Flight Center, based on such optimization, defined this staging velocity as being close to 10,000 feet per second. For Grumman's two-stage fully-reusable configuration, used as a reference for purpose of comparison, it was 9,750 ft/sec.

With the liquid hydrogen now to be carried in lightweight external tankage, a re-calculation of this optimum showed that it would be advantageous to make these tanks larger, allowing them to carry more of this fuel. The orbiter then would have to carry more liquid oxygen, stored within its fuselage. This, however, would be easy to accommodate. Liquid oxygen is dense, denser than water. Hence, the extra quantity would require little additional volume and would not compromise the overall design of the orbiter.

The upshot was that the optimum staging velocity, concomitant with the lowest overall weight, would drop to only 7,000 ft/sec. This would greatly ease the task of designing the booster. The booster now could be considerably

---

6. Report B35-43 RP-11 (Grumman), Section 1; *Aviation Week*, March 29, 1971, pp. 45-46.

## A Shuttle to Fit the Budget

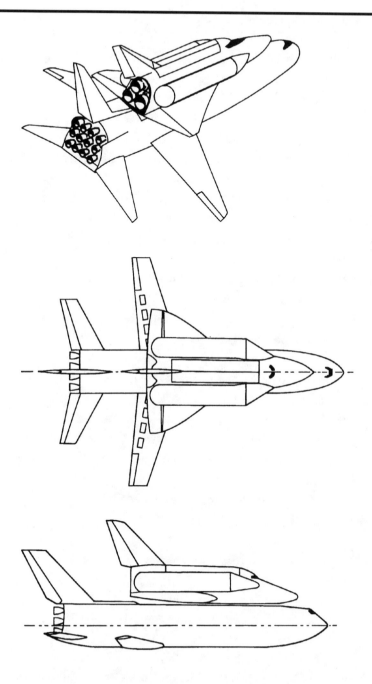

*Grumman's two-stage shuttle, which put the orbiter's liquid hydrogen in expendable tanks. (Art by Dennis Jenkins)*

smaller and lighter in weight, reflecting its reduction in required performance. It also would be much easier to protect thermally.

The orbiter design followed standard practice, with a main structure of titanium. Grumman's report stated that the company would rely on "materials, analysis, manufacturing and test procedures developed for the F-14 aircraft," a Navy fighter that was just entering service. The orbiter's thermal protection was to rely primarily on hot structures, in the fashion of McDonnell Douglas, with carbon composite and silica tiles at the nose and leading edges. The use of external tankage cut the dry or unfueled weight of the complete two-stage shuttle by nearly one-third, from 1.02 million pounds to 692,000 pounds. In the words of the report, this weight saving "means structure we eliminate from design, do not provide tooling for, nor build, maintain, refurbish or otherwise pay for."

This comparison of weights drew on the fact that Grumman had worked to encourage such comparisons deliberately, by carrying through studies of designs with both external and internal orbiter tankage, according to the same ground rules. The report noted that

> *for those who have in the past undertaken to compare configurations from several contractors, the necessity for a dual effort of this nature is readily apparent. There is nothing more frustrating and inaccurate than to attempt to compare weight, performance, and cost from several contractors, using, by definition, their own unique preliminary design groundrules and criteria.*

As a consequence, Grumman's work had to be taken seriously when it pointed to financial advantage:

|  | *Fully Reusable* | *External Hydrogen* |
| --- | --- | --- |
| Development cost, $ M | 7,777 | 6,497 |
| Peak funding, $ B | 2.20 | 1.85 |
| Cost per flight, $ M | 4.29 | 4.22 |

The peak funding level, $1.85 billion, was a long way from the OMB requirement of $1 billion. Nevertheless, it was $350 million closer to this goal than the fully-reusable design. Moreover, in a brilliant example of having one's cake and eating it, Grumman proposed that the expendable tankage

## A Shuttle to Fit the Budget

would actually *reduce* the cost per flight. The tanks per se would cost $740,000 per flight. Other savings, however, would more than offset this, with the largest of them stemming from a substantial cut in the amount of propellants for a flight, and from eliminating the need to refurbish the thermal protection of the now-simpler booster.[7]

While Lockheed's Star-Clipper was widely known for its use of external tankage, it used no booster, relying on a single set of engines in the core to carry it from liftoff to orbit. The use of a reusable booster, in conjunction with expendable tankage on the orbiter, now opened new prospects to explore. An important group of explorers worked at McDonnell Douglas, where they proposed particularly large external tanks that would allow the orbiter to ignite its engines at liftoff, with these engines burning all the way to orbit. Though this shuttle still needed a booster, its staging velocity fell even further, to 6,200 ft/sec. This booster's dry weight, 346,000 pounds, was only three-fifths that of the booster in the fully-reusable system. For the complete two-stage shuttle, cost of development dropped from $9.82 billion to $8.67 billion, yielding a reduction on a par with that projected by Grumman.[8]

Another group of designers worked with Max Faget at NASA's Manned Spacecraft Center. Decisions dating to early 1970 had given that center responsibility for technical direction of orbiter concepts, with NASA Marshall receiving similar responsibility for boosters. Faget had responded by initiating studies of a succession of two-stage, fully-reusable configurations, indulging to the full his taste for straight wings and for lightweight payloads in small payload bays that would suit the mission of space-station resupply. In January 1971, when NASA officials had met with Air Force counterparts, they had agreed that the orbiter should have delta wings along with a payload capacity of 60 x 15 feet and 65,000 pounds. Faget, nevertheless, had gone his merry way, as if this agreement did not apply to him. He continued to pursue his personal preferences in shuttle design; it was well into 1971 before he caught up with the rest of the world.

He knew a good thing when he saw it. Although he had strongly advocated the fully-reusable orbiter with internal tankage, he quickly turned to designs with expendable tanks. The Grumman and McDonnell Douglas con-

---

7. Report B35-43 RP-11 (Grumman); *Aviation Week*, July 12, 1971, pp. 36-39; Powers, *Shuttle*, p. 240.
8. Report MDC E0376-1 (McDonnell Douglas).

# THE SPACE SHUTTLE DECISION

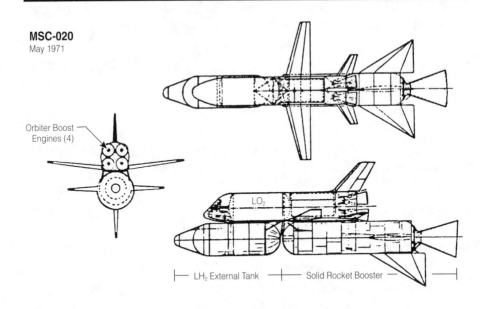

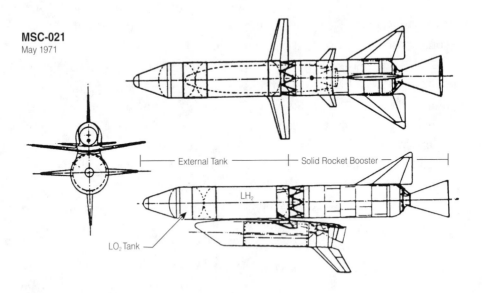

How Max Faget adopted external tankage and moved toward the final orbiter concept, 1971. (Art by Dennis Jenkins)

*A Shuttle to Fit the Budget*

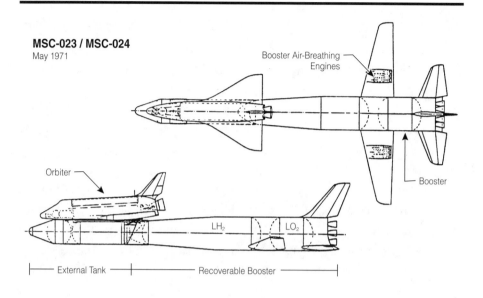

**MSC-023 / MSC-024**
May 1971

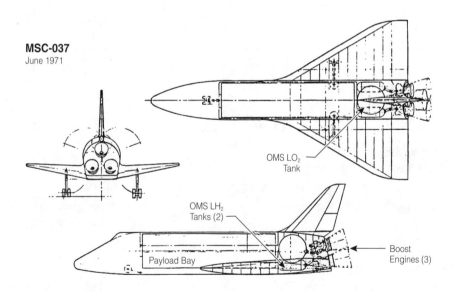

**MSC-037**
June 1971

cepts had carried their hydrogen in long cylinders running alongside the orbiter fuselage, just above the wing. As early as May 1971, Faget prepared a set of drawings designated MSC-020 that put the hydrogen in a single tank, slung beneath the front of the vehicle. This concept followed his preferences by specifying straight wings, a 20,000-pound payload, and a 30 x 15-foot bay. It showed that, at least in this respect, he was willing to change with the times.

The next step was to lengthen this single external tank to allow it to carry liquid oxygen as well. This would reduce the size of the orbiter to a bare minimum. The tank, attached to the orbiter's belly, would demand structural strengthening, for its store of liquid oxygen would be quite heavy. With all propellant removed from the orbiter, that vehicle could achieve a standard design, independent of the tank. The tank could grow to a particularly large size, further lowering the staging velocity of the booster. In turn, this lower staging velocity would further reduce the size of the booster, cutting the cost of the Shuttle program anew.

The first such concept, the MSC-021, came forth during that May. Again, it featured straight wings and the same payload weight as in the MSC-020. Though the bay now had a length of 40 feet, it still was much shorter than what the Air Force would accept. During that same month, Faget also proposed the MSC-023, again with all propellants for the orbiter in a single large underbelly tank. It featured delta wings and a full-size bay, 15 x 60 feet. Here, for the first time, was the outline of a shuttle orbiter that would actually be built.

Even so, Faget was not ready to offer an uncritical embrace of delta wings and large payload bays. Though he now had an external tank that he liked and would stick with, his subsequent designs continued to show small bays and straight or swept wings, and sometimes both. He also examined a number of variations in the arrangement of the orbiter's main engines. Though the MSC-023 had only a single such engine, in June he released the MSC-037, amounting to a variant with three engines and a 40,000-pound payload. It matched the final design in important respects.

The contractors quickly followed this lead, as they launched new studies that assessed its merits. Grumman, which initially put only the hydrogen in external tanks, declared that an orbiter such as the MSC-037 was at least as promising. Lockheed, McDonnell Douglas, and North American found that it was superior, with all three firms giving it a strong endorsement. Indeed, as early as September 1, North American presented its own version of the

## A Shuttle to Fit the Budget

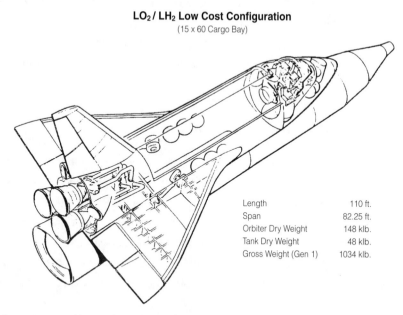

Orbiter concept of September 1, 1971, which foreshadowed the Shuttle that NASA would build. (North American Rockwell)

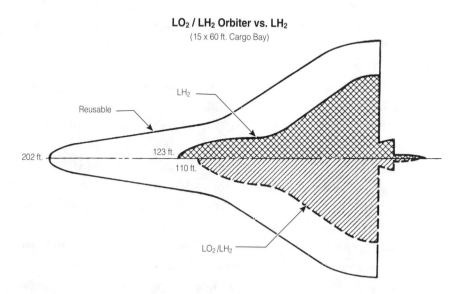

Shrinkage of the shuttle orbiter. Planforms compare the sizes of the fully-reusable orbiter with internal tankage; the orbiter with internal $LO_2$ and with $LH_2$ in external tanks; and the final orbiter, with all propellants carried externally. (North American Rockwell)

MSC-037, singling out this design as the one to pursue. An artist's rendering showed an orbiter closely resembling the one that NASA would build. Eleven days later, NASA formally instructed its contractors to adopt a variant, the MSC-040, and to use it as a basis for comparison within their ongoing studies.[9]

## *The Booster: Confusion and Doubt*

In contrast to the rapid convergence of the orbiter designs, the booster studies fell into disarray. The initial points of reference lay in the Phase B booster concepts of McDonnell Douglas and North American Rockwell. Both contractors proposed vehicles somewhat larger than a Boeing 747, and weighing some five times as much when fueled. The North American concept had two dozen engines: 12 SSMEs and 12 jet engines, the latter serving during flyback to the launch site. McDonnell Douglas's booster, only slightly less ambitious, was to use a dozen SSMEs as well, with ten turbojets.

The North American orbiter had featured a primary structure and skin of titanium, with tiles providing most of the thermal protection. The booster design was more eclectic. Its front section, which enclosed the crew compartment, had frames and skin of René 41, a nickel-chromium alloy that contained cobalt and molybdenum. Though it was strong and oxidation-resistant, it was hard to fabricate. Wings, vertical fin and canards, small forward-mounted winglets used for control, all were of titanium. The fuselage and its tankage were largely of aluminum.

Then came the thermal protection, which avoided the use of tiles in favor of hot structures. The designers might have simplified matters by specifying the wide use of tiles, as on the orbiter. Tiles, however, were in their infancy and North American had to show that it also understood the design of hot structures. As usual, bare titanium skin sufficed for the upper surfaces of the wings and fuselage. The underside, nose and forward fuselage, however, were a metallurgist's delight—or nightmare. The wing leading edges would use coated columbium. Large surface areas would rely on the alloys René 41 and Haynes 188, which were exotic mixtures of nickel, cobalt, chromium, tungsten, and

---

9. Jenkins, *Space Shuttle*, pp. 107-116; Hallion, *Hypersonic*, pp. 1018, 1048-1068; Reports SV 71-40 (North American Rockwell); (Grumman); LMSC A-995931 (Lockheed), Section 1, *Interim Report* (McDonnell Douglas).

## A Shuttle to Fit the Budget

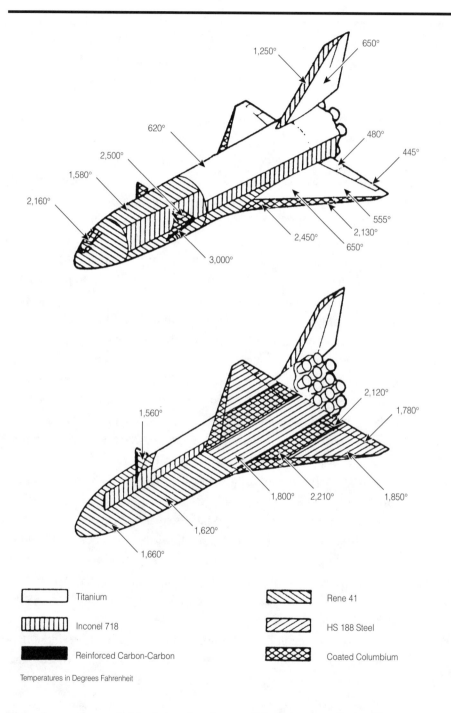

Thermal protection on the North American booster. (North American Rockwell)

molybdenum. Because these metals would expand when hot, designers proposed to build the thermal protection in the form of shells, free to slide over the underlying structure. On paper, at least, the shells would not come loose.

Such a design certainly was not in keeping with the preferences of Max Faget, who always sought the most conservative approach. It was an engineering efflorescence to match the economic efflorescence of the cost-benefit studies at Mathematica, and was likely to raise just as many eyebrows. Moreover, it showed that the exuberance of 1969, when Tom Paine wanted to go to Mars, was still alive. By contrast, much engineering experience is summed up in the acronym KISS: Keep It Simple, Stupid. This is what John Yardley's designers had done at McDonnell Douglas, in crafting their own booster concept.

One can describe it succinctly: aluminum primary structure throughout, including wings, canards, and tail, with tiles of varying thickness to provide the thermal protection. Though metallic shingles would protect the leading edges, they would see use only in the limited areas where tiles would fail to suffice.[10]

These design exercises gave both contractors the opportunity to conduct detailed investigations using a range of approaches: hot structures, aluminum and titanium primary structures, tiles, and metallic thermal protection. Neither contractor offered anything so simple as an aluminum orbiter protected with tiles, though McDonnell Douglas used this approach in its booster. Despite their complexities, however, the concepts were not obviously infeasible, and some people believed they could actually be built.

Along came the Grumman design studies, which put the orbiter's liquid hydrogen in external tanks and lowered the booster's staging velocity from 9,750 to 7,000 ft/sec. Boeing, which had built the first stage of the Saturn V, was teamed with Grumman and had responsibility for the booster. Its engineers determined that this reduction in velocity would bring a disproportionate reduction in the size and weight of the booster, which now would require much less propellant. In addition to this, the lower velocity meant that the booster would be much closer to the launch site when it released the orbiter and began its return. Being closer, it would need less fuel for its jet engines, further reducing its size. The upshot was that whereas the

---

10. Brady and Clauser, *Materials*, pp. 241, 272; Lynch, *Handbook*, pp. 552, 822; *Aviation Week*, January 18, 1971, p. 37; June 7, 1971, pp. 55-61; AIAA Papers 71-804, 71-805; Reports LE 71-7, SD 71-114-1 (North American Rockwell); MDC E0308 (McDonnell Douglas)

## A Shuttle to Fit the Budget

fully-reusable orbiter's booster would have a dry weight of 798,500 pounds, the new orbiter would cut this booster weight to 494,900 pounds.

Better still, the lower staging velocity virtually eliminated the need for thermal protection. The booster now would need neither tiles nor exotic metals. Instead, it would use its structure as a heat sink, just as with the X-15. During reentry, it would experience a sharp but brief pulse of heat, which a conventional aircraft structure could accept and absorb without exceeding temperature limits. Hot areas would continue to demand a titanium hot structure, which would cover some one-eighth of the booster. The rest of this vehicle, however, would make considerable use of aluminum.

How could bare aluminum, without thermal protection, serve in a shuttle booster? It was common understanding that aluminum airframes would lose strength due to aerodynamic heating at speeds beyond Mach 2; higher speeds required titanium, with its greater temperature resistance. These principles, however, dealt with aircraft in cruise, which would face their temperatures continually. The Boeing booster would reenter at Mach 7. Its thermal environment, however, would resemble a fire that does not burn your hand when you whisk it through quickly. Across part of the underside, the vehicle would protect itself by the simple method of using metal with more than usual thickness, to cope with the heat. Even these areas would be limited, with the contractors noting that "the material gauges [thicknesses] required for strength exceed the minimum heat sink gauges over the majority of the vehicle."[11]

In 1954, amid the early feasibility studies that led to the X-15, investigators found, to their pleasant surprise, that they could follow standard aircraft design practice in crafting this research airplane. As a bonus, they also discovered that its metal skin, designed for strength rather than for heat resistance, nevertheless would have enough thickness to serve effectively as a heat sink. Now, in 1971, engineers at Boeing were learning that their booster could offer the same bonus, while providing the convenience of aluminum, the most familiar of metals and the easiest to use.

When McDonnell Douglas went further, introducing an external-tank orbiter that lowered the staging velocity to 6200 ft/sec, its engineers designed a winged booster that was 82 percent aluminum heat sink. Though these

---

11. Report B35-43 RP-11 (Grumman); *Aviation Week*, July 12, 1971, pp. 36-39.

designers could have lowered the staging velocity still further, by putting more propellant in the orbiter's tanks, their selected configuration brought the largest savings in the weight of thermal protection.[12]

Unfortunately, while the move to external tankage brought a welcome reduction in peak annual funding, it took NASA less than one-third of the distance from the $2.2 billion peak funding of the fully-reusables to the $1 billion of the OMB. This move had addressed the easiest part of the problem; the rest of the solution would prove considerably more elusive.

The next step came during June 1971, as the new NASA Administrator, James Fletcher, embraced what his agency had previously rejected: a phased approach to shuttle development. This called for the extensive use of interim systems that would make it possible to build and fly an initial shuttle orbiter, with development of the final systems being delayed for several years. Such an approach would be wasteful, for the interim arrangements would cost money and yet would serve merely as a stopgap. Phased development also carried political risks, for in Washington, few things are so permanent as a temporary solution. If NASA could get any kind of shuttle into space, even one of interim design, it might face strong opposition and long delays before it could win permission to build the shuttle it truly wanted. A phased approach, however, would spread the program over a term of years, reducing the all-important peak funding level.

Fletcher had formally taken over his office on May 1. On June 16, in his first major decision, he made an announcement:

> *The preferred configuration which is emerging from these studies is a two-stage delta wing reusable system in which the orbiter has external tanks that can be jettisoned.*
>
> *Although our studies to date have mostly been based on a concurrent approach in which development and testing of both the orbiter and booster stages would proceed at the same time, we have been studying in parallel, the idea of sequencing the development, test and verification of critical new technology features of the system. We now believe a "phased approach" is feasible and may offer significant advantages.*[13]

---

12. Report MDC E0376-1 (McDonnell Douglas).
13. Fletcher, statement, June 16, 1971; *Aviation Week*, June 21, 1971, p. 19; June 28, 1971, p. 16; Gomersall and Wilcox, *Working Paper*.

## A Shuttle to Fit the Budget

The SSME represented a potentially important element of this approach. It had been slated to receive early attention and funding, for it was to power both Shuttle stages. Its development therefore would pace the entire program. An alternative engine did exist: Rocketdyne's J-2S was an uprated variant of the J-2 engine used in Apollo. It would lack the performance of the SSME, while delivering considerably less thrust. A cluster of these engines nevertheless would push a shuttle orbiter into space, though with a greatly reduced payload weight. Also, use of the J-2S would permit a delay (potentially a long one) in proceeding with the SSME. Indeed, Max Faget was already incorporating the J-2S in a number of his orbiter designs.[14]

Most of the cuts in peak funding, however, would come by putting off plans to develop a fully-reusable booster, even one of smaller size and simplified design such as one of the new heat-sink versions. NASA instead would fly its orbiter atop an existing rocket stage such as the S-IC, the first stage of the Saturn V. By delaying development of a fully-reusable booster for three years, the agency could cut the peak funding to $1.3 billion per year. This would represent a reduction of nearly a billion dollars from the proposed peak funding levels of the Phase B studies, and would put NASA within hailing distance of the OMB requirement.

While the move to external orbiter tankage represented ingenuity, Fletcher's decision amounted to desperation. Dale Myers, the Associate Administrator for Manned Space Flight, admitted to *Aviation Week* that phased development might prove to be impractical. Fletcher himself wrote a letter to a leading shuttle critic, the space scientist James Van Allen: "The political cards are so heavily stacked against this program...that no opposition from the scientific community is necessary. I think you are shooting at a dead horse."[15]

Nevertheless, the prospect of an interim booster spurred hope among the contractors that had products ready to offer. Boeing's situation now was advantageous, for that company's S-IC was the only immediately-available stage with the power to carry a full-size shuttle orbiter. For Boeing, NASA's new interest in that stage also represented a reversal of fortune. Only a few months earlier, the firm was looking ahead to an imminent shutdown of its

---

14. Rocketdyne, *Expendable Launch Vehicle Engines*; Jenkins, *Space Shuttle*, pp. 110, 113-115.
15. Letter, Fletcher to Van Allen, July 12, 1971; *Aviation Week*, June 21, 1971, p. 19.

*Fully-reusable shuttle orbiter with the S-IC used as an interim first stage. (Grumman)*

Saturn production facility. Now it could cherish the hope that production of this rocket stage might continue after all.

Martin Marietta was also in the picture. This firm had actively promoted the Titan III-M, in an era when NASA had been willing to consider a small shuttle orbiter that would amount to an enlarged Dyna-Soar. Though the company had done little of significance for the Shuttle after 1969, now, in mid-1971, it came forward with a new concept, the Titan III-L ("large"). This would use a new liquid-fueled core with a diameter of 16 feet, compared to 10 feet for the standard Titan III, and with up to six solid-fuel boosters of 120-inch diameter.

David LeVine, Martin's vice president for launch vehicles, noted that the solid boosters were already in production at United Technology Corp., and could readily be lengthened to yield more thrust. Though the core would be new, it too would use existing engines from Aerojet General. Though the Titan III-L would be expendable, it would cost no more than $30 million per launch, compared to $73 million for the S-IC. The Shuttle orbiter riding atop the new Titan would certainly not resemble Dyna-Soar; it would be a full-size orbiter, carrying up to 65,000 pounds of payload.[16]

The principal builders of solid rocket motors—Thiokol, Aerojet, United Technology, Lockheed Propulsion Co.—were prepared to make a pitch for their own units as well, arguing that suitable clusters could also provide a good interim shuttle booster. A few years earlier, these firms had built and test-fired rockets with diameters of 156 and even 260 inches. This technology had subsequently been set aside as having no immediate application. Now, however, with the Shuttle needing an interim booster, these big solid motors might see their day.

A fourth approach came from NASA Marshall: the pressure-fed expendable booster. It amounted to reinventing the liquid-fueled rocket, using approaches selected for their simplicity. Conventional rocket stages used structures of thin-gauge aluminum that saved weight. Their engines relied on turbopumps that pumped the propellants to high pressures, with the engines operating under internal pressures that were similarly high. Such engines offered good performance and strong thrust from compact and lightweight units. The turbopumps, however, were costly and difficult to develop, and at times were prone to failure.

Marshall's new concept did away with the turbopumps, relying on gas pressure within the propellant tanks to push the fuel and oxidizer into the engines. Though these engines would have to be of new design, their lack of turbopumps would greatly ease the problem of development. They would be larger and heavier than their pump-fed counterparts and would give less performance. Moreover, the booster would gain weight, for it would require heavy thick-walled tanks to contain the pressure. Though this approach was inelegant, flying in the face of the quest for high performance that marked the SSME, it offered one more route toward reducing the peak annual funding.[17]

This plethora of possibilities reflected the use of external orbiter tankage, which reduced the staging velocity and made it easier for a booster to do its

---

16. *Aviation Week*, June 28, 1971, p. 16; July 12, 1971, p. 38; August 2, 1971, pp. 40-41; Report MCR-71-309 (Martin Marietta).
17. Jenkins, *Space Shuttle*, pp. 122-123; *Aviation Week*, June 28, 1971, p. 16.

# THE SPACE SHUTTLE DECISION

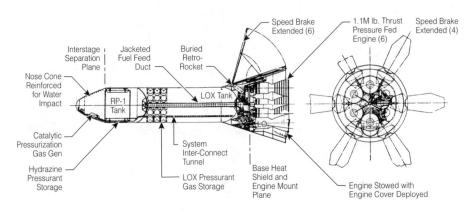

*Recoverable pressure-fed booster. (McDonnell Douglas)*

job. This diversity of boosters meant that there now was no clear reason to choose any of them. The wide range of alternatives recalled the era of the late 1960s, when a hundred flowers had bloomed and when neither NASA nor the Air Force had yet developed a convincing idea of how a shuttle should look.

Though the contractors might have helped by settling on a preferred type of booster, this happened only in part. Early in September, North American declared that all of them appeared acceptable, and recommended deferring a choice. Grumman also found little reason to prefer any of them—but noted that within an interim program of 30 flights or fewer, the S-IC, built by its partner Boeing, would offer lower costs because it was already in hand and would not demand up-front spending for development. Lockheed gave the nod to a cluster of 156-inch solid boosters. McDonnell Douglas did the same, but also had nice words for the Titan III-L.[18]

Having thoroughly muddied the waters with these expendable boosters, NASA officials proceeded to do an about-face as they learned that they might indeed build a reusable booster after all. This happened as further studies of external-tank orbiters showed that they could cut the staging velocity to 5,000

---

18. Reports SV 71-40 (North American Rockwell); LMSC-A995931 (Lockheed); B35-43 RP-21 (Grumman); *Interim Report* (McDonnell Douglas).

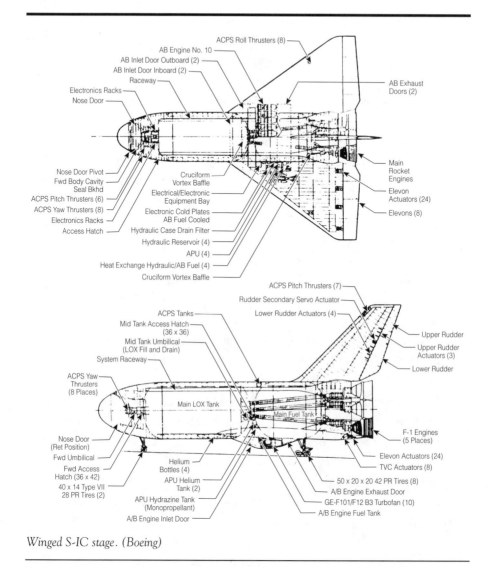

*Winged S-IC stage. (Boeing)*

and even below 4,000 ft/sec. This further enhanced the prospects for reducing the booster size and easing the problem of its thermal protection.

Though it lacked wings, the pressure-fed booster became a candidate for reusability because its thick aluminum skin would easily serve as a heat sink. This same robust skin would allow the vehicle to come down by parachute and land in the ocean, surviving the impact and the subsequent perils of the sea. It would enter the water 200 miles offshore; a boat would then bring it back. No one was prepared to describe this as graceful; with this form of makeshift added to those of its design, the concept was unofficially called the Big Dumb Booster. Such dumbness seemed, however, to promise a new path to wisdom.

A more elegant reusable booster concept came from Boeing, which proposed to remodel its S-IC by turning it into a big airplane. It would receive wings, a tail, a nose with a flight deck, and 10 jet engines for the return to its launch site. Though the standard S-IC had never been built for reentry and reuse, thermal protection would not be a problem; modest thickenings of its aluminum skin now would provide heat sink. To emphasize its near-term feasibility, Boeing's technical artists presented top and side views in lavish detail, even specifying the location of the onboard power units and the choice of tires for the landing gear. At a time when NASA still expected to defer building a flyback booster for several years, one member of the study team emphasized that "our proposal *is* the reusable booster."

By late September, NASA was ready to abandon phased development, returning to the original plan of simultaneously building the booster and orbiter. The booster would be either the winged S-IC that used standard F-1 engines, or the Big Dumb Booster with pressure-fed propulsion. In turn, the orbiter would use phased technology. It would have a full-size payload bay, but would use four J-2S engines; the SSME program would go on the shelf. For thermal protection, the orbiter would rely on an ablative heat shield, even though this would demand costly replacement after every flight. In time, however, the SSME could be resurrected for use in an upgraded orbiter. This orbiter would also use tiles or hot structures, as reusable and low-cost thermal protection.

The development costs and peak annual funding for the complete Shuttle, in billions of dollars, now looked highly encouraging:[19]

|  | Pressure-Fed Reusable | | Winged S-IC | |
|---|---|---|---|---|
|  | Development | Peak | Development | Peak |
| Grumman | $ 4.08 | $ 1.02 | $ 4.5 | $ 1.11 |
| North American | 5.12 | 0.94 | 5.79 | 1.21 |
| McDonnell Douglas | 5.16 | 0.81 | 7.51 | 1.30 |
| Lockheed | (Not given) | | 4.41 | 0.99 |

---

19. *Aviation Week*, September 20, 1971, pp. 16-17; October 11, 1971, p. 23; October 25, 1971, pp. 12-13; Reports LMSC-A995931 (Lockheed); MDC E-0497 (McDonnell Douglas); B35-43 RP-30 (Grumman); SV 71-50 (North American Rockwell); *Ballistic Recoverable* (Boeing).

## A Shuttle to Fit the Budget

These costs compared with estimates prepared by the Aerospace Corp. for a two-stage fully-reusable shuttle: development, $9.92 billion, with a peak of $2.34 billion.[20] The pressure-fed booster yielded a development cost less than half that value, with a peak comfortably under the OMB ceiling of $1 billion. The winged S-IC was only slightly more costly, and it preserved all the operational advantages of the fully-reusable boosters of Phase B, on which NASA had placed so much hope.

The new booster concepts did even more, for they addressed major technical deficiencies of the Phase B boosters. Thermal protection had appeared particularly difficult, at a time when tiles were too new to trust. Charles Donlan, the Shuttle program director, put it this way:

> *Phase B was the first really extensive effort to put together studies related to the completely reusable vehicle. As we went along, it became increasingly evident that there were some problems. You had to develop two hypersonic aircraft configurations, simultaneously. The engine problem was compromised. Trying to use common engines made them not optimized for either case. We ran into a problem of pilot escape from the booster in the event of an abort. We never could quite figure out what to do about it. And then as we looked at the development problems, they became pretty expensive.*
>
> *We learned also that the metallic heat shield, of which the wings were to be made, was by no means ready for use. The slightest scratch and you are in trouble. It became increasingly evident that you want to have the cheapest possible configuration, but you put all this time and effort on a vehicle, the biggest part of which (the first stage), its only role was to get the orbiter up high enough for it to fly itself. So here you're spending all this effort on a part of the system that had no basic payoff. The important thing is the orbiter—that is the payoff.*
>
> *So you see, it was easy to say, "Why do we break our necks putting twelve engines in this damn big reusable booster? You cannot get the pilots away, the metallic heat shield is going to give you all kinds of problems, to do it with ablative stuff makes it too heavy; drop it entirely. That started the ball rolling into other types of boosters for the orbiter.*

---

20. Morgenstern and Heiss, *Analysis*, May 31, 1971, p. 0-15.

## THE SPACE SHUTTLE DECISION

The merits of an aluminum heat-sink booster appeared stronger yet when one recalls that the most plausible Phase B alternative, which used hot structures in the fashion of the North American Rockwell design, amounted to covering much of the surface of a craft the size of a Boeing 747 with alloys that had previously seen use only in turbine blades. By comparison, the simplicity of aluminum heat sink was overwhelming. It was as much as to say that instead of attempting to leap with one swoop into the metallurgy and launch vehicle design of the twenty-first century, engineers instead could build the booster using design methods that dated to World War II.

The Phase B work had also pointed to demanding issues of safety. To reduce weight to a minimum, the internal tanks had to carry part of the Shuttle's weights and loads. The resulting stresses would tend to produce leaks. Hydrogen leaks are difficult to detect, raising the prospect that hydrogen could build up beneath the skin of an orbiter, to produce a damaging fire or explosion. External tankage solved this problem, at least for the orbiter. A leak of propellant now might indeed be detected, since it would occur in plain view. The leak would also be far less dangerous, for the propellant would waft away on the winds rather than build up to form a dangerous concentration.

The Phase B shuttles also promised severe difficulty in technical development. Airplanes and spacecraft, as a rule, tend to gain weight in the course of development. A two-stage shuttle could accommodate such weight growth by enlarging the tanks, to allow them to carry more propellant. With the tankage being internal in both stages, this would demand extensive redesign of one or both stages—which would introduce opportunities for further weight growth. External tankage addressed this problem as well. A simple lengthening of the tanks would do, leaving both booster and orbiter untouched.[21]

It thus appeared that in mere months, NASA had scored an impressive coup, addressing these technical issues while simultaneously meeting the OMB's cost limits. As with the Mathematica study, however, OMB officials again found reason aplenty to view this work with skepticism.

The new configurations closely suited NASA's institutional arrangements, which dated to the time of Apollo. Those arrangements had given NASA Marshall responsibility for the Saturn launch vehicles and their engines, with

---

21. John Mauer interview, Charles Donlan, Washington, October 19, 1983, pp. 19-20; Loftus et al., *Evolution*, pp. 15-18.

the Manned Spacecraft Center holding responsibility for the Apollo moonship, with its piloted spacecraft and lunar lander. This division of authority had carried over to the Shuttle, with Marshall in charge of the booster and MSC dealing with the orbiter. The Shuttle certainly needed an orbiter; hence MSC's prospects were clear. It was less obvious, however, that the Shuttle needed a booster, particularly if it was to be complex and costly. At Mathematica, Klaus Heiss had managed the Shuttle studies and now was recommending simplified designs within NASA. He met resistance, and as he told the *National Journal*, "For a long time some people over there kept seriously telling us, 'We can't go that route, because we've got to have something for the Marshall Space Center as well as something for the Houston Space Center.'"[22]

NASA Marshall had taken shape as a reflected image of Wernher von Braun, its founder and longtime director. He and his fellow veterans of the wartime V-2 effort, who still held senior positions, had long since nailed their flag to the mast of liquid-fuel rocketry, and had left the development of large solid rockets to the Air Force. The recent Shuttle studies reflected this strong Marshall preference. Thus, early in September, the prospect had emerged that the Shuttle might use a cluster of 156-inch solids as its booster. Lockheed and McDonnell Douglas had both made outright selections of this choice, while North American and Grumman had found no reason to reject it.

NASA thus had a clear opportunity to seek a phased-development program with these solids as the interim booster. Instead, in a directive dated September 12, 1971, NASA had instructed its contractors to set aside their studies of phased development and to return to assessments of concurrent development, using two new reusable booster concepts: the pressure-fed and the winged S-IC. Both had NASA Marshall written all over them. Of course, Marshall was the world's leading center for rocket development; its voice would certainly be heard. NASA's decision, however, had amounted to a peremptory dismissal of solids. The results of the new studies were a little too good to be true.[23]

The winged S-IC, for one, promised the fully-reusable booster of NASA's hopes. Its shuttle would do nearly everything that the Phase B fully reusables were to accomplish, with a billion dollars less in peak annual funding. Yet

---

22. *National Journal*, August 12, 1972, p. 1299.
23. NASA Tech. Directive GAC-3, September 16, 1971; Reports B35-43 RP-28 (Grumman), p. 11; SV 71-50 (North American Rockwell), pp. 2, 3, 5.

because this booster was to grow out of the standard S-IC and would use the F-1 engine, it would continue to demand work aplenty from both Marshall and its contractors. The budget analysts at OMB had been trained to spot flimflam. They had reason for doubt when they saw that in an astonishingly brief time, NASA had succeeded in devising a new shuttle design with a flyback booster that would make the agency's wishes come true—while still purporting to meet the cost goals of the OMB.

Projected cost savings also raised questions, for they seemed disproportionately large when contrasted with the relatively modest engineering changes that had led to the shuttle with the winged S-IC. Were the Phase B cost estimates valid? They were certainly high, and if NASA could receive largesse on such a scale, it could continue to maintain itself in the style to which it had become accustomed. On the other hand, one could equally question the low estimates of the recent exercises. Lockheed's invited particular scrutiny, for it was the lowest—and its peak level, interestingly, was just a hair under the OMB mark of $1 billion. Everyone knew that Lockheed had won the C-5A contract with an unrealistically low bid; everyone also knew that the man who had presided over that bid, Dan Haughton, was now the firm's chairman. Was Lockheed attempting again to buy in? Were the other contractors very far behind?

Even if one cared to accept everyone's cost estimates as reflecting good faith, the trend of these estimates invited further questions. On its face, this trend meant that in only six months, NASA had cut the planned cost of development in half, from $10 billion to $5 billion, while sacrificing little in the Shuttle's capability. These cuts had been won at a price, for the cost per flight was now on the rise—while still remaining low enough, at least in the published estimates, to make the Shuttle attractive to users. Still, it was appropriate to ask if NASA had any other rabbits it might pull from a hat. Thus, late in November two OMB staff members, Daniel Taft and the economist John Sullivan, sent a memo to Donald Rice, an OMB assistant director:

> In the light of the innovative Shuttle designs which have been forthcoming over the past several months, we believe that the best procedure would be to provide NASA with a constraint in terms of total investment cost (say, $3-4 B) rather than have us try to define a preferred configuration. If NASA's resourcefulness to date in changing the Shuttle's design is any guide, we have

## A Shuttle to Fit the Budget

*not yet begun to see what they could achieve if they really tried to optimize a system for $3-4 B.*[24]

Rice's domain included the Economics, Science, and Technology Programs Division, which covered NASA. He reported directly to the OMB deputy director, Caspar Weinberger. Rice also had noted that NASA's recent exercises in redesign had brought the Shuttle very close to what NASA had wanted originally. Four years later, he noted particularly that NASA's basic assumptions had seemed to be set in concrete:

*I guess what sticks in my mind more than anything else about it was the difficulty of getting any solid attention paid to alternative designs. I don't mean alternative in the technical detail sense, but alternative in terms of mission requirements and why that mattered. How hard it was to get an examination of alternative specifications of what you would like to accomplish, and the systems designs that reasonably derived from that will lead to each different specification of what you wanted to do.*

He added that within the Pentagon, "they do a hell of a lot better job of looking through the alternatives before they head down one of those roads."[25]

Even so, there had been change within the Shuttle program, at least in terms of its engineering design. The two-stage fully reusables were dead, and Charles Donlan would not miss them, later declaring, "It wasn't till the phase B's came along and we had a hard look at the reality of what we meant by fully reusable that we shook our heads saying, 'No way you're going to build this thing in this century.' As I say, 'Thank God for all the pressures that were brought to bear to not go that route.'"[26]

The winged S-IC soon would die as well, for it appeared more costly than the pressure-fed reusable booster. That one might look and fly like an ugly duckling, but it was a graceful swan in the realm of budgets, and would survive into the next round of designs. This round would resurrect the solid-propellant booster, and would determine the shape of the Shuttle in the form that would actually be built.

---

24. Memo, Taft and Sullivan to Rice, November 29, 1971.
25. John Logsdon interview, Donald Rice, November 13, 1975, p. 1.
26. John Mauer interview, Charles Donlan, Washington, October 19, 1983, pp. 23-24.

THE SPACE SHUTTLE DECISION

## *End Game in the Shuttle Debate*

Within the internal debates of the Executive Branch, the end game, during the second half of 1971, had much in common with the opening gambits of early 1969. Those gambits had featured opposition to NASA's plans from the Budget Bureau, along with the high-level review panel of Charles Townes that had recommended much less than Tom Paine had hoped to pursue. Now, two and a half years later, a similar review would introduce a concept for a mini-shuttle that would win support from NASA's critics—and that this agency could not accept.

Congress was not a significant player within these debates, as it remained generally supportive. The flurry of design changes, in the wake of Phase B, did not dent this support. Congressman Don Fuqua, a leader within the House space committee, notes that these changes initially produced "great consternation" among his colleagues:

> *We had just finished defending one configuration on the floor and then suddenly they announced they were going to change it. Tiger Teague got the top brass from NASA over here and raked them over the coals.*
>
> *We all wanted to know how long they had known they were going to change and how much of this kind of thing was going on behind the committee's back. They explained the reasons behind the changes, and everybody calmed down. After that, though events moved pretty fast, they did try to keep us reasonably well informed.*[27]

The President's Science Advisory Committee, however, wanted its own sources of information. Nixon's first science advisor, Lee DuBridge, had retired in August 1970, citing his wish to step down "well in advance of my 70th birthday in 1971." His successor, Edward David, had been executive director of communications research at Bell Labs. David inherited the existing staff of the Office of Science and Technology. Within this staff, Russell Drew soon proposed that a new PSAC panel should review the Shuttle's prospects and offer views on how NASA should proceed. Daniel Taft, an OMB staffer with responsibility for the NASA budget, warmly endorsed this review and urged his management to support it as well.

---

27. *National Journal*, August 12, 1972, p. 1294.

## A Shuttle to Fit the Budget

The resulting panel took shape in mid-1971, with its chairman being Alexander Flax, president of the Institute for Defense Analyses, a Pentagon think tank. In July, at the outset, Rice presented David with a list of questions he hoped that Flax would address. The first meeting of this Flax Committee was a three-day affair in mid-August, at Woods Hole, Massachusetts, far from the heat and humidity of Washington, with Martha's Vineyard visible across the ocean. The group proceeded to meet about once a month, holding discussions with NASA officials and Shuttle contractors. During the summer and autumn, as these meetings proceeded, NASA replayed anew its familiar struggle with the OMB.[28]

On June 16, when Fletcher announced that he was extending the ongoing contractor studies to consider phased development, he had formally advised the OMB of this in a letter. A month later, in mid-July, Daniel Taft drafted a letter of reply for Don Rice, which asked NASA to "identify an orbiter with minimum performance characteristics." These would include low crossrange, along with substantial reductions in payload size and weight. Rice's letter urged Fletcher to place emphasis on "defining approaches which will substantially reduce the overall investment cost" of the Shuttle, citing a number of design approaches that would cover "the full range of alternatives." Rice, however, did not have to cite a preferred configuration that the OMB would support.

On August 2 the OMB deputy director, Caspar Weinberger, sent a letter to Fletcher that set forth budgetary ceilings for use in preparation of NASA's budget request for FY 1973, which was due at the end of September:

*The planning ceilings established for your agency for the 1973 budget are:*

*(in millions)*
*Net budget authority . . . . . . . . . . . . . . . . . . . . . . . . . . . $ 2,835*
*Net outlays . . . . . . . . . . . . . . . . . . . . . . . . . . . . . . . . . . $ 2,975*

*The above figures provide the basis for developing your 1973 budget submission. The President's budget decisions require that you submit your*

---

28. *Science*, August 28, 1970, pp. 843-844; September 18, 1970, p. 1185; October 23, 1970, pp. 417-419; *National Journal*, August 12, 1972, p. 1295; memo, Taft to Young, January 22, 1971; letters, Rice to David, July 14, 1971; David to Fletcher, July 26, 1971; memo, Drew to PSAC Space Shuttle panel members, August 4, 1971; Myers, Memo for Record, August 19, 1971; NASA, "Documentation of the Space Shuttle Decision Process," February 4, 1972.

## THE SPACE SHUTTLE DECISION

*budget **at or below those figures**. [Emphasis in original].... It should also be understood that subsequent developments may necessitate reducing these planning ceiling amounts; thus, there is no assurance that your final budget allowance will remain at this level.*

These budget marks, if enacted, would represent a further and substantial cut from recent levels: $3,268.7 million appropriated in FY 1971, $3,298 million in FY 1972. With such a budget, NASA certainly would not be able to start the Shuttle during the upcoming fiscal year.[29]

Weinberger was well aware of this. He also knew that these budget marks reflected staff recommendations that had reached his desk by making their way upward through the OMB chain of command. He was willing to listen to Fletcher's viewpoint as well, and three days later he met with Fletcher and with John Young, the head of OMB's Economics, Science, and Technology Programs Division. In notes from that meeting, Fletcher presented several conclusions:

1. Come in with budget that meets spec.
2. Present several alternatives (incl. shuttle) which bring us back to 3.2 outlay & maybe 3.27 authority.
3. Cap...didn't realize manned would be out of business if we made no new starts a la shuttle. Jack Young concurred.

When Weinberger heard that a $2.8-billion budget would mean the end of piloted space flight, he suggested that NASA might be able to stay at the FY 1972 level. On August 12, he wrote a memorandum to Nixon:

*Present tentative plans call for major reductions or change in NASA, by eliminating the last two Apollo flights (16 and 17), and eliminating or sharply reducing the balance of the Manned Space Program (Skylab and Space Shuttle) and many remaining NASA programs.*

*I believe this would be a mistake.*

*1. The real reason for sharp reductions in the NASA budget is that NASA is entirely in the 28% of the budget that is controllable. In short we cut it*

---

29. NASA SP-4012, vol. III, p. 12; letter, Fletcher to Rice, June 16, 1971; memo, Taft to Rice, July 15, 1971; letter, Rice to Fletcher, July 20, 1971; letter, Weinberger to Fletcher, August 2, 1971.

# A Shuttle to Fit the Budget

*Caspar Weinberger with Nixon. (National Archives)*

*because it is cuttable, not because it is doing a bad job or an unnecessary one.*

*2. We are being driven, by the uncontrollable items, to spend more and more on programs that offer no real hope for the future: Model Cities, OEO, Welfare, interest on the National Debt, unemployment compensation, Medicare, etc. Of course, some of these have to be continued, in one form or another, but essentially they are programs, not of our choice, designed to repair mistakes of the past, not of our making.*

*3. We do need to reduce the budget, in my opinion, but we should not make all our reduction decisions on the basis of what is reducible, rather than on the merits of individual programs.*

*4. There is real merit to the future of NASA, and its proposed programs. The Space Shuttle and NERVA particularly offer the opportunity, among other things, to secure substantial scientific fall-out for the civilian economy at the same time that large numbers of valuable (and hard-to-employ-elsewhere) scientists and engineers are kept at work on projects that increase our knowl-*

## THE SPACE SHUTTLE DECISION

*edge of space, our ability to develop for lower cost space exploration, travel, and to secure, through NERVA, twice the existing propulsion efficiency for our rockets.*

He warned against canceling Apollo 16 and 17, noting that such action...

*...would have a very bad effect, coming so soon after Apollo 15's triumph. It would be confirming, in some respects, a belief that I fear is gaining credence at home and abroad: That our best years are behind us, that we are turning inward, reducing our defense commitments, and voluntarily starting to give up our super-power status, and our desire to maintain our world superiority.*

*America should be able to afford something besides increased welfare, programs to repair our cities, or Appalachian relief and the like....*

*7. I believe I can find enough reductions in other programs to pay for continuing NASA at generally the $3.3 - $3.4 billion level I propose here. This figure is about $400 - $500 million more than the present planning targets.*

Here was a milestone. For the first time since the heyday of Apollo, NASA now had an advocate who had real clout within the budget and policymaking process. What was more, Weinberger's memo brought a response, as Nixon read it and wrote in the margin, "I agree with Cap."[30]

By mid-August George Low, the deputy NASA administrator, had reason to believe that piloted flight might survive, if only at a bare-bones level. He wrote, "My own view is that we might be able to bring the 1973 budget back to the 1972 level, but that our chances of bringing it above that level are essentially nonexistent." The 1972 budget had been too small to launch the Shuttle as a new start, and if future budgets would continue at that level, a shuttle would be out of the question. Taking the bull by the horns, he added that in this situation, "we should drop the Shuttle right now":

*My view is that we should assume that this is a permanent situation, that we should drop the shuttle, and that we come up with a new manned space flight program. In my view, this program should be based on an evolutionary space*

---

30. Fletcher, meeting notes, August 5, 1971; Weinberger, Memorandum for the President, August 12, 1971. Reprinted in part in NASA SP-4407, volume I, pp. 546-547. For Nixon's comment see attachment to memo, Huntsman to Shultz, September 13, 1971.

## A Shuttle to Fit the Budget

> *station development, leading from Skylab through a series of research and applications modules to a distant goal of a permanent space station.... The transportation system for this manned space flight program would consist of Apollo hardware for Skylab; a glider launched on an expendable booster for the research and applications modules; and finally, the shuttle but delayed 5 to 10 years beyond our present thinking. The new element in this plan is the expendable booster-launched glider. The whole program ties together in that none of it is dead-ended. The glider would be both an up and down logistics system for the research and applications modules, and, at the same time, lead toward the development of a shuttle in the future.*

The glider would lack engines; it thus would resemble, once again, an enlarged Dyna-Soar. Martin Marietta had advocated such vehicles as recently as 1969; now, with its proposed Titan III-L, it had a launch vehicle suited to this task. Low's notes amounted to admitting that with piloted flight *in extremis*, this glider might form the basis of a fallback position that NASA might accept if continued budget reductions were to force the agency to abandon its plans for a Shuttle.

Two days later, in a review of Shuttle configurations and alternatives, Low described the glider more fully:

> *The glider itself would look somewhat like the shuttle, but would be smaller. It would carry a payload of 12 feet by 40 feet and a payload weight of about 30,000 pounds. It would have sufficient propulsion for on-orbit maneuvering but would not have the engines or propellant tanks required to propel itself into orbit. It would make use of current technology in avionics and other on-board systems. The glider would be placed into orbit with a two-stage vehicle of the Titan IIIL class. The glider and its payload would be reusable but both booster stages would be expended. The requirement for a 15 ft. by 60 ft., 65,000-lb. payload, could be met with the same expendable launch vehicle.*[31]

Needless to say, this was not an approach that NASA would embrace willingly. The agency was still a gung ho outfit, deeply engaged even at that

---

31. Low, Personal Notes No. 52, August 15, 1971; Low, "Space Transportation System Planning," August 17, 1971.

## THE SPACE SHUTTLE DECISION

moment in sending astronauts to the Moon, and its senior officials were not about to go back to Dyna-Soar if they could avoid it. Dale Myers, head of the Office of Manned Space Flight, wrote a memo to Low in which he scoffed at the glider, declaring that its advocates wanted no more than to send an astronaut "'whirling about the Earth' with no evident use for him in space. In the meantime, these people can get back to doing things as they do them now, with various sized ballistic systems, a relatively constant budget, and a relatively 'status quo' sort of operation."

LeRoy Day, deputy director of the Shuttle program, would criticize the glider more pointedly:

> *You had to put this thing on top of an enormous booster which you had to throw away each time. And so you had an operating cost that was getting to be kind of ridiculous. The vehicle size and everything—it didn't have much utility. It was kind of a demonstration. It would certainly have been a research vehicle that you could have studied re-entry with. When you got all through with that then you would have said, "Gee, that would be nice if it was big enough to really do something." Then you would have to turn around and build another vehicle. And with the way the budget climate looked, we were pretty sure that we'd be shut out. We'd never be able to say, "Okay, now let's start up a real program and build another one that will be an enlarged version and have more capability." The OMB and Congress would never support it; it would be like two different programs, and we said, "That'll be the death of it for sure."* [32]

Then in mid-September, just when it counted, George Shultz, director of the OMB, received a staff memo:

> *The President read with interest and agreed with Mr. Weinberger's memorandum of August 12, 1971 on the subject of the future of NASA.*
>
> *Further, the President approved Mr. Weinberger's plan to find enough reductions in other programs to pay for continuing NASA at generally the 3.3–3.4 billion dollar level, or about 400 to 500 million more than the present planning targets.*

---

32. Memo, Myers to Low, September 29, 1971; John Mauer interview, Leroy Day, October 17, 1983, p. 29.

## A Shuttle to Fit the Budget

Significantly, this approval did not embrace the Space Shuttle itself. Low writes that during the next two weeks, "Fletcher and I debated whether we should not forego the Shuttle entirely and develop instead some alternative manned space flight program." Late in September, however, in presenting NASA's formal budget request for FY 1973, Fletcher screwed his courage to the sticking point and indeed sought funds for a Shuttle.

This budget request took the form of a 13-page letter to Shultz. Fletcher stated, "My minimum recommended program requires budgetary authority of $3,385 million and outlays of $3,225 million for FY 1973." He placed the Space Shuttle at the top of his agenda, and asked for funding of $228 million: $200 million for research and development, and $28 million for facilities. Three weeks later, in a follow-up letter to Weinberger, he emphasized the Shuttle anew:

*Space shuttle development should be started no later than the summer of 1972 [emphasis in original]. Our studies have led to a shuttle concept that minimizes peak funding requirements, and at the same time drastically reduces development cost required for the first manned orbital flight. Yet this concept still leads to a productive space transportation system to meet the needs of U.S. civilian and military space programs.*[33]

Fletcher, however, was not the only one with ideas on how to build a piloted spacecraft. Within the Flax Committee, reviewing the Shuttle program on behalf of the White House, a member named Eugene Fubini had taken a leading role. He had been the Pentagon's deputy director of research and engineering, and he had willingly met with the same officials of Martin Marietta who had been recommending the Titan III-L and its glider in lieu of a true Space Shuttle.

Influenced by Fubini, Flax sent an interim report to Edward David on October 19. This report presented a set of alternatives in which the glider, far from ranking lower than the least acceptable form of piloted space vehicle, was actually the most ambitious option that the Flax Committee was willing to endorse. Moreover, the committee's glider would carry only 10,000 pounds

---

33. Memo, Huntsman to Shultz, September 13, 1971; letter, Fletcher to Shultz, September 30, 1971; Low, Personal Notes No. 55, October 2, 1971; letter, Fletcher to Weinberger, October 19, 1971.

## THE SPACE SHUTTLE DECISION

of payload, one-third as much as the version hesitantly considered by NASA's George Low.

What could be less ambitious than this glider? There was the possibility of modifying Apollo spacecraft to make them refurbishable and continuing to fly them using the existing Saturn I-B, on rare occasions. Another option called for developing the Titan III-M and its Big Gemini, a variant of this existing spacecraft that would grow to carry as many as nine people. Such alternatives might keep piloted space flight alive—but it would resemble a patient on life support.

Flax's report also addressed the subject of shuttle economics. He dealt specifically with NASA's preferred concept, which called for concurrent development of booster and orbiter, with the orbiter using phased technology. People described this orbiter as "Mark I/Mark II," referring to an initial version that would later be upgraded with SSME engines, reusable thermal protection, and advanced onboard electronics.

NASA's estimated cost per flight was $9.0 million for Mark I, falling to $5.5 million for the improved Mark II. Significantly, and like the OMB, Flax did not challenge these estimates. He merely denied that they promised advantage. As his report stated,

> *Considering all of the technological and operational unknowns involved in the shuttle development and the fact that no vehicles of similar function have ever been designed before or have ever operated over the range of flight regimes required for the shuttle, prudent extrapolation of prior experience would indicate that estimated development costs may be 30 to 50 percent on the low side. Thus, the estimates of $6.5 billion in RDT&E[34] for the Mk I/Mk II shuttle program may range between $8.5 to $10 billion, reflecting increased program costs of $2.5 to $3.5 billion. Similar uncertainties must be considered to apply to other non-recurring costs such as production and facilities (amounting to about $4 billion). Thus a possible cost uncertainty of about $5 billion for total program costs might be envisioned giving a high estimate of total non-recurring cost of about $15 billion.*
>
> *At a launch rate of about 40 per year (DOD, NASA and other) over the 13 years used in the NASA cost model and an average payload cost of $30*

---

34. Cost of development: Research, Development, Test and Engineering.

## A Shuttle to Fit the Budget

*million (not unrepresentative of the mix of current unmanned payloads), the total payload costs would be $15.5 billion. Thus, even if the total payload cost were saved (including those launched to Mars, Venus, etc.) over a 13-year period by recovery and reuse at zero refurbishment cost, it would, in the case of the high-end cost estimate, barely offset the cost of the shuttle program without discounting. A more realistic (although probably generous) estimate of the savings possible through payload recovery might be 50 percent of payload costs which could account for only $7.5 billion.*

*The other area of savings which is offered by the shuttle is its launch cost. Average launch cost with current expendable boosters is $12 million (projected into the 1978-90 era in the NASA cost model). Thus, with current expendable boosters, the annual launch cost will be $500 million. The cost of Mk II shuttle operation per flight is usually cited at $5.5 million; thus the cost for 40 flights per year will be $220 million. The saving of $280 million annually for 13 years amounts to $3.6 billion. However, a doubling of the operational cost would reduce the saving to $60 million annually or $780 million....*

*The operating cost estimates of $5.5 million per flight for the shuttle, within narrow limits, must be considered to be a very rough estimate at this time, particularly for the early years of shuttle operation. The actual value will depend upon the time between overhaul of equipment not yet designed, refurbishability of thermal protection system materials not yet out of the laboratory, and on the feasibility of operating in the shuttle in an "airline" mode radically different from all past experience in space operations.*[35]

These few paragraphs delivered a body blow to the Shuttle's economic prospects, for although their conclusions were highly unfavorable, they actually carried a strong bias in *favor* of the Shuttle. This was because this analysis used current or undiscounted dollars. The Mathematica study had moved heaven and earth to try to justify the Shuttle in dollars discounted at ten percent per year, and members of the Flax Committee had little use for that study's findings. In the words of a committee staff member, "No one believed all the fancy economics and no one believed the mission model. I think they were on

---

35. Note, Frank Williams to Von Braun, August 23, 1971; Low, Personal Notes No. 56, October 17, 1971; letter, Flax to David, October 19, 1971.

hemp when they were talking about sixty flights per year." Yet it had seemed easy to show that the Shuttle could be cost-effective in undiscounted money.

Now, however, Flax was saying that the Space Shuttle would fail this test as well. NASA might have moved mountains to try to cut the Shuttle's cost of development, but the complete nonrecurring cost, taking account of a plausible overrun, could easily approach $15 billion. The undiscounted savings, with a reasonable mission model and an optimistic cost per flight, would barely top $11 billion and could be less. The OMB economist John Sullivan, who had made shuttle studies his specialty, summarized the matter within a memo to Don Rice: "A Shuttle cannot be justified when using cost-effectiveness as the criterion."

Nevertheless, no one wanted to shut down the piloted-flight effort. In early October, Low wrote of a meeting with the OMB's John Young:

> *I took Jack Young to lunch about a week ago, largely because I had heard that he was the one most negative toward manned space flight within OMB. During our discussions, he agreed that the manned space program and the unmanned planetary programs were the big swing factors in the NASA budget. He indicated, however, that he understood that only the level of manned space flight was in question and not whether to have a manned space flight program at all. He agreed that the President would not, and could not, stop the nation's manned space flight effort.*[36]

Yet if piloted space flight was to involve more than infrequent missions in which astronauts would show the flag, then NASA would need help, and in a hurry. The help came from Mathematica, which now gave a strong endorsement to a preferred shuttle configuration.

## *TAOS: A New Alternative*

The Mathematica analysis of May 31, 1971, had drawn criticism because it had dealt only with the two-stage fully-reusable approach of Phase B, paying no heed to design alternatives. Yet a comparison of such alternatives, on eco-

---

36. Memo, Sullivan to Rice, October 19, 1971; Low, Personal Notes No. 55, October 2, 1971; John Logsdon interview, Dave Elliot, p. 3.

372

## A Shuttle to Fit the Budget

nomic grounds, promised insights that even the Flax Committee was willing to pursue. Flax's interim report stated that while direct cost-benefit analysis lacked "sufficient credibility to serve as a primary basis for deciding to undertake such an expensive and high-risk program," economic comparisons "are undoubtedly extremely valuable in making cost tradeoffs and in considering alternatives in design and program planning."

In studying such alternatives, designers traded reduction in development cost against an increase in cost per flight. This tradeoff had to stay within bounds; if the cost per flight was too high, the Shuttle would fail to capture traffic from existing expendables. Yet the Mark I orbiter, as discussed in Flax's report, would cost $9 million per flight, comfortably under the $12 million of expendables. This was enough of a margin to give this design a strong advantage.

Klaus Heiss, who was continuing to direct the Mathematica studies, was not in a position to introduce new design concepts of his own. Instead he proceeded by receiving the contractors' concepts, comparing them using methods of economics. As he pursued such comparisons, he concluded that NASA had not pushed its tradeoffs far enough. The opportunity existed to push the cost per flight as high as $10 million, in exchange for a further reduction in development cost and in peak annual funding. Even at $10 million per flight, all but five percent of the Shuttle's planned missions would remain cost-effective when compared with similar missions using expendable launch vehicles.

Heiss saw that a specific class of Shuttle designs would do this. He called this class TAOS (Thrust Assisted Orbiter Shuttle). It would use a standard orbiter, possibly of the Mark I/Mark II type, with an external tank large enough to permit the orbiter's engines to operate with "parallel staging," burning from liftoff to orbit. Two booster rockets would flank this tank, giving added thrust after liftoff, then falling away at staging velocity. The boosters might use pressure-fed liquid-fuel engines. Alternately, they could use solid propellant.

"If you could go to $10 million," Heiss later told *National Journal*, "then some kind of thrust-assisted orbiter shuttle beat out all other systems. It had the lowest development costs of any system capable of sustaining continuous manned flight."[37]

---

37. Letter, Flax to David, October 19, 1971, p. 6; *National Journal*, August 12, 1972, pp. 1298-1299.

## THE SPACE SHUTTLE DECISION

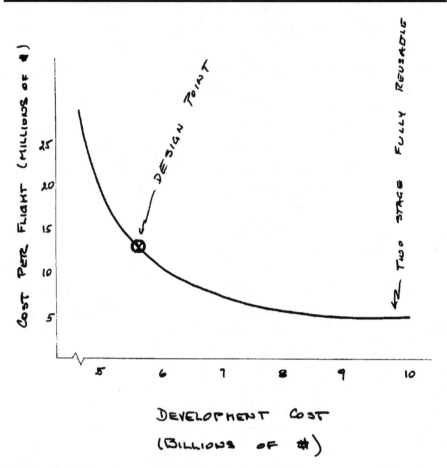

Sketch made by George Low in 1979, illustrating the trade-off between cost per flight and development cost. (Drawing by George Low courtesy of John Logsdon)

Such designs already existed. McDonnell Douglas had one called RATO, Rocket Assisted Take-Off. It put the orbiter's propellants in two external tanks, with a 180-inch solid rocket mounted to the belly. Grumman had TAHO, Thrust Assisted Hydrogen-Oxygen, with a single external tank flanked by twin pressure-fed liquid boosters. This background meant that key NASA contractors already had people who had introduced TAOS concepts and had studied some of their engineering issues.

For TAOS, the 260-inch solid motor would not do. It would be difficult to develop, hard to handle due to its enormous size, and yet it would offer no

*A Shuttle to Fit the Budget*

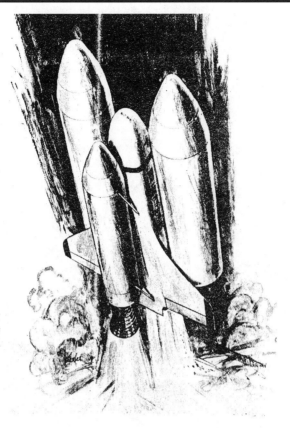

Example of TAOS. Grumman concept of a shuttle with a single large solid-propellant booster and propellants in external tankage. McDonnell Douglas had a similar concept called RATO (Rocket Assisted Take-Off). (Grumman)

cost advantage. The TAOS approach, however, specifically included the selected Shuttle configuration of 1972, with twin solids, smaller in size, for the booster. Ironically, this specific configuration appears to have originated within the Institute for Defense Analyses, the think tank whose president, Alexander Flax, was challenging NASA's approaches so effectively.

The IDA had close ties to the Air Force, which had been using solid boosters for years on the Titan III, and NASA's Charles Donlan gives credit for the twin-solid TAOS to two IDA staffers, Reinald Finke and George Brady. In 1986, Donlan declared that "Brady came up with the configuration that's almost identical" to the one that NASA built. "And so a year or so ago,

## THE SPACE SHUTTLE DECISION

I recommended that NASA give him one of their awards, when they were recognizing Shuttle contributors."

During 1971, however, TAOS had a strong disadvantage: those booster rockets, whether with liquid or solid propellant, would be unpiloted. NASA still was pinning its hopes on a piloted booster, quite possibly the winged S-IC, and its officials paid little heed to Heiss's proposals. Heiss knew why: "In the first place, there was the irresistible urge to go for the most advanced design and technology possible. And then there was a deep-seated bureaucratic bias for two manned vehicles." There also were the institutional prerogatives of NASA Marshall, which might accept pressure-fed liquid boosters in a pinch, but had little background in solids. The OMB's Don Rice would note that NASA "pushed so hard for the liquid fuel thing in the first place, because it was hard to find something for Marshall to do. A reason to keep Marshall around if they didn't have one of those big booster development programs underway."[38]

Heiss nevertheless pressed the case for TAOS within NASA, working with Robert Lindley, the director of engineering and operations within the Office of Manned Space Flight. Lindley had carried out initial cost-benefit studies of the Shuttle, as an in-house NASA effort, and had worked closely with Heiss when Mathematica pursued its subsequent efforts. Heiss would describe Lindley as "one of the few people over there to grasp clearly the real economic and design tradeoffs."

Through September and October, Heiss and Lindley exhorted McDonnell Douglas and Grumman to include TAOS designs in their presentations to NASA. Says Heiss, however, "some NASA officials kept telling them to forget it, the configuration had no chance." As late as October 15, in a presentation to the Flax Committee, NASA failed to include TAOS. Toward the end of the month, Heiss adds, "we thought the whole program was on a catastrophic course."

"There were still many people in NASA who believed they could sell the Administration an $8-billion to $10-billion two-stage flyback system," which might be even more ambitious than the phased-technology orbiter with winged S-IC. "At the other end of the spectrum, the OST [White House Office of

---

38. Report B35-43 RP-28 (Grumman); *Interim Report* (McDonnell Douglas), pp. 13, 25; John Mauer interview, Charles Donlan, October 19, 1983, pp. 28-29; John Logsdon interview, Donald Rice, November 13, 1975, p. 2; *National Journal*, August 12, 1972, p. 1299.

## A Shuttle to Fit the Budget

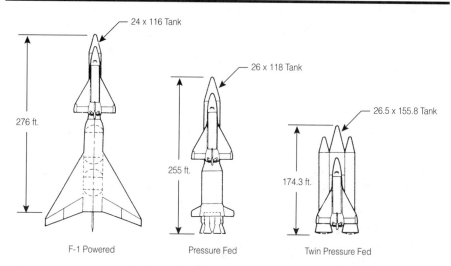

*Booster concepts of autumn 1971, with a winged S-IC at left. Configuration at right is a TAOS. All three alternatives were to boost the same orbiter, which would carry all its propellant in a single expendable tank. (McDonnell Douglas)*

Science and Technology] and OMB seemed to be drawing back toward some kind of advanced expendable system," based on the Titan III-L or the even less capable Titan III-M. A major shuttle design review loomed during November, and in Heiss's words, "We decided to try and get to Fletcher."

Heiss and his boss, Oskar Morgenstern, wrote a 15-page report that called strongly for TAOS. Under standard procedures, they would have submitted it through channels; senior officials with axes to grind could have sat on it or downplayed it before the report could reach the policymaking level. Instead, Heiss and Morgenstern sent it to Fletcher directly. "Those who say Mathematica was a kept child of NASA really don't know what they're talking about," Donlan later said. "Morgenstern was nobody's kept child. In fact, we had no way of controlling what he did."[39]

The conclusions of his report dealt not only with the Shuttle but addressed the Space Tug, a reusable rocket stage intended to carry spacecraft to and from high orbit:

---

39. *National Journal*, August 12, 1972, p. 1299; John Mauer interview, Charles Donlan, October 19, 1983, p. 28.

*The TAOS concept foregoes the development of a Two Stage Shuttle System. With the use of thrust assists of either solid rocket motors or high pressure feed systems—which can be made in part reusable for low staging velocities—the TAOS concepts promise a reduction of the non-recurring costs (RDT&E and initial fleet investment) from about $9 billion or more (two stage systems, including reusable S-IC) to about $6 billion or less, with a minimal operating cost increase, if any, in the operating phase of the TAOS system.*

*The detailed economic justifications of the TAOS concept—when compared to any two stage reusable system are:*

1. *The non-recurring costs of TAOS are estimated by industry to be $6 billion or less over the period to 1979 or to 1984-85, depending on the objectives and choices of NASA.*
2. *The risks in the TAOS development are in balance lower but still substantial. Intact abort with external hydrogen and oxygen tanks is feasible; lagging performance in the engine area can be made up by added external tank capacity. A large reusable manned booster is not needed.*
3. *The TAOS's that were analyzed promise the **same capabilities** as the original two stage shuttle, including a 40,000 pound lift capability into polar orbit and a 60 x 15 feet payload bay.*
4. *The TAOS can carry the **Space Tug** and capture high energy missions from 1979 on.*
5. *The most economic TAOS would use the **advanced orbiter engines** immediately. Our calculations indicate that among the alternative TAOS configurations an early full operational capability (i.e., high performance engines on the orbiter) is economically most advantageous, and feasible, within budget constraints of $1 billion peak funding.*
6. *The TAOS **can** use J-2S engines on the orbiter for an interim period.*
7. *The TAOS **abolishes** completely the immediate need to decide on a **reusable booster** and allows postponement of that decision without blocking later transition to that system if still desired. Thereby, TAOS eliminates or lowers the risk and potential cost overruns in booster development.*
8. *The TAOS can use **"parallel burn"** concepts, which, if feasible, may change the reusable booster decision.*

*A Shuttle to Fit the Budget*

9. *Technological progress may make **tank costs**, and thrust assisted rocket costs less expensive, thus further aiding TAOS concepts when compared to two stage concepts.*
10. *TAOS assures NASA an **early program definition**, and a purpose to the agency. An agreement on TAOS will allow NASA Headquarters a quick and clear reorganization of major NASA centers to meet the TAOS development requirements economically.*
11. *The TAOS funding schedule makes an early Space Tug development possible. The Space Tug is an important part of the Space Shuttle System. A 1979 Space Tug should recover its complete development costs before 1985 even with the stretched build up of Space Shuttle missions from 1979 to 1985.*
12. *A clear policy on TAOS development will give an incentive to European countries to undertake and fund the Space Tug development—thereby possibly even eliminating Space Tug funding from NASA budget considerations.*
13. *The cost per launch of TAOS can be as low as $6 million or even less on an **incremental cost** basis, with reuse of parts of the thrust assist rockets (either SRM or pressure-fed). With Point 9 realized, the costs of TAOS would practically match the **costs per launch of the two stage fully reusable system**.*
14. *TAOS practically assures NASA of a reusable space transportation system with **major objectives achieved**.*[40]

Like Grumman a few months earlier, Mathematica, with Point 13, was not above assuring Fletcher that he might have his cake and eat it too. This low cost per flight indeed might be attained—but only if it proved possible to dunk a pressure-fed booster into the Atlantic and fish it out none the worse for wear, ready for refurbishment and reuse at minimal cost. Contractor studies, however, soon supported Mathematica's high hopes, showing that TAOS indeed offered low cost—and that solid boosters promised costs that were lower still:[41]

---

40. Heiss and Morgenstern, *Factors*. Reprinted in NASA SP-4407, vol. I, pp. 549-555.
41. Reports MDC E0497 (McDonnell Douglas), p. 1-23; B35-43 RP-30 (Grumman), pp. 2, 17.

## THE SPACE SHUTTLE DECISION

| Concept | Development cost $ B | Peak annual funding $ B | Cost per flight $ M |
|---|---|---|---|
| Grumman: twin pressure-fed | 4.02 | 0.97 | 8.2 |
| McDonnell Douglas: twin pressure-fed | 4.83 | 0.74 | 6.4 |
| McDonnell Douglas: twin solid motors | 4.34 | 0.71 | 9.9 |

TAOS did not take NASA by storm. Its design concepts were already part of the mix of alternatives; they had merely been languishing for want of attention. TAOS, however, proved highly useful as it provided a fallback concept, less costly than the winged S-IC, which NASA would adopt as it continued to yield to pressure from the OMB.

## A Time to Decide

For NASA, it was time to have the OMB fish or cut bait. Fletcher had submitted his budget request on September 30, with a line item of $228 million to initiate Shuttle development. This NASA request was one of many budget proposals from federal agencies and cabinet departments; the OMB would modify them and then assemble them into the President's budget, which would go to Congress early in 1972. An early milestone in this budget process, the Director's Review, included a session titled "Space and General Research." Weinberger ran those reviews in 1971. They represented meetings at which OMB staff would recommend actions on agencies' requests, with Weinberger being free to respond. Following standard rules, NASA officials would not be present.

OMB staff members were well prepared for this session. A staff paper, prepared in early October, admitted that the use of a single pressure-fed booster would be more cost-effective than use of the winged S-IC. This represented a response to NASA's booster alternatives of September; the pressure-fed option was a rocket stage of the usual type, and the Shuttle was not a TAOS. While this configuration looked better than the alternative, it still was not good enough. The paper concluded that the most cost-effective option of all, more so than any shuttle, was an uprated Titan III with Big Gemini.

## A Shuttle to Fit the Budget

By contrast, Fletcher wanted his shuttle and he wanted it very soon. In a letter to Weinberger of October 19, he pressed his point:

*The aerospace industry will be hurt by continuing indecision and further delay in the shuttle program. A firm go-ahead, on the other hand, will quickly create jobs in the industry.*

*It will not be possible to sustain the momentum now built up in the shuttle program much longer. A loss in momentum will have serious and costly consequences, and may even be irreversible.*[42]

It would not be possible to maintain the Shuttle effort indefinitely as an exercise in design study and analysis. The aerospace industry was set up to build, not to dither endlessly. Its preliminary-design groups, which had carried forward the Shuttle work to date, practiced a highly skilled and specialized trade. If their efforts were not to bear fruit, their companies would be weakened unless senior management could reassign these people to the preparation of other proposals.

NASA's Director's Review session took place on October 22, and the OMB staff recommendation was blunt: cancel the Shuttle program. Staff members also proposed that if this was not feasible, the decision should be held off for another year, when the OMB would deal with the budget for FY 1974. Weinberger was not eager to accept such options. Neither was he willing to flatly override his staff, even though this was his prerogative, and endorse NASA's position. NASA wanted the Mark I/Mark II orbiter with winged S-IC booster, and Weinberger was interested in alternatives studied by the Flax Committee.

That committee had held a key meeting only a week earlier, with the OMB staff economist John Sullivan attending as an observer, and Flax had reviewed shuttle alternatives only three days earlier, in his interim report to the White House's Edward David. These alternatives included the glider. As Weinberger later described the meeting:

*The staff, in effect, decided they weren't going to do it. The staff then told me that, if we wanted to do it, it could be done less expensively. So I was*

---

42. Letters: Fletcher to Shultz, September 30, 1971; Fletcher to Weinberger, October 19, 1971; NASA, "Documentation of the Space Shuttle Decision," February 4, 1972; *Space Policy*, May 1986, p. 112.

# THE SPACE SHUTTLE DECISION

*delighted to hear that; and so they went back and worked with NASA to work out a different configuration.*

*I could have cut it off at the Director's Review and insist that we are going to do it the way NASA wants it. But the opportunity to do it at a lower cost on additional analysis appealed to me. I never had any doubt in my own mind but, one way or another, I wanted to do it. I thought it was the proper thing for the government to do at that time, and that we needed some forward-looking new activities geared toward the future instead of the past.*

With Solomonic precision, Weinberger then split the difference between the views of NASA and of his staff. He told Don Rice that some sort of shuttle indeed would be approved; NASA would not have to settle for Big Gemini. Rice and his staff, however, would have a free hand to seek lower costs. If this meant that NASA would wind up with nothing more than the glider—well, no one in Washington expected the OMB to come forward as a bearer of gifts.[43]

NASA had some support on the White House staff, but its critics were both numerous and well-placed. In mid-November, George Low gave a summary view of the players that reflected the assessments of William Anders, a former astronaut who now was running the interagency National Aeronautics and Space Council:

> **Weinberger:** *is a real space buff. The only one in OMB really positive toward the NASA program. Causes Rice to over-balance in the opposite direction. Everybody lower in OMB is negative.*
> 
> **Rice:** *the most knowledgeable opposition comes from Rice. Feels that NASA is out of control; however, he will probably support a glider on a TITAN IIIL.*
> 
> **Ed David:** *was noticeably quiet, measuring his words, and repeatedly saying that he only represented science and that other factors are also involved.... Not really plugged into the President.*
> 
> **Flax:** *Fubini is really running the Flax Committee. Flax apparently states that no program as large as the Shuttle will gain continuing support. We need a less costly program.... Anders feels that Flax is driving David toward the glider and not vice versa. Anders believes David will support the Orbiter*

---

43. NASA, "Documentation of the Space Shuttle Decision," February 4, 1972; John Logsdon interview, Caspar Weinberger, San Francisco, August 23, 1977, pp. 5, 13; *Space Policy*, May 1986, p. 113; *Science*, May 30, 1986, pp. 1102-1103.

*with the parallel staged pressure-fed booster if Flax so recommends.*

**Whitehead:** *Whitehead could be helpful in making Flanigan a meaningful communications link to the President, which Anders believes Flanigan needs to be. Whitehead's main motivation now is to improve the Fletcher/Flanigan communications link. Whitehead can be extremely helpful in selling the NASA desired Shuttle approach.... Believes in a $3.5 billion NASA.*

**Rose:** *is the California unemployment buff in the White House. Tries to be helpful and sees Flanigan all the time. He defers to Whitehead when Whitehead is present.*

**Flanigan:** *states that the Shuttle story is improving; however, he is by no means convinced that there should be a Shuttle. Is strongly influenced by Whitehead, Rose and David.*

**Peterson:** *is the most negative of all about NASA. Perhaps the most dangerous opposition we have within the White House. Believes the space program is the place to take money to stimulate technology. Asked why not take $1 billion out of space and who needs manned space flight.*

**Ehrlichman:** *asked the question, "Given the public attitude on space, why not put the money in aeronautics?" However, he is very much concerned about the aerospace industry and will probably go along with whatever OMB/OST/Flanigan recommend.*[44]

These were not the scoundrels depicted in the subsequent bestseller on Watergate, *All the President's Men*. Though skeptical, they were open to argument and perhaps even to persuasion, and it was up to Fletcher and Low to try to sway them.

On November 22, Fletcher responded to a request from Jonathan Rose, a member of Flanigan's staff, and sent him a 10-page paper that presented NASA's case for the Shuttle. On that same day, Low addressed a request from Rice and offered his own paper, which gave a reply to Flax's interim report of a month earlier. Flax had discussed a range of alternatives that included Big Gemini and a small glider. Low now gave his own range of alternatives, with each receiving a succinct description. Big Gemini was conspicuous by its absence, but in other respects, Low's list covered the most important con-

---

44. Low, Memo for Record, November 15, 1971.

## THE SPACE SHUTTLE DECISION

figurations of 1971.

At the high-cost end of this list were two-stage reusables, both the fully-reusables of Phase B and the variant with external hydrogen tanks on the orbiter. In the middle was the Mark I/Mark II orbiter with a single large tank carrying both propellants, and four choices of booster: winged S-IC, single pressure-fed first stage, TAOS with twin pressure-fed boosters, TAOS with twin solids. Anchoring the low-cost end was the Titan III-L with a glider carrying payload of 30,000 pounds. Low recommended a version of TAOS: "The most promising candidate configuration today is the Mark I/Mark II orbiter with the parallel-staged pressure-fed booster," with all engines burning at liftoff.

Low also went to a meeting of the Flax Committee and drew a diagram on a blackboard, summarizing the points of this paper. Fletcher sent a copy to Edward David at the White House, with a cover letter:

*All of these configurations of the Shuttle can be developed for costs substantially below those we planned six months ago. We have progressed to the point where a decision to proceed with the shuttle in connection with the FY 1973 budget process is definitely in order.*[45]

Needless to say, the OMB staff had its own views of the matter. Low's glider had a 12 x 40-foot payload bay; it was the lowest in his range of options, but was the most ambitious of OMB's. On November 29, OMB staffers Daniel Taft and John Sullivan sent their own memo to Rice:

*Even a small (10' x 20') glider could capture all manned space flight traffic, e.g.: Space station visits and return to Earth (station modules could be launched unmanned by expendable rockets).*

*Since the Shuttle is not an economic system under optimistic assumptions, the importance of whether or not all of the payload benefits are realized becomes less significant. The important factors are really such considerations as national prestige, continuation of a manned space flight program, and advancement of technology. Any of the reusable vehicles discussed in this paper (even a 10' by 20' glider) provide this type of intangible benefit.*

*A 10' x 20' glider would provide virtually the same intangible benefits as*

---

45. Memo, Fletcher to Rose, November 22, 1971; letters: Low to Rice, November 22, 1971; Fletcher to David, November 24, 1971; Low, Personal Notes No. 59, November 28, 1971.

# A Shuttle to Fit the Budget

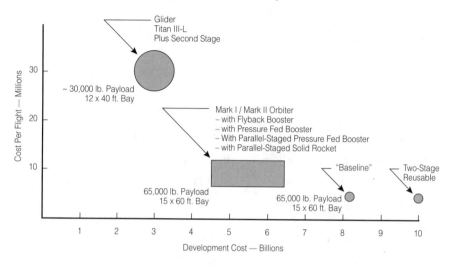

*Alternate concepts of November 1971, showing cost per flight and development cost. Compare with figure on page 374. (NASA)*

*a 15' x 60' orbiter for less than half the investment cost.*

*A 12' x 40' glider or orbiter would provide more operational experience with larger payloads than would a 10' x 20' and hence would make it easier to grow to a larger system should that later be desired.*[46]

The elements of a consensus now appeared to be at hand. Following Low's presentation to the Flax Committee, its members felt that NASA had "finally come around to something reasonable"—with the reasonable thing being the glider, not the TAOS booster and orbiter. Such a glider also appeared on the option lists of both Low and the OMB staff. Within the White House, Flanigan met with two of the top executives of North American Rockwell. He told them that there definitely would be a Shuttle program, that the government was about ready to make the decision, but that some issues still had to be sorted out.

Weinberger had already rejected Big Gemini; Low now argued against the glider. Lacking its own propulsion, it could not make use of designs such as TAOS to keep the cost per flight within bounds. Instead, a glider would

---

46. Memo, Taft and Sullivan to Rice, November 29, 1971.

## THE SPACE SHUTTLE DECISION

require its own two-stage rocket, which would drive the cost per flight as high as $35 million. Low met with Rice, Whitehead, Rose, and Edward David, and wrote that "we demonstrated quite conclusively that a glider would not be cost-effective. Apparently, we also made considerable headway with all the glider advocates in convincing them of this point."

A moment of decision came on December 2, as the OMB sent Nixon a Memorandum for the President. It dealt with space policy, covering a range of issues. The memo included a two-page discussion of the Space Shuttle, and presented the OMB's recommendation: "a smaller reduced cost version of a manned reusable Shuttle with an investment of $4-5 billion over the next eight years—less than one-half NASA's original proposal." With a stroke of his pen, Nixon would grant his consent:

|   | Approve | Disapprove |
|---|---|---|
| 1. Initiate reduced-cost smaller Space Shuttle program | _____ | _____ |
| 2. Conduct Soviet docking mission | _____ | _____ |
| 3. Conduct other manned Earth-orbital missions | _____ | _____ |
| 4. Apollo 16 and 17 | | |
|     Cancel both missions | _____ | _____ |
|     Cancel just Apollo 16 | _____ | _____ |
|     Reschedule Apollo 16 and fly both | _____ | _____ |

For NASA, however, this Shuttle that would fit the budget amounted to a throwback to earlier years, when NASA had considered only its own needs and had not yet introduced the large payload capacity that the Air Force would demand. The new shuttle would not be a glider; it would have its own propulsion. It would, however, carry only 30,000 pounds, in a 10 x 30-foot bay. It offered considerably less than the lowest-cost version of TAOS considered to date.

For months, NASA's senior administrators had been negotiating with the OMB, which time and again had pressed for less costly designs. These same administrators then had to turn around and act as an in-house version of OMB, pressing their engineering managers to pursue such designs. These managers operated at the working level; meetings with Weinberger or Flanigan took place far over their heads. They liked the two-stage fully

## A Shuttle to Fit the Budget

reusable Space Shuttle; they had nurtured it and did not want to give it up. Low and Myers thus had to put a good deal of effort into keeping their troops in line.

This internal NASA debate, paralleling that between NASA and the OMB, brought such contretemps as Max Faget's reluctance to accept Air Force requirements, the resistance that Heiss and Lindley encountered when they advocated TAOS, and insistence at NASA Marshall that the booster had to be liquid-fueled. The December 2 Space Shuttle, which the OMB proposed and Nixon endorsed a week later, fell outside the bounds of what NASA was willing to take. Fletcher, dealing with both the White House and the OMB, would promptly declare that it was unacceptable.[47]

---

47. OMB, Memorandum for the President, December 2, 1971; OMB, "Space Shuttle Program," December 10, 1971; Low, Personal Notes No. 60, December 12, 1971; letter, Low to John Logsdon, January 23, 1979.

# CHAPTER NINE

## Nixon's Decision

Richard Nixon liked space flight. "I can remember Nixon coming off a phone conversation with the astronauts," John Ehrlichman recalls.

> And you know, they are up on the moon, and [Nixon was] as high as a kite. He got a big charge out of them. Then when the astronauts would come to the White House for dinner afterwards, he would always be enormously stimulated by contact with these folks. He liked heroes. He thought it was good for this country to have heroes.

Like other presidents before and since, he basked in the reflected glory of spacefarers. When the crew of Apollo 11 returned from the first landing on the Moon, he was aboard the aircraft carrier *USS Hornet* to greet them. He then used this triumph to gain diplomatic advantage, for after hailing the achievement, he set out on a nine-day world tour that took him to capitals in Southeast Asia, India, Pakistan, and Europe. Significantly, he had planned this tour well in advance of the Apollo 11 flight, anticipating its safe return. "The President had rather daringly pegged his trip to the success of this operation," Tom Paine later remarked.

> Had he gone out to the Pacific to be present at the splashdown and had there been some kind of an accident, it might have harmed considerably his ability then to have the successful trip, which was his first trip abroad as President. I was scared to death that we would have a fiasco or even a

*tragedy. We just wondered whether he knew the odds as well as we did. Well, fortunately Apollo 11 was a success, and in the ensuing world trip, everywhere the President went, the only thing about the United States that anybody wanted to talk about was of course the lunar landing."*[1]

Yet while Nixon willingly embraced Apollo, which he had inherited from Lyndon Johnson, he took his time in committing the nation to new initiatives, whether in space or in other areas of technology. Between 1960 and 1980, such civilian initiatives were largely a province of Democrats: Kennedy and Johnson with Apollo and NASA, Jimmy Carter with his ambitious synthetic-fuels program in the late 1970s. But when George Shultz presented Nixon with NASA's plan for the space shuttle and urged him to accept it, he did.

## *Nixon and Technology*

If Nixon had wished to emulate Kennedy by supporting a new push in space, he could have endorsed the September 1969 report of the Space Task Group, with its recommended focus on a piloted mission to Mars. Nixon did no such thing. He did not even respond to this report in a timely fashion, waiting nearly six months before issuing his own statement on space policy. Nor did this statement come from a senior advisor such as Ehrlichman or Henry Kissinger. Instead, it was the work of two middle-ranking staffers, William Kriegsman and Clay Whitehead, who reported to Peter Flanigan.

John Kennedy, while in the White House, had repeatedly spoken of space flight with the ring of a clarion call, and it is appropriate to note the contrast. Here is JFK, speaking at Rice University in September 1962:

> *The exploration of space will go ahead whether we join in it or not, and it is one of the great adventures of all time, and no nation which expects to be the leader of other nations can expect to stay behind in this race for space.*
>
> *For the eyes of the world now look into space—to the moon and to the planets beyond—and we have vowed that we shall not see it governed by a hostile flag of conquest, but by a banner of freedom and peace.*

---

1. Manchester, *Glory*, pp. 1159-1160; John Logsdon interview, John Ehrlichman, Santa Fe, May 6, 1983, pp. 21, 25-26; Eugene Emme interviews, Thomas Paine, Washington, August 3, 1970, pp. 27-28; September, 1970, pp. 13-14.

# Nixon's Decision

*We set sail on this new sea because there is new knowledge to be gained and new rights to be won, and they must be won and used for the progress of all people.*

*But why, some say, the moon? Why choose this as our goal? And they may well ask, why climb the highest mountain? Why, thirty-five years ago, fly the Atlantic? Why does Rice play Texas?*

*We choose to go to the moon! We choose to go to the moon in this decade, and do the other things, not because they are easy, but because they are hard. Because that goal will serve to organize and measure the best of our energies and skills. Because that challenge is one that we are willing to accept, one we are unwilling to postpone, and one we intend to win.*[2]

Similarly, here is Nixon in his statement of March 1970, which amounted to a most uncertain trumpet:

*Having completed that long stride into the future which has been our objective for the past decade, we now must define new goals which make sense for the Seventies. We must build on the successes of the past, always reaching out for new achievements. But we must also recognize that many critical problems here on this planet make high priority demands on our attention and our resources. By no means should we allow our space program to stagnate. But—with the entire future and the entire universe before us—we should not try to do everything at once. Our approach to space must continue to be bold—but it must also be balanced....*

*We must realize that space activities will be a part of our lives for the rest of time. We must think of them as part of a continuing process—one which will go on day in and day out, year in and year out—and not as a series of separate leaps, each requiring a massive concentration of energy and will and accomplished on a crash timetable.... We must also realize that space expenditures must take their proper place within a rigorous system of national priorities.*

The statement endorsed the activities that were under way at the moment or were well along in preparation: additional Apollo flights, Skylab, planetary

---

2. New York *Times*, September 13, 1962, p. 16.

## THE SPACE SHUTTLE DECISION

missions, and Earth-orbiting spacecraft. It called for further moves toward international cooperation in space. It stopped, however, well short of endorsing the Space Shuttle:

> *We should work to reduce* ***substantially the cost of space operations*** *[emphasis in original]. Our present rocket technology will provide a reliable launch capability for some time. But as we build for the longer-range future, we must devise less costly and less complicated ways of transporting payloads into space. Such a capability—designed so that it will be suitable for a wide range of scientific, defense and commercial uses—can help us realize important economies in all aspects of our space program. We are currently examining in greater detail the feasibility of re-usable space shuttles as one way of achieving this objective.*

This paragraph amounted to a Nixonian blessing for the continued "examining" that would proceed during the next two years. The overall statement, however, endorsed only one new initiative: "Grand Tour," a program that would take advantage of a rare alignment of the outer planets to visit Jupiter, Saturn, Uranus, and Neptune. Moreover, Nixon's statement specifically supported his budget for FY 1971, which continued a policy of cuts in appropriations that dated to 1966. In 1970, NASA was still in retreat, and this statement underscored this march to the rear.[3]

Yet amid this retreat, Nixon maintained strong support for his existing program, with the SST being a prime example. In Ehrlichman's words,

> *Nixon died very hard on the SST, and he had a commitment to that which had to do with chauvinism, I think, is the proper word. We had to be at the leading edge of this kind of applied technological development. And if we weren't, then a great deal of national virtue was lost, and our standing in the world and all that. He was terribly troubled to go to an international conference and have the French president arrive in an SST. That was why that was very hard on him.*

In an attempt to recoup, Nixon borrowed an idea of the Democrats: that the government should use the resources of aerospace to solve domestic

---

3. Nixon, "Statement by the President," March 7, 1970. Reprinted in Launius, *NASA*, pp. 216-221.

problems. During the second half of 1971, several of his senior advisors tried to launch an effort called the New Technology Opportunities Program (NTOP), which sought to define specific projects that might be ripe for federal support. The key man in this effort, William Magruder, had been Nixon's head of the SST program. Ehrlichman recalls that Nixon gave an instruction: "Let's keep in science and technology, and let's find something good for Magruder to do."[4]

The activity began on July 1, when Ehrlichman sent letters to 15 agencies, asking for proposals. The responses went to the Office of Science and Technology, where Edward David's staff carried out initial evaluations. Then in September, Nixon bypassed David as head of this program, even though David had come up within Bell Labs as a technologist, and even though his purview specifically included technology. In an action that effectively downgraded this side of David's domain, Nixon named Magruder as head of NTOP. The OST staff continued its assessments, but Magruder broadened the review to include people from the Treasury Department and the Council of Economic Advisers. Magruder also worked with the OMB's Donald Rice, who assigned a staff member, Hugh Loweth, to work full-time on budgetary aspects.

The effort went forward under tight deadlines, for Ehrlichman wanted to have a finished set of proposals in hand by the end of 1971, in time for inclusion in the FY 1973 budget and the President's State of the Union message. Magruder expanded his reach by seeking ideas from private industry, sending out hundreds of letters to corporations and trade associations. "We were continually in a crisis situation," said one OST manager. "Toward the end, we were killing those guys in the OMB, hitting them with more and more proposals every day. Poor Hugh Loweth was working practically a 24-hour day."

The reviewers quickly discovered, however, that the prospective domestic initiatives carried difficulties that ranged well beyond the merely technical. One important proposal called for full-scale development of high-speed rail transportation in the Northeast, laying new rail and refurbishing passenger stations. The New York Central and Pennsylvania railroads, recently merged into the Penn Central, had allowed this service to deteriorate badly; the Penn Central had gone bankrupt in 1970. Yet its tracks and rights-of-way were still serviceable.[5]

---

4. John Logsdon interview, John Ehrlichman, Santa Fe, May 6, 1983, pp. 2-4.
5. *National Journal*: October 23, 1971, pp. 2114-2124; October 30, 1971, pp. 2156-2163; May 6, 1972, pp. 756-765; *Audacity*, Summer 1993, pp. 52-62.

## THE SPACE SHUTTLE DECISION

We pushed that pretty hard," said Lawrence Goldmuntz, who directed the OST's proposal evaluations.

*And who can argue that it shouldn't be a high-priority item? But in analyzing that proposal the White House also had to take into account the fact that there are several thousand government jurisdictions involved, that the Penn Central is not the most popular railroad in the country today, that it might get athwart union work rules—and well, a number of complicated issues like this came up.*

Another proposal envisioned the development of two-way television, which would foreshadow today's personal computers with e-mail. Two-way TV would allow individual citizens to communicate directly with city social-service agencies, including health, welfare, and police-protection programs. "But we quickly got caught in a crossfire between the Corporation for Public Broadcasting, the Office of Telecommunications Policy, and the cable TV interests," said Goldmuntz. "The policy questions were just too complex."

Another concept promised to develop integrated utilities, which would combine sewage and solid-waste disposal, power, heat, and light within a single system. Such systems, built as modules, offered major savings through lower fuel consumption, with the modules being installed in office towers and apartment buildings. Such integrated services, however, would have raised opposition from unionized municipal workers.

Other proposals ran afoul of recent changes in the national mood. The Atomic Energy Commission had a long-running interest in peaceful uses of nuclear explosives. Its officials endorsed a demonstration project that would use multiple detonations to fracture impermeable rock formations that held natural gas. The concept gained high marks from Magruder's reviewers; the AEC plan even seemed to promise commercial feasibility. However, 1972 was an election year, and with environmentalists showing their strength, the Administration could not touch it.

Still other proposals faced political opposition, such as a plan to build offshore terminals for deep-draft supertankers that drew too much water to enter conventional ports. Such terminals promised to cut shipping costs by eliminating the need to route the supertankers to the Caribbean, where they would transfer their cargoes of oil to smaller tankers of lesser draft. This proposal,

## Nixon's Decision

however, faced strong opposition from governors of states in the Northeast, who feared oil spills. It also would have tended to push the White House toward endorsing expansion of oil imports, a policy that Nixon was reluctant to support.

Four senior White House officials carried out the final review: John Ehrlichman, head of the Domestic Council; OMB director George Shultz; Peter Flanigan, special assistant to the President; and Peter Peterson, director of the Council on International Economic Policy. They declined to endorse any of Magruder's proposals, and the main reason was that in the course of the NTOP exercise, key people had come to realize that they truly knew little about the process of technological innovation. By Christmas, NTOP was moribund.

"We did think in the summer that we could do more and do it quickly," said Peterson. "By December, we were determined to go slow and keep our feet on the ground. I didn't think we should jump into anything before we knew where we were going." Edward David added that "one of the things many of us had driven home more clearly than before was that R and D is not the whole story—you've got to take into account customs, mores, politics, existing structures, and a whole host of other things when you attack a technological issue." Raymond Bisplinghoff, deputy director of the National Science Foundation, concluded that this exercise "verified that we do not know how to make major interventions by the federal government in the R and D sector."

It is a matter of record that as the NTOP fell, the Space Shuttle rose and won Nixon's approval. The latter event, however, did not follow from the former. Edwin Harper, assistant director of the Domestic Council, told *National Journal*, "I was at all the relevant meetings and the two programs were never discussed in terms of a trade-off. The timing of the space-shuttle decision had an independent history." Both NTOP and the shuttle had to stand on their respective merits; in the end, only the Shuttle survived the cut.

NTOP nevertheless was important, for it represented a serious White House attempt to redirect the resources of aerospace toward new domestic priorities. When the attempt faltered, it soon became clear that Nixon would not try to help the beleaguered aerospace industry by having its people work on mass transit or pollution control. Instead, he would give them an election-year gift by keeping that industry's resources within the realm of aerospace.[6]

---

6. *National Journal,* May 6, 1972, pp. 756-765.

THE SPACE SHUTTLE DECISION

## Space Shuttle: The Last Moves

In mid-November, three weeks after the Director's Review and three weeks before the OMB would request Nixon's approval for a downsized shuttle, Low wrote: "The shuttle configuration is beginning to be focused on a considerably smaller orbiter with external hydrogen and oxygen tanks (but with the same payload size and weight) and with a pressure-fed recoverable booster which might be parallel staged." The use of a pressure-fed booster was an option dating to the studies of early summer; the possibility of parallel staging, with booster and orbiter engines all burning at liftoff, showed the influence of TAOS. On November 22, in his report to Rice, Low was more definite: "The most promising candidate configuration today is the Mark I/Mark II orbiter with the parallel-staged pressure-fed booster."

Ten days later, the OMB's Memorandum for the President acknowledged NASA's recent design revisions but called on the agency to accept a version of the shuttle that would be less costly still:

> Last year NASA was proposing a $10-12 B Shuttle. In response to questions from OMB and OST about whether the benefits justified such a large investment, NASA has since designed a $6 B Shuttle which can do all the missions of the larger, more expensive one because it has exactly the same payload capability. (We think both costs are underestimated, perhaps by 50%, i.e., cost overruns are likely on both but more likely on the more expensive version.)
>
> In either case, NASA would plan to replace all of the U.S. expendable booster programs with the Shuttle. Thus, one program, the Shuttle, would dominate NASA for the coming decade, as did Apollo in the 1960's. This would make efforts to reorient NASA to domestic pursuits more difficult, and tend to starve unmanned earth applications missions for resources.
>
> The Shuttle alternative that is chosen must balance costs, benefits and subjective considerations.
>
> What are the Options? NASA, NASA contractors, OST, PSAC and OMB have all given consideration to alternatives to NASA's large Space Shuttle proposal. In summary these alternatives run the gamut from:
> – large systems with both reusable powered orbiters and boosters ($12 B) to
> – small systems with an unpowered reusable orbiter and a non-reusable launch vehicle ($3 B).

## Nixon's Decision

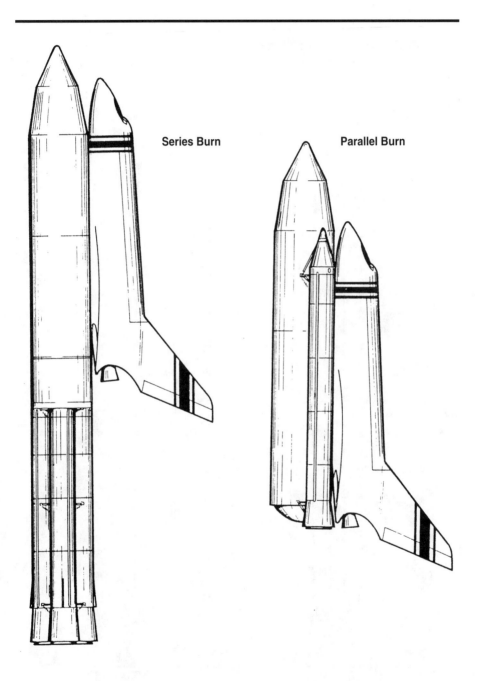

*Series vs. parallel burn. Series burn calls for conventional staging, with the orbiter engines igniting following cutoff of the first stage. Parallel burn ignites both booster and orbiter engines prior to liftoff. TAOS concepts called for parallel burn. (Thiokol)*

# THE SPACE SHUTTLE DECISION

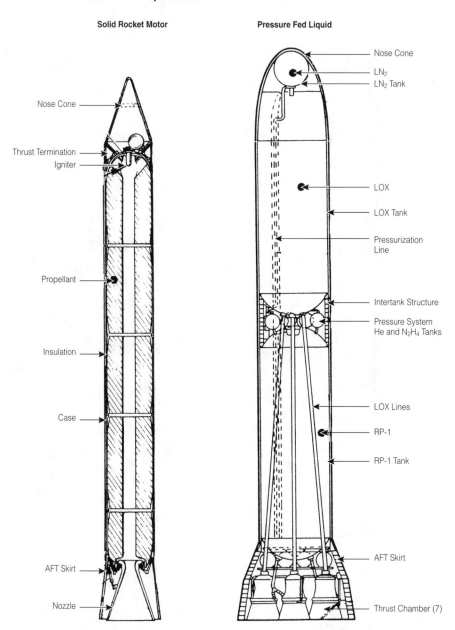

TAOS alternatives: solid rocket motors and pressure-fed liquid boosters, to fly as strap-ons. (Thiokol)

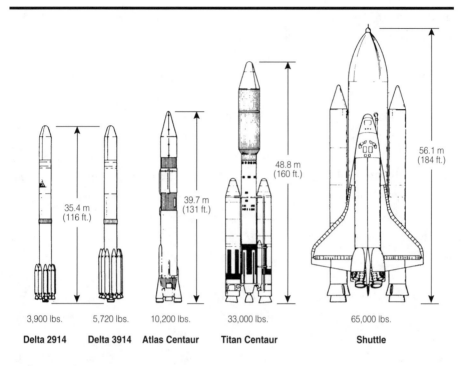

*Expendable launch vehicles of the 1970s, intended for replacement by the Shuttle. (NASA)*

> *Operating costs vary from a high of $30 million per launch for the lowest investment cost option to a low of $5 million per launch for the highest investment cost option.*
>
> *The revised program proposed in this memorandum would develop a smaller reduced cost version of a manned reusable Shuttle with an investment of $4-5 billion over the next 8 years—less than one-half NASA's original proposal.*

The OMB proposal was not the result of a thoughtless exercise aimed at pressing NASA until the pips squeaked; it represented a serious alternative. With its 10 x 30-foot payload bay and 30,000-pound capacity, the OMB shuttle would still capture some 80 percent of the payloads of the larger designs. Similarly, the OMB would not permit its shuttle to be all things to all people; its memo to Nixon stated that "for national security purposes, we may not want all our eggs in one basket." The OMB stated explicitly that the

## THE SPACE SHUTTLE DECISION

nation was to "retain the reliable Titan III expendable booster to launch the few largest payloads that would not fit the smaller Shuttle. These include space telescopes and large intelligence satellites."[7]

Nixon took about a week before he read and accepted the OMB memo. On Saturday, December 11, Fletcher and Low met with Rice, David, and Flanigan. These NASA officials learned that Nixon indeed had decided to go ahead with the shuttle—provided that it was the downsized version of the OMB. This brought considerable and heated discussion. Fletcher finally declared that he could not accept such a decision, and that he wanted to see the President. This meeting did not take place, for Fletcher subsequently decided that he would not help his cause by staging a confrontation with Nixon.

Fletcher nevertheless would fight for a larger shuttle, and in doing this, he would stand for NASA in the fashion of his predecessors. Paine had been the visionary, pushing toward Mars. Low had succeeded Paine; as Acting Administrator, he pushed strongly for two-stage fully-reusable designs. He also sought OMB approval to initiate shuttle development during FY 1972, and won permission to proceed with the SSME on such a schedule. By contrast, Fletcher had acquiesced to OMB pressure from the outset, abandoning the high-cost alternatives as he struggled to meet the OMB's stringent cost limits. He nevertheless had a limited amount of wiggle room, and he would make the most of it.

Though the OMB memo to Nixon specified a development cost, it did not state a payload size or capacity. The OMB presented its numbers separately in a paper for NASA: 10 x 30 feet, 30,000 pounds. Low responded quickly; on Monday, December 13, he sent a memo to Dale Myers that asked for an in-house assessment of this OMB shuttle, to be completed by the end of the month. In this assessment, Myers was to try once more for the full-capacity Shuttle of 15 x 60 feet and 65,000 pounds, by comparing its merits with those of the OMB's configuration.

Flanigan also proved helpful, as he sent a memo to Fletcher from the White House: "None of the figures in the paper given to you are set in concrete. Rather, they should be viewed as a new way to approach the problem, against which an initial estimate will be made within a couple of weeks. Obviously,

---

7. Low, Personal Notes No. 58, November 14, 1971; letter, Low to Rice, November 22, 1971; OMB, Memorandum for the President, December 2, 1971.

those figures will have to be refined in succeeding weeks." He added pointedly, "There is no written directive from the President on this subject."[8]

Fletcher, however, had to prepare once again to give ground. The OMB was willing to permit NASA to build its shuttle with the SSME; it was willing to bypass the interim Phase I orbiter with its J-2S engines. The OMB also would allow a booster "of the reusable pressure-fed type," which Low had recommended. Nevertheless, while rejecting the OMB's small payload size and weight, Fletcher and Low had to find new ways to save money and to cut the developmental cost anew.

With OMB ready to toss the Air Force's payload bay out the window, and with the OMB calling for continued use of the Titan III for large Air Force payloads, it was appropriate to take a fresh look at another Air Force requirement, which had demanded a crossrange of 1,100 nautical miles. This had been a prime reason for the choice of a delta wing for the orbiter, which drove up the weight of the Shuttle and increased its cost by requiring more thermal protection. With the OMB now pressing NASA to return to the payload capacity recommended by Max Faget, the agency had to consider whether it might cut costs by also returning to Faget's straight-wing orbiter design.

Charles Donlan, acting director of the Shuttle Program Office, ruled this out. In a memo to Low, he emphasized that high crossrange would be "fundamental to the operation of the orbiter." It would make the Shuttle maneuverable, greatly broadening the opportunities to abort a mission and perhaps save the lives of astronauts. A high crossrange would also afford frequent opportunities to return to Cape Canaveral in the course of a normal mission, following launch from that site.

Delta wings also promised advantages that were entirely separate from crossrange. A delta orbiter would be stable in flight from hypersonic to subsonic speeds, throughout a wide range of nose-high attitudes. The aerodynamic flow over such an orbiter would be smooth and predictable, with the delta wing thus permitting accurate predictions of heating during reentry and giving confidence in the design of the thermal protection system. In addition to this, the delta vehicle would experience relatively low temperatures of 600 to 800 °F over its sides and upper surfaces.

---

8. OMB, "Space Shuttle Program," December 10, 1971; Low, Personal Notes No. 60, December 12, 1971; memos: Low to Myers, December 13, 1971; Flanigan to Fletcher, December 17, 1971; *Science*, May 30, 1986, p. 1103.

## THE SPACE SHUTTLE DECISION

By contrast, straight-wing configurations would produce complex hypersonic flow fields, with high local temperatures and severe changes in temperature on the wing, body and tail. Temperatures on the sides of the fuselage would run from 900 to 1,300 degrees, making the analysis and design of thermal protection more complex. During transition from supersonic to subsonic speeds, the straight-wing orbiter would experience unsteady flow and buffeting, making it harder to fly.

Because of this combination of aerodynamic and operational advantages, Donlan favored the delta wing for reasons that were entirely separate from those of the Air Force. Fletcher, however, still could back off on the issue of payload-bay capacity. A smaller bay would give a smaller and somewhat less costly orbiter. Reducing the payload weight would trim the size and mass of the entire Shuttle, cutting costs even further. Moreover, for the past several weeks Fletcher had been holding in reserve the possibility of such cuts.

He had gone to lunch on October 19, 1971, with David Packard, the Deputy Secretary of Defense, and Packard had stated that he felt very uneasy over the Air Force's design requirements for the shuttle. As Fletcher wrote in a memo to Low,

> *The requirements which he was particularly concerned about were the cross-range requirement, and payload requirement, and the size requirement. He felt that the cross-range requirement might have been an artificial one and although he didn't fully understand the implications of it felt that if it were causing difficulties, it could easily be modified. I assured him that the diameter requirement came primarily from NASA and not from the Air Force, but that the length probably came from the Air Force. He knew quite well which program caused the length difficulty and although it can't be discussed in this memorandum, it is clear to both of us that something could be done in this regard also. We both agreed that the payload requirement was somewhat arbitrary at this point.*

Fletcher did not discuss this with Rice or Weinberger, for he was about to engage them in the high-stakes game that followed the Director's Review, and he did not want to tip his hand. Two months later, however, with Fletcher preparing to play this card of shuttle capacity, Low's memo of December 13 directed Myers to compare four cases: the OMB shuttle, the full-size NASA shuttle, a design of 12 x 40 feet and 30,000 pounds that NASA had studied previously, and a new case of 14 x 50 feet and 65,000 pounds. They all knew, however, that they could not start with a smaller bay and then enlarge it. Donlan had written that such a course "is not considered practical. The most cost effective system is one sized properly at the outset for its intended use."[9]

In addition to this, with Low presenting new alternatives to the NASA staff for evaluation, Don Rice found himself in a position to add a few alternatives of his own. He had formal channels for receiving information from NASA, but now he opened up some back-channels that gave him access to additional sources. "Some of my information came from the Defense Department, but not very much of it," he later remarked. "Some of it came from industry. There were clearly some people in industry who were concerned that NASA was going to lead them down the road of another C-5A debacle and that they would end up with nothing."

Fletcher saw what was happening:

*He would come up with his own drawings on what it should look like. He had private sources that he was turning to. Contractors wanting their particular version. Contractors do that. And there is no reason why they shouldn't. I hate to say this about Don but he really didn't understand aerospace contractors. Tell him to design a cheaper system, he'd design a cheaper system. He went that far; he came up with his own design.*

Fletcher took it in stride: "We kept dealing with him. After all, you have to be polite, because you never know who's going to be your boss next." Low also joined in the exchanges. At one point, Rice presented a developmental cost of $4 billion and claimed that it came from a contractor whose name he could not divulge. Low asked him to disclose the contractor, so that NASA could critique his claim. Rice refused, but based on the questions he had asked, Low concluded that the contractor was North American Rockwell. That company had given far more information to NASA than to Rice, and Low determined that Rice's $4 billion left out company profit and the cost of NASA program support. Indeed, it was the equivalent of numbers that NASA itself had shown.

---

9. Memos: Fletcher to Low, October 20, 1971; Donlan to Low, December 5, 1971; Low to Myers, December 13, 1971.

## THE SPACE SHUTTLE DECISION

Every time Rice got a bright idea, he would send it over to NASA for assessment by its staff. The OMB's William Niskanen, head of its Evaluation Division, chimed in with similar requests of his own. Within NASA, William Lilly was the chief budget officer and was close to the working level where people had to take the time to deal with these matters.

Lilly tried to come between Rice and his staff:

*It was some dirty warfare going on at that time. Our approach was to, in essence, to create doubt in their boss's mind of the value of the work they had done. We did our studies and in many ways in order just to prove that they were biased, and they were wrong and they didn't know what they were doing. And any way we could embarrass them, we did it.*

Lilly took care to control the flow of information from NASA to the OMB, for he knew that a small leak can sink a great ship:

*I reached the point, to hell with you, back of my hand to you. I mean, you can't do your job if I don't let you in the agency [NASA]. And in essence, I probably went about shutting them off from information throughout the agency. I just told my people that, in a way that they understood, but not really telling them, they weren't to respond. I wanted to see everything that went over. There was to be no answers to questions given throughout the agency to OMB without it coming through me personally.*

"They kept throwing road blocks on us," Lilly continues. "Every time we would come up with a study and do what they wanted in terms of comparative analysis. They couldn't challenge what we had done in many ways, they would turn around and come with studying a new concept." In time, Niskanen sent a request that went too far: "To be brutally frank about it, the reaction we had or my attitude on it, and what, in essence, I told them? I told them to go shove it up your ass. I wasn't going to do any such thing. It didn't make any goddamn sense. I wasn't about to do such a study, and we never did."

Fletcher, by then, was using channels of his own: "We were kind of mean to Don in going over his head all the time. But I figured that was the only way we could get this thing done." He met with Caspar Weinberger "quite frequently," with Peter Flanigan "almost as often," and with George Shultz "two or three times." He also arranged a meeting with John Ehrlichman.[10]

These exchanges continued through Christmas. On Monday, December 27, having reviewed the latest contractor studies, Fletcher prepared to play his poker card of shuttle capacity and to make an offer that would counter the OMB shuttle of two weeks earlier. "We met all day to discuss the various options," Low wrote. Low continued those meetings on Tuesday morning, talking with Myers, with Lilly, and with others as well. Following those meetings, Low and Fletcher agreed on the terms of their counteroffer. Fletcher then presented these terms to Weinberger in a December 29 letter:

> We have concluded that the full capability 15 x 60' - 65,000# shuttle still represents a "best buy", and in ordinary times should be developed. However, in recognition of the extremely severe near-term budgetary problems, we are recommending a somewhat smaller vehicle—one with a 14 x 45' - 45,000# payload capability, at a somewhat reduced overall cost.
>
> This is the smallest vehicle that we can still consider to be useful for manned flight as well as a variety of unmanned payloads. However, it will not accommodate many DOD payloads and some planetary payloads. Also, it will not accommodate a space tug together with a payload, and will therefore not provide an effective capability to return payloads or propulsive stages from high "synchronous" orbits, where most applications payloads are placed.

An accompanying table compared costs for five options:

| CASE | 1 | 2 | 2A | 3 | 4 |
|---|---|---|---|---|---|
| Payload bay (ft.) | 10 x 30 | 12 x 40 | 14 x 45 | 14 x 50 | 15 x 60 |
| Payload weight (lbs.) | 30,000 | 30,000 | 45,000 | 65,000 | 65,000 |
| Development cost ($billions) | 4.7 | 4.9 | 5.0 | 5.2 | 5.5 |
| Operating cost ($millions/flight) | 6.6 | 7.0 | 7.5 | 7.6 | 7.7 |
| Payload costs ($/pound) | 220 | 223 | 167 | 115 | 118 |

---

10. John Logsdon interviews: Donald Rice, November 13, 1975, p. 2; James Fletcher, September 21, 1977, pp. 20-21, 23; John Mauer interview, William Lilly, Washington, October 20, 1983, pp. 28-33; Roger Launius interview, James Fletcher, September 19, 1991, p. 14; Low, Personal Notes No. 61, January 2, 1972.

## THE SPACE SHUTTLE DECISION

This table included the four cases cited by Low in his memo to Myers of December 13, and reviewed by Myers. Case 1 was the OMB shuttle; Case 4 was the shuttle NASA wanted and now apparently would not have. Following standard custom, the option Fletcher proposed was right in the middle as Case 2A.[11]

Fletcher privately knew that he could go still lower. Talking with Low, he decided that they could accept something as small as 14 x 40 feet with 40,000 pounds. The two men then went to an afternoon meeting with White House and OMB officials: Shultz, Weinberger, Flanigan, David, Rice and Rose. Shultz was now the key man; he headed the OMB, he was Rice's boss, and he had Nixon's ear.

Shultz looked at NASA's presentation and decided that the only thing that made any sense, as NASA had said all along, would be the full-size version, Case 4. Shultz, however, did not press this point for Rice objected vigorously. Rice's staff was still active; only one day earlier, his economist John Sullivan had sent him a memo arguing anew that the most cost-effective system was still the Titan III. The meeting broke up with no decision. Fletcher and Low, however, came away fairly confident that they would at least get Case 2A, which they had recommended. Indeed, Shultz's support, however tentative, allowed them to hope that they might even win the full-capability Case 4.

Rice again prevailed, as he talked further with Weinberger. In a phone conversation with Fletcher, Weinberger stated that he wanted NASA to look at a 14 x 45-foot shuttle—with 30,000 pounds of payload, only two-thirds that of Case 2A. In Low's words, "Fletcher came close to telling Weinberger to go to hell but restrained himself perhaps better than I could." Fletcher then phoned Shultz and talked with him at length. Shultz remained unwilling to make a decision, but recommended that NASA should take one more look at Rice's request.

Although Rice was holding firm on a weight of 30,000 pounds, he now was willing to budge slightly on payload size, for Sullivan's memo had discussed a 12 x 40-foot shuttle with twin solid boosters. Though this configuration would carry no more weight than a Titan III, it could fly with the boosters of a Titan III: 120-inch solid motors that were in production and had known costs. Such a shuttle still would not match the cost-effectiveness of the Titan III itself, but it would come close. In Sullivan's own analysis, that

---

11. Letter, Fletcher to Weinberger, December 29, 1971; Low, Personal Notes No. 61, January 2, 1972.

Titan would save only $100 million when compared to that shuttle. Within the OMB, this was as near to an endorsement as any type of shuttle was likely to receive.

Low phoned Rice and asked him to put his questions in writing. Rice replied that he might have further questions subsequently, but he drew up a set of queries and sent it over to Low late on Friday evening, which was New Year's Eve. Low discussed them with Fletcher and Myers over the weekend; on Monday, January 3, they completed their response. A sampling will illustrate the exchanges:

1. *If future budgets for NASA were constrained to $3.2 - $3.3 B, would you want to do a large shuttle?*
   **Answer:** *The answer is yes. The NASA budget is not committed to exceed the FY 1973 level even if the maximum credible cost overrun occurred.*

2. *Why should a relatively few space station modules for the mid-1980's determine the size and weight capabilities of the shuttle? What other missions are really driving the payload size and weight requirements?*
   **Answer:** *In addition to "a few space station modules," the payload length is driven above 40 feet by most NASA planetary payloads, most DOD payloads, and a few of the NASA science payloads.... The payload diameter is driven above 13 feet by manned payloads, some NASA science payloads, and some NASA planetary payloads. The payload weight is driven above 40,000 lbs. not only by space station modules but by space station resupply logistic vehicles, as well as sortie cans;*[12] *a 40,000-lb. payload capability is also exceeded by many DOD payloads, as well as 13 different science, applications and planetary payloads.*

4. *What capabilities and dollar benefits would be lost by going to a 12 x 40' (30 - 35,000 lb.) shuttle launched by SRM's?*
   **Answer:** *At this size and weight we lose most DOD payloads, all manned payloads (including resupply logistics vehicles), most planetary payloads, and many science and applications payloads.*

---

12. Small laboratories that would not fly freely in space but would remain within the payload bay and return with the Shuttle orbiter.

*THE SPACE SHUTTLE DECISION*

These exchanges with Rice would feed into a letter that Fletcher intended to prepare for Weinberger on Monday, January 3. On Sunday, however, with no resolution in sight, Low confided privately that "there is nobody in the White House willing to make any decisions."

He later wrote,

*The single most significant factor was that there was no top-level leadership in the White House. Nixon was unwilling to deal with his agency heads and dealt solely with his staff. This placed a great deal of the decision-making responsibility with the OMB, and by definition the OMB is far more interested in short-range budgetary problems than in the long-range future of the nation.*[13]

## *The Hinge of Decision*

Low did not know it, but the commitment he wanted already was imminent. At a recent budget session held in Key Biscayne near Miami, a decision had been made to include funds for the Shuttle. A subsequent decision called for Nixon to release an announcement at the Western White House in San Clemente, California. He would also meet there with Fletcher, who would hold a press conference. Fletcher learned of this during that New Year weekend. He and Jonathan Rose responded by asking William Anders of the National Aeronautics and Space Council to prepare a draft of this presidential statement.

Monday dawned, with the day's action items prominently featuring Fletcher's letter to Weinberger and NASA's responses to Rice's question set. People at the Manned Spacecraft Center had been working over the weekend to come up with the answers, but when these responses reached Washington, early that afternoon, they were quite inconsistent. It quickly developed that some of the specialists in Houston had misread some key tables of data. With time pressing, NASA Headquarters gave them only one hour to come up with the right answers, which they did. These answers made it possible for Fletcher to finish his letter:

---

13. Memo, Sullivan to Rice, December 28, 1971; "Space Shuttle Questions Provided by OMB on 12/31/71"; Low, Personal Notes No. 61, January 2, 1972; letter, Low to John Logsdon, January 23, 1979.

*Nixon's Decision*

> *Our view that the shuttle with the 14x45 - 45,000# capability is the minimum acceptable option is still valid.*
>
> *The OMB proposed option of a 14x45 - 30,000# shuttle is not acceptable because it will not handle manned space station modules, manned sortie flights or manned resupply missions in a standard space station orbit: in short, it would not provide a manned spaceflight capability for the United States. Also, this shuttle would not handle 28 different science, applications and planetary payloads that could be carried with a 45,000# payload capability.*

At 6 p.m., Fletcher and Low met in George Shultz's office with a group of participants that again included Shultz, Weinberger, Flanigan, Rice, and David. The opponents' positions remained the same, with Rice calling for a smaller design and David proposing that they put off a commitment for several months, to give NASA time to refine its most recent estimates. Shultz responded differently. Unlike those critics, he had been at Key Biscayne.

He had given Rice full opportunity to raise his objections. Now, in this meeting, he had seen NASA respond to Rice's questions, with answers that militated strongly against the OMB shuttle. What happened next proved to be the hinge of decision.

Only Low wrote of this moment soon afterward, and in a terse manner: "Shultz agreed with our position that the 15 x 60' 65,000 lb. Shuttle should be developed." Weinberger, Rice, and Fletcher all gave interviews during subsequent years, but did not discuss the events of this meeting. Fletcher and Low, however, spoke of the matter with senior NASA officials, who would present their own recollections:[14]

> Lilly: *Rice got a little bit confused. There was a feeling with Low and Fletcher that Rice got too carried away, moving toward misstatements, trying to exaggerate some things. George Shultz picked up the phone and called Morgenstern during the meeting and asked him about it. Now, NASA had prepared for this kind of thing—to be sure Morgenstern was fully knowledgeable. And Shultz got off the phone and made the decision—we'll go this way, and to prepare the papers for the approval.*

---

14. Memos: David N. Parker (White House staff), December 31, 1971; Rice to Shultz, January 3, 1972; letter, Fletcher to Weinberger, January 3, 1972; Low, Personal Notes No. 62, January 15, 1972.

## THE SPACE SHUTTLE DECISION

> Donlan: *I get this directly from Dr. Fletcher. He was telling his wife the story of how he went over there with the information that I supplied him from our shuttle program studies that showed that there wasn't all that much to be gained by going 45-foot length and, furthermore, it invalidated a lot of missions that the Air Force claimed they needed for the sizing of the satellites. So Shultz said, "Well, what are you fooling around with that 45-foot configuration for? It doesn't cost that much more. Why don't you get the one you want—take the 60-foot one." And Fletcher came back with that message. That's how it was settled.*

> Willis Shapley, NASA Associate Deputy Administrator: *George Shultz—he had personally spent almost as many hours going through all the planning studies, and especially this famous economic analysis that Klaus Heiss at Mathematica had done. And he personally called up Oskar Morgenstern and other people there, and he was satisfied, finally, that it was a reasonable proposition. So when it was clear that all the boys made their case, Shultz said, 'If we're going to do it, let's do it right; let's do the big shuttle and forget about the Bureau of the Budget shuttle.' So that's how we ended up with what it was."*

It was Christmas in January, for whereas Fletcher and Low had come in hoping for approval of the downsized version that they really did not want, Shultz now was ready to recommend that they receive the full-size version that had not been in play for over a month. Similarly, the views of Rice and David would carry no weight. Shultz, after all, was Rice's boss, and it was Shultz who would meet with Nixon. David also reported to Nixon, but he had little clout. He had been bypassed in favor of Magruder during the recent NTOP exercise. A year later, amid a post-election White House reorganization, Nixon would abolish his post of presidential science advisor—and would also abolish the OST in the bargain.[15]

The decision to proceed with the Shuttle became firm during the meeting in Shultz's office, with Shultz confirming his assent to NASA's request for $200 million as startup funds for the Shuttle within Nixon's FY 1973 budget.

---

15 John Mauer interviews: Charles Donlan, Washington, October 19, 1983, pp. 33-34; William Lilly, Washington, October 20, 1983, p. 39; Willis Shapley, October 26, 1984, p. 26; *Science*, January 19, 1973, p. 233; February 2, 1973, pp. 455, 458-459; February 16, 1973, p. 641; March 30, 1973, p. 1311.

*Nixon's Decision*

*Nixon meets with Fletcher and approves the Shuttle as a program, January 5, 1972. (National Archives)*

Flanigan now asked Low and Fletcher to prepare a draft of the presidential statement—which Anders was writing already. In turn, Anders' statement formed the bulk of the material used in Nixon's release.

As recently as November, Flanigan had anticipated that any White House announcement would be low-key. At that time, with the $3 billion glider as the likely new initiative, Flanigan expected to see the main coverage limited to the aerospace trade press, thereby reassuring this industry of Nixon's support while avoiding the high visibility that would draw fire from critics. The Space Shuttle, however, had metamorphosed now into a $5.5 billion program. As early as the previous Friday, prior to the meeting in Shultz's office, a White House staffer had already laid out the high-profile announcement that now was scheduled for Wednesday, January 5, 1972.

Fletcher and Low flew out to California, editing two NASA statements along the way. Nixon greeted them at the Western White House, as did John

## THE SPACE SHUTTLE DECISION

Ehrlichman. Though the President had planned to spend only 15 minutes with his visitors, the meeting ran well beyond a half-hour as he showed strong interest in the Shuttle and the space program. Fletcher had brought a model of a TAOS, and Ehrlichman would remember "Nixon's fascination with the model. And he held it and, in fact, I wasn't sure that Fletcher was going to be able to get it away from him when the thing was over."

Nixon stated that NASA should stress civilian applications but should not hesitate to note the military uses as well. He showed interest in the possibility of routine operations and quick reaction times, for he saw that these could allow the Shuttle to help in disasters such as earthquakes or floods. He also liked the idea of using the Shuttle to dispose of nuclear waste by launching it into space. Fletcher mentioned that it might become possible to collect solar power in orbit and beam it to Earth in the form of electricity. Nixon replied that such developments tend to happen much more quickly than people expect, and that they should not hesitate to talk about them.

He liked the fact that ordinary people, who would not be highly-trained astronauts, would be able to fly in the Shuttle. He asked if the Shuttle was a good investment, and agreed that it was indeed, for it promised a tenfold reduction in the cost of space flight. He added that even if it was not a good investment, the nation would have to do it anyway, because space flight was here to stay. Fletcher came away from the meeting saying, "The President thinks about space just like McCurdy does," referring to a colleague within NASA's upper management.

Although his formal statement largely reflected NASA's views, Nixon edited the draft in his own hand. The final version showed a firmness and sense of direction that had been utterly lacking in his March 1970 statement on space policy. It also featured a grace note that might have suited John Kennedy:

> *I have decided today that the United States should proceed at once with the development of an entirely new type of space transportation system designed to help transform the space frontier of the 1970s into familiar territory, easily accessible for human endeavor in the 1980s and '90s.*
>
> *This system will center on a space vehicle that can shuttle repeatedly from earth to orbit and back. It will revolutionize transportation into near space, by routinizing it. It will take the astronomical costs out of astronautics. In short, it will go a long way toward delivering the rich benefits of practical*

*space utilization and the valuable spinoffs from space efforts into the daily lives of Americans and all people....*

*Views of the earth from space have shown us how small and fragile our home planet truly is. We are learning the imperatives of universal brotherhood and global ecology—learning to think and act as guardians of one tiny blue and green island in the trackless oceans of the universe. This new program will give more people more access to the liberating perspectives of space....*

*"We must sail sometimes with the wind and sometimes against it," said Oliver Wendell Holmes, "but we must sail, and not drift, nor lie at anchor." So with man's epic voyage into space—a voyage the United States of America has led and still shall lead.*[16]

It was appropriate to give a name to this new ship of space, and of state. Fletcher, Shapley, and Low had prepared a list that included Pegasus, Hermes, Astroplane, and Skylark. Flanigan passed this list to White House staffers, who picked the name Space Clipper. A draft of Nixon's statement used this name, which resembled Lockheed's Star-Clipper. Nixon himself, however, decided that it would be better to refer to the vehicle in the usual fashion, as the "Space Shuttle." Earlier piloted spacecraft had carried names such as Mercury, Gemini, and Apollo, but the new one would break with this practice.

Criticism of Nixon's decision came from the usual sources, with Senator Proxmire issuing a press release: "The President has clearly reordered our priorities. But he has reordered them the wrong way." Senator William Fulbright played the same note: "I believe the shuttle simply cannot rank high on our list of priorities in view of the critical social and economic problems we face." They, however, were merely outriders within a Democratic Party that had long commanded the center of American politics but now was slipping to the left, and whose presidential candidate, George McGovern, would shortly receive one of the worst electoral drubbings ever administered. Though Congress would perform its constitutional role by voting the funding, it would not become an important source of opposition.

---

16. Low: Personal Notes No. 59, November 28, 1971; No. 62, January 15, 1972; letter, Fletcher to Weinberger, January 4, 1972; memos: David Parker, December 31, 1971; Flanigan to Nixon, January 4, 1972; Nixon, "Statement by the President," January 5, 1972; John Logsdon interview, John Ehrlichman, Santa Fe, May 6, 1983, p. 1; Low, Memo for Record, January 12, 1972. Reprinted in NASA SP-4407, vol. I, pp. 558-559.

## THE SPACE SHUTTLE DECISION

Certainly there was politics aplenty in Nixon's decision. He wanted to help the aerospace industry during the upcoming election year, and the staffer Jonathan Rose, reporting to Flanigan, had been monitoring that industry's unemployment. Fletcher, in a letter to Rose of November 22, had written that an early start on the Shuttle "would lead to a direct employment of 8,800 by the end of 1972, and 24,000 by the end of 1973." Ehrlichman would recall that this was

*a very important consideration in Nixon's mind. There are what we call battleground states; they are the pivotal states that control big blocks of electoral votes. So when you look at unemployment numbers, and you key them to the battleground states, the space program has an importance out of proportion to its budget.*

The politics also reached a much higher level, touching the matter of presidential decisions that could stand as a legacy, with consequences that would reverberate through coming decades. Theodore White, a chronicler of presidents, had written in 1965 that "on the far edge of the plateau lie problems which we in this decade cannot conceive of as political." These included "the Moon and space. How large a part of American energy should be invested in this exploration with no definable certainty except the certainty that it will change the lives of all our children?"

Weinberger had made an important point in his memo to Nixon of August 12: the United States, as the world's great reserve of strength, could do more than merely add bricks to the welfare state. Such policies might suit the British, of whom the former secretary of state Dean Acheson had said, "Great Britain has lost an Empire and has not yet found a role." They would not suit America.

One could view Nixon's decision as a straightforward exercise in daily work at the White House. Nixon had a strong interest in management; he had set up a staff system, which included a strengthened OMB, that could weigh policy alternatives with considerable effectiveness and present him with well-researched options for his decision. As one Flax Committee staffer put it, "Once they decided to do it, Nixon and Ehrlichman weren't going to argue with NASA whether it ought to be a 60 or a 45-foot-long bay, or 12 or 15 feet. If the head of NASA is telling them that it has to be this size and they want to go ahead with the project, they are going to say go ahead." The demise of

NTOP also helped. While the Space Shuttle stood on its merits and did not simply replace NTOP as a backup, Nixon and Shultz had been prepared to include funding for NTOP initiatives in the FY 1973 budget. When no suitable proposals came forth, that made it easier to shift the putative NTOP funds over to the Shuttle, and to approve a larger Shuttle as well.

Yet while the Shuttle could not match the significance of Nixon's opening to China, it drew on more than good staff work. In Ehrlichman's words, "There wasn't anybody who made those final decisions except Nixon, in this kind of area. Defense, space, certain kinds of domestic problems—he was the final arbiter." The Shuttle carried Nixon's personal stamp because it carried his personal decision.[17]

## Loose Ends I: A Final Configuration

Now that Shultz had handed NASA its Space Shuttle on a silver platter, the agency had to decide how it would look. The question of choosing a booster was still up in the air, and it was far from clear that the Shuttle indeed would be a TAOS; liquid-fueled boosters designed as conventional first stages were making a strong comeback. Similarly, the agency could not simply walk away from Fletcher's alternative of 14 x 45 feet and 45,000 pounds; NASA itself had proposed it, and it merited additional attention because it offered the potential advantage of being able to use existing 120-inch solid rocket motors. Further study of this design would also discourage the OMB from complaining that NASA once again was abandoning a good possibility with unseemly haste.

Though the issue of payload size and weight was still open, the basic design of the orbiter was approaching a definitive form. During the fall of 1971, when the Mark I/Mark II approach was still in the forefront, the contractors had worked from a Max Faget configuration known as MSC-040A. It elaborated the earlier MSC-040 by adding small liquid-fuel engines for orbital maneuvers, along with thrusters for attitude control that were mounted at the tips of the wings and tail.

The Mark I/Mark II concept, however, with its phased technology, had never been more than an artificial stratagem to reduce peak funding by stretch-

---

17. *National Journal*, August 19, 1972, p. 1329; John Logsdon interviews: Dave Elliot, p. 4; John Ehrlichman, Santa Fe, May 6, 1983, pp. 9, 32-33; White, *1964*, pp. 476, 478; *Oxford*, p. 1; letter, Fletcher to Rose, November 22, 1971. Reprinted in NASA SP-407, vol. I, pp. 555-558.

ing out the development, while accepting serious compromises in design. With Shultz's support, NASA now was free to build an honest orbiter, one that would be right the first time. MSC-040A had called for four J-2S engines; a variant of January 1972, MSC-040C, replaced them with three SSMEs.

Other decisions shaped the orbiter's structure and thermal protection. Though hot structures now looked like an open invitation to a cost overrun, everyone knew how to build an aluminum airplane, and the orbiter indeed would take shape as an aircraft built largely of this metal. It then needed thermal protection, and NASA now placed its hope in the still-unproven tiles. Though recent research had increased confidence that they indeed would serve, what made tiles more attractive yet was that NASA could count with reasonable assurance on using ablative heat shields as a backup. Ongoing work with ablatives had cut their cost dramatically while reducing their weight to 15 pounds per square foot, matching the weight of the tiles. A year later, Eugene Love, a director at NASA Langley, would write, "Ablators are baselined as a confident fall-back solution (temporary) for both leading edges and large surface areas, should development of the baseline approaches lag."[18]

Though the choice of booster was still unsettled, during the early weeks of 1972 there was excellent reason to believe that NASA's eventual selection would take good care of the Marshall Space Flight Center. The winged S-IC would have done splendidly, but even Boeing, which had built the basic S-IC and knew this concept best, had been unable to drive its development cost low enough to compete with alternatives such as TAOS. It fell by the wayside around the end of 1971, amid criticism even within the shuttle community. John Yardley, who headed the work at McDonnell Douglas, told *Aviation Week*, "You just could not build the world's largest airplane without all the problems that would go with a 700,000-pound craft. And it doesn't buy you much flying it back if you can do the same job in a cheaper way."

Boeing and NASA Marshall, however, would not be denied, as they proposed a new alternative: a pump-fed booster. Though this again was to be an S-IC variant, it would be without wings, tail, jet engines, landing gear, or crew compartment. Instead it amounted to the standard S-IC, fitted out to land in the ocean by using parachutes. A retro-rocket was to cushion its impact in the

---

18. AIAA Paper 73-31; Jenkins, *Space Shuttle*, p. 115; memo, Taft to Rice, January 27, 1972; letter, Low to Rice, January 11, 1972.

sea; it then would float like a ditched airplane as it awaited rescue. After refurbishment, it would fly again.

TAOS concepts were still very much in the running, with solid motors receiving a share of attention. The liquid-fueled TAOS, however, with twin pressure-fed boosters flanking the external tank, had lost favor. A key group of shuttle design reviewers, at the Manned Spacecraft Center, had come around instead to recommending a single pressure-fed booster that would take the form of a conventional first stage. This, too, would provide grist for the mill of NASA Marshall, for that center would manage development of both the booster and its engines.

Within the OMB, Daniel Taft, who worked with the NASA budget, saw an opportunity—and smelled a rat. The opportunity existed because NASA's own estimates proposed that a suitable solid rocket motor would cost up to a billion dollars less to develop than a pressure-fed booster. In addition to this, the Air Force had already developed the 120-inch solids of the Titan III, thus providing a base of experience along with confidence in the validity of the new cost estimates for solids. Pressure-fed versions carried no such experience and no such confidence, for they had never been built before.

NASA, however, wanted a pressure-fed booster, and Taft knew that to lead it to solids would not be easy. In a memo to Rice, late in January, he laid out the issues. He wrote that "NASA's schedule for the selection of the final configuration...is extremely tight (March 1)." Drawing on Rice's back-channels to the contractors, Taft noted that one such source had recommended that pressure-fed designs should be studied for six to twelve months. Taft also asked "whether NASA can overcome its instinctive dislike" of solid rocket motors. He added:

> *NASA has recently let contracts with the four major solid rocket contractors ($150 K each) for quick (1 month) studies of development and production costs and technical aspects of SRMs. This is truly a hasty last minute effort by NASA which can hardly be expected to make up for NASA's failure to study SRMs seriously in the past.*
>
> *Of course, the requirement for Marshall's involvement in the shuttle program would be quite weak if SRMs were selected. Ironically, Marshall, which has little understanding of SRMs and much to lose by their selection, is managing the SRM contractor effort....*

*THE SPACE SHUTTLE DECISION*

> *If left to their own desires, NASA would probably select the 15 x 60' orbiter with the pressure-fed booster. This is regrettable because we consider the pressure-fed booster to be a high risk option from the standpoint of both investment cost and operating cost....*
>
> *At this time I believe that we should lay our cards on the table and explain frankly to NASA our concerns about the risks involved in the pressure-fed booster....*[19]

Meanwhile, there was the irksome matter of the budget. Nixon's message to Congress included NASA's full amount of $228 million for the Shuttle: $200 million for the program per se, $28 million for construction of facilities. NASA's FY 1973 request totaled $3.379 billion in new obligational authority, $3.192 billion in outlays. In a letter to Fletcher dated February 9, Weinberger, however, emphasized that the second of these numbers was the one that counted: "For planning purposes an annual spending level of $3.2 billion should be assumed for the foreseeable future"—that is, through FY 1978. A week later, Shultz repeated this and added, "We also fully expect NASA to develop a shuttle system within the $5.5 billion estimate." The billion-dollar difference in development costs, pressure-fed versus solid, came to nearly one-fifth of this total. Choosing solids thus would give much needed leeway.[20]

During the subsequent week, contractors presented briefings and gave their recommendations concerning the choice of booster. Low wrote that these briefings "yielded the recommendations for each contractor that were most predictable based on vested interests." They also were predictable based on the contractors' choices during a similar exercise six months earlier, when they had compared expendable boosters for interim use. On both these occasions, Boeing's recommendations had been particularly egregious.

Boeing was home to the S-IC and was teamed with Grumman. In their report of September 1, they had proposed that the standard S-IC, which needed no development, would give lower costs in an interim program of up

---

19. *Aviation Week*, January 10, 1972, pp. 15-16; January 24, 1972, pp. 36-37, 40-41; Low, Personal Notes No. 61, January 2, 1972; letter, Low to Rice, January 11, 1972; memo, Taft to Rice, January 27, 1972; Report B35-43 RP-30 (Grumman).
20. *Aviation Week*, January 31, 1972, pp. 24-25; letters: Weinberger to Fletcher, February 9, 1972; Shultz to Fletcher, February 16, 1972; memo, Low to Fletcher, March 22, 1972.

to 30 shuttle flights. (Boeing had built only 15 S-ICs for the whole of Apollo.) Now, in February, Boeing continued to root for the home team by coming out in favor of its pump-fed booster. This report also came out strongly against solid rockets, urging that they "should be eliminated from further consideration."

By contrast, Lockheed was a major builder of solids. In September, it merely had weighed the merits of competing sizes and arrangements of solids, drawing on this in-house expertise. Now, however, it compared a range of alternatives that included liquid boosters—and found again that solids were best. The February briefing from North American Rockwell was also in character. In September, that company had found no reason to choose between the alternatives of the day. Now it hedged anew, stating that one could choose either solids or a conventional liquid first stage, depending on what cost goals were most important.

McDonnell Douglas also liked solids. It certainly had long experience with liquids, having built the Thor missile, the Delta launch vehicle, and the S-IVB, the third stage of the Saturn V. It also was familiar with solids, being accustomed to use them to provide the widely-used Delta with extra thrust at launch. It had endorsed solids in September; it now did so again. In addition to this, its report carried a lengthy review of their safety.

The review covered 2,128 solid-motor firings, as Delta strap-ons, Titan III boosters, Minuteman ICBMs, and the small four-stage Scout. Thirteen had failed, in ways that were pertinent to the Shuttle, and McDonnell Douglas took care to note both the causes of the failures and the changes in design or in quality-control procedures that could prevent them from recurring. The report noted particularly that in the event of such recurrences, it would usually be possible to safely abort a Shuttle launch. However, there was an exception.

This would happen if the hot, high-pressure gas within a solid motor succeeded in burning through its casing. Large solids were built in segments, pinned together at their joints, and such joints posed particular hazard of a burnthrough. The report noted:[21]

---

21. Reports: B35-43 RP-21; *Evaluation* (quote, p. 211) (Grumman); LMSC-A995931, -D157302 (Lockheed); *Interim Report*; *Design Review* (briefing chart, p. 60) (McDonnell Douglas); SV 71-40, SV 72-14 (North American Rockwell); Low, Personal Notes No. 65, February 27, 1972.

## THE SPACE SHUTTLE DECISION

| Problem | Remedy | Shuttle Abort Consideration |
|---|---|---|
| Case burnthrough | Increase case insulation thickness; use two "O" rings between segments | If burnthrough occurs adjacent to HO tank or orbiter, timing sensing may not be feasible and abort not possible |

Like a distant flash of lightning on a pleasant summer day, this briefing chart clearly foreshadowed the loss of the *Challenger*, 14 years later.

On the whole, contractor studies found no advantage in the smaller orbiter, no offset to its compromise of NASA's ability to carry DoD payloads. This removed the last questions as to whether NASA would get the full-size version that it wanted. But with the booster recommendations ranging over the map, NASA once again had no clear way to proceed. Six months earlier, a similar confusion over choice of boosters had worked to NASA's advantage, by opening the door to new possibilities that included the winged S-IC. The situation now was different; the agency wanted to narrow its alternatives, not expand them. Yet within the contractors' reports, data on costs gave an overriding basis for a decision.

Between January and March, while the development cost of a pressure-fed booster stayed virtually constant, the SSME escalated sharply and the orbiter went higher still. An internal OMB memo, from the economist Sullivan to Rice, summarized NASA's own estimate of the changes, in millions of dollars:

|  | 1/3/72 Estimate (Pressure-fed) | Pressure-fed | Current Estimates Pump-fed | Solids |
|---|---|---|---|---|
| Orbiter | 3,058 | 3,660 | 3,660 | 3,750 |
| Main Engine | 450 | 580 | 580 | 580 |
| Booster | 1,390 | 1,400 | 1,080 | 350 |
| Program Support | 602 | 570 | 560 | 470 |
| Total | 5,500 | 6,210 | 5,880 | 5,150 |
| Cost/flight | 7.7 | 9.3 | 8.6 | 10.4 |

The situation was not completely bleak; the increased cost of the orbiter, as much as $700 million, included $260 million for specialized solid rockets that might be useful in abort. Sullivan noted that the justification for this abort system "seems weak," and it soon was dropped from the design. As a result, these estimates could be reduced.

But when NASA officials confirmed these estimates, they threw in the towel. Both Weinberger and Shultz had insisted that NASA stay within a $5.5 billion development cost, and it would be most unseemly if the program were to start with a projected cost overrun that would violate this limit. This could happen quite severely with the pressure-fed option; it would also happen with the less costly pump-fed booster, even with the abort system deleted. Moreover, while the projected costs for the solids drew on Air Force experience and were both low and firm, estimates for either of the liquid-fueled options were dodgy and likely to increase by the next design review.

The strong case of a solid motor also gave a strong case for choosing the solid motor. No one had previously tried to recover and reuse a solid booster; those of the Titan III had simply plopped into the deep, to provide homes for fishes. Early in January, a NASA official had said, "It is not contemplated at this time that a solid-rocket booster would be recoverable." Yet the modest staging velocity of the solids, as low as 4000 ft/sec, meant that their heavy casings could easily serve as a heat sink. They also could withstand the stress of dropping by parachute into the ocean. NASA Marshall and its contractors found that reusability of these solids would cut the cost per flight to around $10 million, allowing the Shuttle to maintain its advantage and to capture its traffic from expendables.[22]

NASA also had to consider the danger of the sea, for inevitably, some boosters would be lost. The high cost of a liquid booster meant that losing even one of them would be quite expensive. Moreover, although the pump-fed booster would save on development costs through its use of the existing F-1 engine, its thin-walled structure would easily sustain damage while afloat. The casing of a solid booster would cost much less. It would be relatively impervious to damage, and the occasional loss of such a casing would not compromise the program's overall economics.

Low wrote that at the end of February 1972, "Dale Myers presented the OMSF and OMSF/Center recommendation, which was to go with...solid

---

22. *Aviation Week*, January 10, 1972, p. 15; March 20, 1972, pp. 14-16; memo, Sullivan to Rice, March 13, 1972; letter, Low to Rice, January 11, 1972.

## THE SPACE SHUTTLE DECISION

rocket motor boosters." Fletcher took the news to a meeting with David, Anders, Flanigan, and Rose. In Fletcher's words, "There was uniform agreement in the group that we had made the right choice." He also met with Weinberger, who "seemed quite receptive" to NASA's decision. Weinberger also agreed that there was no need to make a new decision on the size of the orbiter; that choice had already been made in early January. Low met with Rice and wrote that "he appeared to accept our conclusions, almost as though he had invented them himself."[23]

The director of NASA Marshall, Eberhard Rees, had expressed hope that the shuttle would use a liquid booster, because that would provide more work for his center. Although he had been deeply involved with liquid rockets since the wartime V-2 effort, he now would have to change with the times. The Shuttle would use two 156-inch boosters,[24] which were as large as could travel on American railroads. Only nine such solids had been test fired—five by Thiokol, four by Lockheed—and Marshall would have plenty to do in bringing them to a level of reliability that would allow them to carry astronauts. Marshall also would manage the development and production of the SSME and the external tank. This center thus would not wither on the vine.

Budget officer William Lilly went over the estimates and came up with a development cost of $5.15 billion, well below the target of $5.5 billion. Myers objected, insisting that he could accept no lower figure than $5.34 billion, but Low sided with Lilly and persuaded Myers to accept his number. Naturally, this was the one that went to the OMB. It did, however, include the $260 million for the abort system that later was discarded, and thus carried a margin for further reduction.[25]

With this, the Shuttle took form in the shape that NASA would build and that flies to this day. Ironically, though it was a NASA project from the start, its main design features reflected pressures from outside that agency. The Air Force had pushed for the large payload capacity and the high crossrange that called for a delta wing; while NASA later accepted these features and made

---

23. Memos: Fletcher to Low, March 3, 1972; Rice to Shultz and Weinberger, March 13, 1972; Fletcher, Memo for Record, March 3, 1972; letters: Fletcher to Weinberger, March 6, 1972; Weinberger to Fletcher, March 13, 1972; Shultz to Fletcher, March 17, 1972; Low: Memo for Record, March 8, 1972; Personal Notes No. 66, March 12, 1972; Donlan, *Space Shuttle*.
24. A year later this specification changed to a diameter of 142 inches, due to a reduction in the design weight of the orbiter. Loftus et al., *Evolution*, p. 26; *Astronautics & Aeronautics*, January, 1974, p. 72.
25. *Aviation Week*, March 20, 1972, pp. 14-16; Low, Personal Notes No. 66, March 12, 1972.

them its own, the initial impetus had come from the Pentagon. Similarly, the solid boosters came from the OMB. Left to its own devices, NASA surely would have picked a liquid booster such as the fully-reusable winged heat-sink type that flourished during the second half of 1971. In this fashion, the Air Force and OMB crafted a design that NASA would construct and operate.

## Loose Ends II: NERVA and Cape Canaveral

NASA now was not only ready to proceed with the orbiter it wanted; it also could look ahead to having most of the principal elements of George Mueller's integrated space program of 1969, which he had planned to culminate in piloted missions to Mars. That program envisioned a Space Shuttle, approval for which was in hand. It called for a space tug, which was on the agenda as part of the development of the operational Shuttle. Mueller had looked ahead to astronaut-tended spacecraft, including a Large Space Telescope that would take shape as today's orbiting Hubble instrument. These, too, were in prospect in 1972. Space station modules, launched by the Saturn V, were key elements in his scenario, with variants of these modules evolving into portions of Mueller's eventual Mars ship. Though the Saturn V would fly no more, Fletcher had pushed successfully for a shuttle that could serve to build a future space station. This effort would come to the forefront during the 1980s.

NASA needed one more element to make this framework complete. It needed the NERVA nuclear rocket, which would power the Mars mission. Work in Nevada with experimental nuclear engines had brought this technology to an advanced level that was ready for mainstream development, and NERVA held influential support in Congress. If this program could go forward, NASA might yet be able to set sail for the Red Planet.

During FY 1970, NERVA moved into a phase of detailed design and hardware fabrication. The goal now was not further research, but rather the development of a flight-qualified engine with 75,000 pounds of thrust, at a cost of $860 million over a period of eight to nine years. The program received $88 million in FY 1970 and $85 million in FY 1971, with the funds coming jointly from NASA and the Atomic Energy Commission.

On Capitol Hill, its political support was unassailable. The program had a research center at Los Alamos, New Mexico. The *National Journal* described that state's Senator Clinton Anderson as "NERVA's most zealous and watchful guardian over the years." Anderson was one of the most senior Democrats in the Senate, and chaired its space committee. The test area was in Nevada; hence the program also held strong support from that state's Senator Howard Cannon, another influential Democrat who was also a member of the space committee. Indeed, support for NERVA was bipartisan. Westinghouse was building its reactors in Pennsylvania, home to another key supporter: the leader of the Senate's Republicans, Hugh Scott. In the House, the Pennsylvania congressman James Fulton was the most senior Republican within that chamber's space committee. He took a strong stand for NERVA as well.

Nevertheless, NERVA faced opposition within the OMB. As one of its officials said, "Here we had a high technology program that was expensive as hell, averaging $100 million a year. It had a long-term potential, but NASA didn't know for a long time what they were going to do with it. It was a logical place in the budget to raise questions."

Late in 1970, amid preparation of the FY 1972 budget, NASA and the AEC requested a total of $88 million. By then, however, NASA was abandoning Mueller's bold Mars plan. There was no need for rapid development of NERVA, which was likely to be ready long before any mission could use it. NASA decided to slow it down by lowering its priority. The OMB responded by treating it as a splendid opportunity to save money by canceling it outright. When the smoke cleared, the budget request for NERVA was down by four-fifths, to only $17.4 million.

Late in January 1971, at a NASA budget briefing, George Low went out of his way to deny that his agency still was looking ahead to a piloted mission to Mars, saying, "We have in our program today no plans for a manned Mars landing." He actually had plans aplenty, but he was telling the truth; they were not in the program. NERVA stood at their core, and NERVA by then was dying.

Congress, exercising its power of the purse, appropriated $69 million for the program. Nixon, however, held the right to impound funds and refuse to spend them. A few years later, that power would be curtailed by the Impoundment Control Act, but in this fashion he released only $29 million for FY 1972, withholding the rest. That did it; when Fletcher sent his FY 1973 budget request to Shultz, in September 1971, he abandoned hope of building a nuclear-powered engine suitable for Mars.

This decision ended a longstanding NASA policy of developing advanced

## Nixon's Decision

engines well before there was need for them. The agency had contracted with Rocketdyne to build the F-1 as early as January 1959, over two years before Kennedy called for Apollo. Development of the J-2 dated to September 1960. The demise of NERVA meant that nobody would be flying to Mars, perhaps not even within our lifetimes.[26]

There also was a good deal of politics in another issue: where to launch the Shuttle. Though NASA had an obvious interest in using the Apollo launch facilities at Cape Canaveral, the Shuttle was not the Saturn V and plenty of people in other areas of the country were ready to propose that they could offer sites having unique advantages. All these people demanded careful attention, for they had congressmen and senators as well as access to local newspapers.

The merits of Cape Canaveral were undeniable. In addition to its Apollo facilities, it was on the Intracoastal Waterway; hence barges could bring oversize rocket stages from such locations as NASA's Michoud Assembly Facility, a government-owned plant near New Orleans that had built the S-IC. The southerly latitude of the Cape also meant that rockets launched from this site would benefit from the Earth's rotation, which would impart a velocity of up to 914 mph while a vehicle sat on its pad.

The Cape, however, had disadvantages as well. Many military launches demanded a polar orbit, but Canaveral could not accommodate rocket flights to due north or south; they might drop spent stages on the Carolinas or Cuba. The Air Force had built its own separate facilities at California's Vandenberg Air Force Base to get around this. The Cape was subject to corrosion from salt air. It needed a 10,000-foot runway to land the Shuttle, and while this was feasible, the Cape often had cloudy or rainy weather. Lying at sea level, it required rockets to blast their way through the entire thickness of the atmosphere.

As early as April 1970, NASA's Tom Paine set up a 14-member Space Shuttle Facilities Group, which went on to evaluate both the Cape and the alternatives. NASA also contracted with the Ralph M. Parsons Co., a major heavy-construction firm, to provide independent outside advice. Some 20 states went on to propose over a hundred possible locations, though many bidders had little idea of what NASA needed. Mill Creek, Oklahoma, invited

---

26. *National Journal*, March 13, 1971, p. 541; May 29, 1971, pp. 1156-1165; May 6, 1972, p. 787; Data Sheets: F-1 Rocket Engine; J-2 Rocket Engine (Rocketdyne).

## THE SPACE SHUTTLE DECISION

NASA to use its town airfield; Brownsville, Texas put in a word for the nearby Cameron County Airport. Another bid came from Michigan's upper peninsula. The source was an unemployed truck driver who told the New York *Times*, "Some of my friends and I were drinking it up a little at the town bar, and this guy came in who had just read about the space base competition..."

Nevertheless, there was at least one serious alternative: the Army's White Sands Missile Range with its adjacent Holloman Air Force Base. Holloman offered existing runways along with a 4,000-foot elevation to complement its southerly latitude, thus giving a double boost to a shuttle. Located amid high desert, its weather was virtually cloudless and its flying conditions nearly ideal. An arid climate discouraged corrosion. The remoteness of White Sands also stimulated thoughts of all-azimuth launches, whereby this single facility would fire the Shuttle into both NASA and Air Force orbits. A community of missile specialists had worked here since World War II. This location had another important resource: the powerful Senator Clinton Anderson of New Mexico, its home state, who was ready to fight vigorously for its selection.

Anderson, however, was not the only Washington baron with eyes on this prize. Another was Congressman Olin Teague, chairman of the Subcommittee on Manned Space Flight within the House space committee. A committee staffer noted that early in 1970,

> *NASA and industry spokesmen suddenly began pointing out that the shuttle potentially could be launched from almost anywhere in the United States. At that juncture, a number of congressmen discovered that they had phased-down or abandoned facilities or Air Force bases in their districts. So Teague and the subcommittee decided they had better make their position—and that of the full committee—clear from the outset.*

In December, he came out strongly in favor of Cape Canaveral, reacting with vigor against the suggestion that the Shuttle might go anywhere else. He warned, "Unless I am convinced that NASA is making maximum use of existing facilities, I intend to oppose any money for the shuttle in every way, form or fashion." He later added that NASA had better put the base at Canaveral, "or come up with a goddamned good case for its removal."

NASA thus had to do a lot of stroking. In October 1970, George Low assured Senator Anderson that White Sands would receive close scrutiny. The

*Nixon's Decision*

following March, Dale Myers reaffirmed that the competition was far from over, and described White Sands as the closest challenger to the Cape. But when NASA announced its decision, in April 1972, it stated clearly that it was not about to build a national shuttle site in the New Mexico desert. It awarded the palm to Cape Canaveral, with Vandenberg AFB to serve for military launches.

In announcing the decision, George Low said that it would cost $150 million to modify the existing Apollo facilities, which included the Vehicle Assembly Building and Pads 39-A and -B. Though Vandenberg had nothing resembling the VAB that it could convert, it did have a big Titan III launch facility, Space Launch Complex 6, known as Slick-Six. This site would require $500 million, with the Air Force paying the bill. That service had no intention of sharing the cost of shuttle development with NASA; NERVA had received its budget partly from NASA and partly from the AEC, but the Shuttle would be entirely a NASA project even though the Air Force was to receive many of its benefits. By promising to pay for its own launch facility at Vandenberg, this service nevertheless showed that it too would become actively involved.

The rejection of both NERVA and White Sands, at nearly the same time, was a double defeat for Senator Anderson. However, he was 76 years old and in poor health. In 1972, he announced that he would retire from the Senate and would not run for re-election. As chairman of that chamber's space committee, he had done much for NASA. Even so, he would not be receiving any major base or program in New Mexico, to serve as his monument.[27]

## *Awarding the Contracts*

The presidential campaign was in full swing in mid-1972, and on July 31 Jean Westwood, chairing the Democratic National Committee, issued a statement:[28]

> *Three days ago, the Nixon Administration awarded the $2.6 billion space shuttle contract to the North American Rockwell Corporation in California.*
> 
> *I regard this decision as the latest, and perhaps most blatant, example of President Nixon's calculated use of the American taxpayers' dollars for his*

---

27. *National Journal*, April 24, 1971, pp. 869-876; April 24, 1972, pp. 706-707.
28. Westwood, "Statement," July 31, 1972.

## THE SPACE SHUTTLE DECISION

> *own re-election purposes....*
>
> *The award of this contract to North American Rockwell also raises questions of ethics. Why is it that five of the current directors of the corporation happened also to be major contributors to Richard Nixon's election in 1968?*
>
> *I ask Democratic members of Congress for a full airing of this contract award....*

Westwood's press statement included a list of the "major contributors," whose 1968 donations had mostly come to $1,000. One wealthy man, Henry Mudd, had given all of $2,000. Yet while stating explicitly that Nixon could be bought and sold for the price of a Volkswagen Beetle, she missed a potentially more significant story: Dale Myers, who had spent his career at North American, had put together the selection board that had picked this company as the winner of the contract.

Myers was NASA's Associate Administrator for Manned Space Flight. He was clearly aware of the potential for conflict of interest, for in Washington, it is well understood that the person who picks the membership of a review board can often determine its decisions in advance. George Low also saw the potential for conflict, and discussed the matter with colleagues. He later wrote that Myers "had fully divested himself of all his connections with North American and since this activity is so closely tied to all that he is going to do over the next several years, it was necessary that he should be involved."

Myers put together this board in January, while the choice of booster was still open. He picked its members based on their positions within NASA and their responsibilities within the shuttle program; the only non-NASA members were from the Air Force. A senior attorney provided legal counsel. Low wrote that he "concurred fully and formally in the selection" of the members. He and Myers also emphasized that the board "will conduct its business in strict accordance with the provisions of the Source Evaluation Board Manual."

The Request for Proposals went out on March 17, shortly after NASA had wrapped up the choice of the booster. Responses were due on May 12, and four companies replied: North American Rockwell, McDonnell Douglas, Lockheed, and Grumman. Fletcher and Low—but not Myers—then reviewed the findings of the board during July, and met with the bidders' corporate officers. With them was another NASA official, Richard McCurdy. He had just

*Nixon's Decision*

*The shuttle concept that won the contract, July 1972. (Rockwell International)*

returned from vacation; his secretary had flown to Spain to give him documents so he could read them on the way back. These three men then made the final decision.

Lockheed ranked fourth in the competition. Though its bid was only $40 million more than North American's, some suspected that the proposal looked like another attempt at buying in. Its shuttle was heavy, and Fletcher wrote that its design had "unnecessary complexity." Lockheed left a 65-second gap during ascent with no provision for abort. It proposed an overly high landing speed, a structure that could accumulate moisture, and a program management that would rely on subcontractors to do much of the detailed design. "We didn't see how they were going to drive all those horses and keep the costs down," said McCurdy. "We ended up not believing their proposal on costs."

Lockheed also was the only bidder with no experience in building piloted spacecraft. By contrast, McDonnell Douglas had been a mainstay in this area, for McDonnell Aircraft had built the Mercury and Gemini capsules while Douglas Aircraft constructed the Skylab space station. This very breadth of experience, however, worked against this merged company, for the concepts within the proposal came partly from Douglas in Huntington Beach, California and partly from McDonnell in St. Louis. "Their proposal was almost like two company proposals," one source confided. "They gave the impression of never having consummated their marriage, and we couldn't live with that."

Moreover, the proposal projected a relatively high cost, and showed technical weaknesses. It divided the flight testing between Edwards Air Force Base and Cape Canaveral in a way that required full data-handling capability at both sites. Its discussion of ground operations did not reflect recent company experience in the Apollo program. In addition to this, although Douglas had designed its DC-10 airliner for easy maintenance, the provisions in the proposal for maintainability of the Shuttle failed to adequately make use of this background.

The company might have strengthened its bid with a good presentation to Fletcher, Low, and McCurdy, but this did not go well. Low wrote that this briefing

> *really did not answer any questions. For example, we still don't know who the Chief Engineer on the Shuttle will be. Answers in general were broad, generalized, and weak. The attitude displayed was "we are a great company and you had better give us the job because we will do the best job for you."*

McDonnell Douglas came away ranked third in the competition.

That left Grumman and North American, which together had built the piloted spacecraft of Apollo. "Theirs is the most recent, the most intense and the most demanding experience with manned space flight," McCurdy said. "There's no question it helped them formulate their proposals."

Grumman actually gained the highest score in the technical areas. Its configuration was not perfect, with Fletcher noting "complex designs" in guidance, control, navigation, and data processing. He, however, wrote that "Grumman's design went to a greater depth of detail than those of the other companies. Its detailed weight estimates were substantiated by the design details." Its structural layouts showed "a thorough understanding of potential problems and positive solutions," and were simple and straightforward.

"Grumman did a very good job in proposing design features to enhance maintainability," he added. "The provisions it made for access throughout the vehicle were outstanding."

Grumman, however, was less outstanding in cost and management. Its proposed cost was higher than NASA liked. Fletcher saw why: the firm planned "to build up its work force rapidly to an early manpower peak. This poses the risk of premature hardening of the specifications and premature commitment of resources during the course of the program." Grumman came in a strong second overall; its excellent design did not outweigh its shortcomings in these other areas.

North American's concept showed weaknesses, such as a crew cabin that would be difficult to build. Ironically, its overall strength stemmed in part from a near-disaster: the flight of Apollo 13, which had an onboard explosion and barely survived. Fletcher wrote that this company's "good understanding of all electric power subsystems reflected the very thorough studies that North American made following the Apollo 13 accident, which had its origin in an electrical subsystem."

This firm's proposal "provided the lightest dry weight of any of the designs submitted." It "presented an excellent analysis of maintainability from the standpoint of design criteria and goals to achieve optimum turnaround conditions and timing between flights. It designed its orbiter vehicle with very good overall accessibility for maintenance." For the critical functions of guidance, control, and navigation, North American provided a "simple design with minimum interfaces."

In addition to this, the proposal was particularly strong in cost and management. Its projected cost was the lowest of the four, and Low noted "the universal opinion that North American, indeed, will wind up with the lowest cost." It achieved this in part by proposing a more measured approach to buildup of employment. "We were impressed by the way North American had thought through the personnel buildup," said McCurdy. "The others attacked this problem a little more like a cavalry charge. North American had the lowest number of man-years in its shuttle proposal. Man-years is where you save money, nowhere else."

This company also gained an edge through its approach to minority hiring. A confrontation with black employees in 1969 had left North American determined to take the lead in promoting equal opportunity, and in 1972 this firm had more blacks, Hispanics, and Asians than any of the three competitors. NASA viewed this as advantageous, for as McCurdy put it, "We're not crusaders for civil rights. But the fact that North American moved forward on this front tells us something about how the company is thinking ahead."

Fletcher, Low, and McCurdy picked this winner on the afternoon of July 26. In Low's words, "we very quickly determined that all three of us wanted to select North American Rockwell on the basis of the highest score" and "the lowest cost." They then made phone calls to spread the news. When Fletcher phoned North American, just after lunchtime in California, he found that this company's executives already had gotten the word. Their local congressman, Del Clawson, had received his own phone call from NASA earlier that afternoon, and beat Fletcher to the punch with a call of his own.

Had Nixon's hand steered the choice? "The lack of White House interest in this selection had been remarkable," Low wrote. Prior to that week's round of meetings, he had asked Fletcher "whether he had any commitment to inform the White House what was going on and whether they wanted to get into the act in any way whatsoever. His answer was an emphatic no to both questions."

At the company briefings, Low asked the bidders to comment on the fairness of the competition. In Low's words, three of these firms "all indicated that this had been the best and fairest competition they had ever participated in." Sanford McDonnell, president of his firm, reserved the right to lodge a protest, and Low noted that "in effect, Sandy McDonnell said that the competition was a fair one if we select McDonnell Douglas, and unfair if we select somebody else." However, he accepted the final decision.[29]

His conduct thus contrasted with that of Pratt & Whitney, which had made just this type of formal protest a year earlier, on losing the SSME competition to Rocketdyne. This complaint had the legal status of a lawsuit, directed not against Rocketdyne but rather against NASA, which allegedly had performed wrongful acts in selecting that contractor. With Pratt as the plaintiff and NASA as defendant, Rocketdyne held the role of a highly interested witness whose testimony could help NASA in seeking to uphold this

---

29. O'Toole and Dash, *Space Shuttle*; memo, Myers to Manned Spacecraft Center Director, February 23, 1972; Source Evaluation Board, *Report of Findings*, June 23, 1972; *General Summary*; Low, Addendum to Personal Notes No. 75, July 29, 1972; Fletcher et al., "Selection of Contractor," September 18, 1972. See also Reports SV 72-19, SSV 72-2 (North American Rockwell).

award. The proceedings did not take place in federal court, but rather went forward under the Comptroller General, Elmer Staats. He issued his decision at the end of March 1972, while preparation of proposals for the complete Shuttle was under way.

Pratt's attorneys, who were highly capable, argued that NASA had "failed to conduct meaningful negotiations." NASA's discussions with the bidders "did not include the pointing out of deficiencies or weaknesses and did not afford offerors an opportunity to improve their proposals." Staats would have none of this:

*"It is also unfair, we think, to help one proposer through successive rounds of discussions to bring his original inadequate proposal up to the level of other adequate proposals by pointing out those weaknesses which were the result of his own lack of diligence, competence, or inventiveness.*

Had NASA "erroneously and illegally accepted a nonresponsive proposal?" Though this introduced engineering issues, Staats found "no basis to object to the technical judgment reached." Was it true that "NASA's determination of Pratt & Whitney's deficiencies was arbitrary and capricious?" Pratt charged that NASA had failed to respond to requests for information, had unfairly penalized certain technical deficiencies in its proposal while giving Rocketdyne something of a free ride, and had failed "to read and fully comprehend" this proposal. Staats wrote, "The administrative report contains a detailed rebuttal of these contentions." NASA's evaluations "provided a sound basis for selecting the most advantageous proposal."

Pratt also complained that "selection of Rocketdyne wastes eleven years" of its experience. Though Pratt indeed had devoted much effort to its XLR-129 rocket engine, Staats concluded that NASA had fairly weighed the merits of this engine and had not overlooked its advantages in determining that "Rocketdyne offered the superior technical approach." NASA had been obliged to give due weight to the experience of Pratt, and had done this.

Staats disagreed with another contention: that Rocketdyne's design invited a cost overrun. He concluded that Rocketdyne's cost estimates had been well substantiated. Similarly, he did not accept that Rocketdyne had "obtained an unfair competitive advantage by diversion of Saturn funds to the

## THE SPACE SHUTTLE DECISION

SSME proposal effort." He cited an audit by which "Rocketdyne determined the amount involved to be $2,526, and that it has made appropriate adjustments to the respective contracts."[30]

He concluded, "We believe the procurement was conducted in a manner which was consistent with applicable law and regulations and was fair to all proposers." The contract award to Rocketdyne would stand; Pratt's attempt to overturn it had failed. Rocketdyne indeed would design and build the SSME. Because this division was part of North American Rockwell, the subsequent award of the main Shuttle contract to this company's Space Division gave this firm responsibility for the entire Shuttle orbiter, including its engine. This was more than North American had carried during Apollo, more than it had held since the days of Navaho some 20 years earlier.

Now that Rocketdyne had the SSME, it intended to keep it. On receiving the initial contract award the previous July, William Bergen, president of North American, had approached Pratt & Whitney, inviting that company to share in the engine development. Pratt chose instead to pursue its appeal, and Low wrote in April 1972 that "there may be a lot of bad blood between the 2 companies. Certainly it is not our intent to force a marriage at this time between Pratt and Whitney and Rocketdyne."

In building the orbiter, however, the Space Division would generously share the work by awarding important subcontracts to its rivals. In March 1973, North American—now known as Rockwell International—gave Grumman responsibility for the orbiter's delta wing, and granted McDonnell Douglas the right to build the small onboard rocket system that would be used for on-orbit maneuvers. NASA conducted additional contract competitions during 1973, choosing Martin Marietta to build the external tank and selecting Thiokol for the solid rocket boosters.[31]

The events of 1972 brought an end to NASA's search for a post-Apollo future. The search had begun in 1965, when George Mueller had set up his Apollo Applications program office. This effort led to Skylab, but that offered no more than one more year of piloted missions. Characteristically, Mueller responded by seeking a larger space station that could fly atop a Saturn V. The Space Shuttle then grew out of this new pursuit, initially as a logistics vehicle

---

30. Letter B-173677, Staats to Fletcher, March 31, 1972.
31. *Ibid.*; Low, Personal Notes No. 68, April 17, 1972; NASA SP-4012, vol. III, p. 49, 122-123.

but growing to take on a life of its own.

Why, finally, did Nixon decide to build the Shuttle? One must not underestimate the tendency of the federal government to look after its own; few major Washington programs reach an end, to vanish into the night. Nixon had no wish to shut down piloted space flight; he wanted to keep it alive. He also was concerned over aerospace employment. Yet he could have addressed such issues with nothing more than Big Gemini riding atop a Titan III-M, to fly occasionally and show the flag.

The key to the Shuttle was its well-founded prospect of low cost and routine operation. This promise did not rest on the cost-benefit studies of Mathematica, which the Flax Committee largely refuted and the OMB rejected out of hand. Rather, it rested on technical developments: automated onboard checkout, reusable thermal protection, rocket engines with long life. No OMB internal memo or White House report ever denied this promise; only experience would do that, years later. The Space Shuttle thus could find its way to approval, within a nation and government that remained willing to embrace the new.

During 1972, the Shuttle entered a new phase, as a mainstream aerospace program. The debates and arguments were finished. NASA now held its future in its own hands, with responsibility for executing what it had planned and delivering what it had promised.

# Bibliography

Some publications have accession numbers from CASI (Center for Aerospace Information) or from SHHDC (Shuttle History Historical Documents Collection), at NASA-Marshall.

Agnew, Spiro, chairman. *The Post-Apollo Space Program: Directions for the Future* [reprinted in NASA SP-4407, vol. I, pp. 522-543] (Space Task Group, September 1969).

Akridge, Max. *Space Shuttle History* (NASA-MSFC; SHHDC-5013, January 8, 1970).

Allen, H. Julian, and A. J. Eggers. *A Study of the Motion and Aerodynamic Heating of Ballistic Missiles Entering the Earth's Atmosphere at High Supersonic Speeds* (NACA Report 1381, 1958).

Ames, Milton B., chairman. *Report of the Ad Hoc Subpanel on Reusable Launch Vehicle Technology* (Aeronautics and Astronautics Coordinating Board, September 14, 1966.)

Baar, James, and William E. Howard. *Polaris!* (New York: Harcourt, Brace, 1960).

Bathie, William W. *Fundamentals of Gas Turbines* (New York: John Wiley, 1984).

Bender, Marylin, and Selig Altschul. *The Chosen Instrument* (New York: Simon & Schuster, 1982).

Berman, Larry. *The Office of Management and Budget and the Presidency, 1921-1979* (Princeton, New Jersey: Princeton University Press, 1979).

Bernstein, Jeremy. *Three Degrees Above Zero* (New York: Scribner, 1984).

Blaug, Mark. *Great Economists Since Keynes* (Brighton, Sussex: Wheatsheaf Books, Ltd., 1985).

Clauser, Henry R., and George S. Brady. *Materials Handbook* (New York: McGraw-Hill, 1991).

Branch, Taylor. *Parting the Waters* (New York: Simon & Schuster, 1988).

Burnet, Charles. *Three Centuries to Concorde* (London: Mechanical Engineering Publications, 1979).

Burroughs, William E. *Deep Black: Space Espionage and National Security* (New York: Random House, 1986).

Catton, Bruce. *A Stillness at Appomattox* (New York: Pocket Books, 1958).

Chaikin, Andrew. *A Man on the Moon* (New York: Penguin Books, 1994).

Clarke, Arthur C. *The Exploration of Space* (London: Temple Press, Ltd., 1951).

—. *2001: A Space Odyssey* (New York: Signet Books, 1968).

Cornelisse, J. W.; H. F. R. Schoyer and K. F. Wakker. *Rocket Propulsion and Spaceflight Dynamics* (Belmont, California: Fearon-Pitman, 1979).

Costello, John, and Terry Hughes. *Concorde* (London: Angus & Robertson, 1976).

Daley, Robert. *An American Saga: Juan Trippe and His Pan Am Empire* (New York: Random House, 1980).

Davies, R. E. G. *A History of the World's Airlines* (London: Oxford University Press, 1964).

Day, Leroy, manager. *NASA Space Shuttle Summary Report* (Space Shuttle Task Group May 19, 1969). See also revised version, July 31, 1969.

—. *NASA Space Shuttle Task Group Report, Volume II: Desired System Characteristics* (Space Shuttle Task Group, June 12, 1969).

—. *Summary Report of Recoverable versus Expendable Booster, Space Shuttle Studies* (Space Shuttle Task Group, December 10, 1969).

Donlan, Charles J. *Space Shuttle Systems Definition Evolution.* Internal NASA paper (July 11, 1972).

Dornberger, Walter. *V-2* (New York: Bantam Books, 1979).

DuBridge, Lee, chairman. *The Post-Apollo Space Program: Directions for the Future* President's Science Advisory Committee (September 1969).

—. *The Next Decade in Space* (President's Science Advisory Committee, CASI 71N-10973, March 1970).

Durant, Will. *The Reformation* (New York: Simon and Schuster, 1957).

Dwiggins, Don. *The SST: Here It Comes Ready or Not* (Garden City, New York: Doubleday, 1968).

Eddy, Paul; Elaine Porter, and Bruce Page. *Destination Disaster* (New York: Quadrangle/New York Times Book Company, 1976).

Ehrlichman, John. *Witness to Power: The Nixon Years* (New York: Simon and Schuster, 1982).

*Eight Decades of Progress* (Lynn, Massachusetts: General Electric Co., 1990).

Emme, Eugene M., editor. *The History of Rocket Technology* (Detroit: Wayne State University Press, 1964).

Faget, Maxime, and Milton Silveira. *Fundamental Design Considerations for an Earth-Surface to Orbit Shuttle* (CASI 70A-44618, October 1970).

Fahrney, Rear Admiral D. S. *The History of Pilotless Aircraft and Guided Missiles* (Washington, D.C.: Archives, Federation of American Scientists, undated).

Fall, Bernard B. *Hell in a Very Small Place: The Siege of Dien Bien Phu* (New York: J. B. Lippincott, 1966).

Grey, Jerry. *Enterprise* (New York: William Morrow, 1979).

Gomersall, Edward W., and Darrell E. Wilcox. *Working Paper: Evolutionary Concepts for a National Space Transportation System (Shuttle)* (NASA-OART, Report MO-71-1, May 28, 1971).

Gunston, Bill. *Fighters of the Fifties* (Osceola, Wisconsin: Specialty Press, 1981).

Halliday, David, and Robert Resnick. *Physics for Students of Science and Engineering* (New York: John Wiley, 1962).

Hallion, Richard P. *The Path to the Space Shuttle: The Evolution of Lifting Reentry Technology* (Edwards Air Force Base, California: History Office, Air Force Flight Test Center, November 1983).

—, ed. *The Hypersonic Revolution: Eight Case Studies in the History of Hypersonic Technology* (Wright-Patterson Air Force Base, Ohio: Special Staff Office, Aeronautical Systems Division, 1987).

Hansen, James R. *Transition to Space: A History of 'Space Plane' Concepts at Langley Aeronautical Laboratory, 1952-1957* (1985).

Heiss, Klaus P., and Oskar Morgenstern. *Factors for a Decision on a New Reusable Space Transportation System.* Memo to James Fletcher [reprinted in NASA SP-4407, vol. I, pp. 549-555] (October 28, 1971).

Heppenheimer, T. A. *Colonies in Space* (Harrisburg, Pennsylvania: Stackpole Books, 1977).

—. *Hypersonic Technologies and the National Aerospace Plane* (Arlington, Virginia: Pasha Publications, 1990).

# Bibliography

—. *Turbulent Skies: The History of Commercial Aviation* (New York: John Wiley, 1995).

—. *Countdown: A History of Space Flight* (New York: John Wiley, 1997).

Horwitch, Mel. *Clipped Wings: The American SST Conflict* (Cambridge, Massachusetts: MIT Press, 1982).

Huggett, Clayton; C. E. Bartley, and Mark M. Mills. *Solid Propellant Rockets* (Princeton, New Jersey: Princeton University Press, 1960).

Hunter, Maxwell W. II. *The Origins of the Shuttle (According to Hunter)* [reprinted in part in *Earth/Space News*, November 1976, pp. 5-7] (Sunnyvale, California: Lockheed, September, 1972).

Huzel, Dieter K. *Peenemunde to Canaveral* (Englewood Cliffs, New Jersey: Prentice-Hall, 1962).

Irving, Clive. *Wide-Body: The Triumph of the 747* (New York: William Morrow, 1993).

Jenkins, Dennis. *Space Shuttle: The History of Developing the National Space Transportation System* (Marceline, Missouri: Walsworth Publishing Co., 1996).

Johnson, Paul. *Modern Times: The World from the Twenties to the Eighties* (New York: Harper & Row, 1983).

Kent, Richard J., Jr. *Safe, Separated and Soaring: A History of Federal Civil Aviation Policy 1961-1972* (Washington: U.S. Government Printing Office, 1980).

Killian, James R. *Sputnik, Scientists and Eisenhower* (Cambridge, Massachusetts: MIT Press, 1977).

Klass, Philip J. *Secret Sentries in Space* (New York: Random House, 1971).

Knight, Geoffrey. *Concorde: The Inside Story* (London: Weidenfeld & Nicolson, 1976).

Koppes, Clayton R. *JPL and the American Space Program* (New Haven, Connecticut: Yale University Press, 1982).

Kuter, Laurence S. *The Great Gamble: The Boeing 747* (University, Alabama: University of Alabama Press, 1973).

Launius, Roger D. *NASA: A History of the U.S. Civil Space Program* (Malabar, Florida: Krieger Publishing Co., 1994).

Lehman, Milton. *This High Man: The Life of Robert Goddard* (New York: Farrar, Straus, 1963).

Levine, Arthur L. *The Future of the U.S. Space Program* (New York: Praeger, 1975).

Ley, Willy. *Rockets, Missiles and Space Travel* (New York: Viking Press, 1957).

Loftus, J. P., Jr.; S. M. Andrich, M. G. Goodhart, and R. C. Kennedy. *Evolution of the Space Shuttle Design*. Report to the Presidential Commission on the Space Shuttle Challenger (March 6, 1986).

Logsdon, John. *The Decision to Go to the Moon* (Cambridge, Massachusetts: MIT Press, 1970).

—. *From Apollo to Shuttle: Policy Making in the Post-Apollo Era*. Unpublished manuscript (Washington: NASA History Office Archives, NASA Headquarters, Document HHR-46, Spring 1983).

Long, Franklin A., chairman. *The Space Program in the Post-Apollo Period* (Joint Space Panels, President's Science Advisory Committee, February 1967).

Lord, Walter. *A Night to Remember* (New York: Bantam, 1963).

Lynch, Charles T., ed. *Handbook of Materials Science. Vol. II: Metals, Composites and Refractory Materials* (Cleveland: CRC Press, 1975).

Manchester, William. *The Glory and the Dream* (Boston: Little, Brown, 1974).

McDonald, Robert A., ed. *Corona Between the Sun and the Earth: The First NRO Reconnaissance Eye in Space* (Bethesda, Maryland: American Society for Photogrammetry and Remote Sensing, 1997).

McDougall, Walter. *...The Heavens and the Earth: A Political History of the Space Age* (New York: Basic Books, 1985).

Miller, Jay. *The X-Planes, X-1 to X-29* (Marine on St. Croix, Minnesota: Specialty Press, 1983).

Miller, Ron, and Frederick C. Durant III. *Worlds Beyond: The Art of Chesley Bonestell* (Norfolk, Virginia: Donning Co., 1983)

Morgenstern, Oskar, and Klaus P. Heiss. *Economic Analysis of New Space Transportation Systems* (Princeton, New Jersey: Mathematica, Inc. CASI 75N-22191, May 31, 1971).

Mosley, Leonard. *Dulles: A Biography of Eleanor, Allen and John Foster Dulles and Their Family Network* (New York: Dial Press, 1978).

Mueller, George. *Address by Dr. George E. Mueller Before the British Interplanetary Society* (London: University College, August 10, 1968).

—. *Space Shuttle Contractors Briefing* (Washington: NASA, May 5, 1969).

Murray, Russ. *Lee Atwood...Dean of Aerospace* (Rockwell International Corp., 1980).

Nau, Richard A. *A Comparison of Fixed Wing Reusable Booster Concepts*. Space Technology Conference. (New York: Society of Automotive Engineers, May, 1967).

Nayler, J. L., and E. Ower. *Aviation: Its Technical Development* (Philadelphia: Dufour Editions, 1965).

Neal, J. A. *The Development of the Navaho Guided Missile, 1945-1953* (Wright-Patterson Air Force Base, Ohio: Wright Air Development Center Historical Branch, January, 1956).

Neufeld, Jacob. *The Development of Ballistic Missiles in the United States Air Force 1945-1960* (Washington: Office of Air Force History, 1990).

Neufeld, Michael. *The Rocket and the Reich: Peenemunde and the Coming of the Ballistic Missile Era* (New York: Free Press, 1994).

Newell, Homer, chairman. *America's Next Decades in Space: A Report for the Space Task Group* (NASA Planning Steering Group, CASI 71N-12420, September, 1969).

Newhouse, John. *The Sporty Game* (New York: Alfred A. Knopf, 1982).

O'Toole, Thomas, and Leon Dash. *How the Space Shuttle Contract Was Won* [newspaper clipping] (October, 1972).

Owen, Kenneth. *Concorde: New Shape in the Sky* (London: Jane's Publishing Co., 1982).

*Oxford Dictionary of Quotations* (New York: Oxford University Press, 1980).

Pace, Scott. *Engineering Design and Political Choice: The Space Shuttle 1969-1972*. M.S. thesis (Cambridge, Massachusetts: MIT, May, 1982).

Pace, Steve. *North American XB-70 Valkyrie* (Blue Ridge Summit, Pennsylvania: Tab Books, 1990).

*Pedigree of Champions: Boeing Since 1916* (Seattle: Boeing, 1985).

Perkins, Rosalie L. *History of the Air Force Plant Representative Office, North American Rockwell Corp., Rocketdyne Division, Canoga Park, California* (Maxwell Air Force Base, Alabama: Air Force Archives, 1967-1972).

Pierce, J. R. *The Beginnings of Satellite Communication* (San Francisco: San Francisco Press, 1968).

Powers, Robert M. *Shuttle: The World's First Spaceship* (Harrisburg, Pennsylvania: Stackpole Books, 1979).

Powers, Thomas. *The Man Who Kept the Secrets: Richard Helms & the CIA* (New York: Alfred A. Knopf, 1979).

Prados, John. *The Soviet Estimate: U.S. Intelligence Analysis and Soviet Strategic Forces* (Princeton, New Jersey: Princeton University Press, 1986).

Proceedings, NASA Space Shuttle Symposium, October 16-19, 1969 (Washington: NASA History Office Archives, NASA Headquarters).

Ramo, Simon. *The Business of Science* (New York: Hill and Wang, 1988).

Ranelagh, John. *The Agency: The Rise and Decline of the CIA* (London: Weidenfeld and Nicolson, 1986).

Rhodes, Richard. *Dark Sun: The Making of the Hydrogen Bomb* (New York: Simon & Schuster, 1995).

Rice, Berkeley. *The C-5A Scandal* (Boston: Houghton Mifflin, 1971).

Richelson, Jeffrey. *American Espionage and the Soviet Target* (New York: William Morrow, 1987).

—. *America's Secret Eyes in Space: The U.S. Keyhole Spy Satellite Program* (New York: Harper & Row, 1990).

Ruffner, Kevin C., ed. *Corona: America's First Satellite Program* (Washington: Central Intelligence Agency, 1995).

Schnyer, A. Dan, and R. G. Voss. *Review of Orbital Transportation Concepts—Low Cost Operations* (Washington: NASA-OMSF, December 18, 1968).

Serling, Robert J. *Legend and Legacy: The Story of Boeing and Its People* (New York: St. Martin's Press, 1992).

Shurcliff, William. *S/S/T and Sonic Boom Handbook* (New York: Ballantine Books, 1970).

Smith, Francis B., manager. *Summary Report: Future Programs Task Group* [reprinted in NASA SP-4407, vol. I, pp. 473-490] (January, 1965).

Solberg, Carl. *Conquest of the Skies* (Boston: Little, Brown, 1979).

Steiner, John E. *Problems and Challenges: A Path to the Future* (London: Royal Aeronautical Society, October 10, 1974).

—. *Jet Aviation Development: One Company's Perspective* (Seattle: Boeing, 1989).

Sutton, George P. *Rocket Propulsion Elements* (New York: John Wiley, 1986).

Thompson, Tina D., ed. *TRW Space Log*, vol. 27 (1991), vol. 31 (1995).

Townes, Charles, chairman. *Report of the Task Force on Space* [reprinted in NASA SP-4407, vol. I, pp. 499-512] (January 8, 1969).

Tuchman, Barbara. *The March of Folly* (New York: Alfred A. Knopf, 1984).

Wattenberg, Ben J. *The Real America* (New York: Doubleday, 1974).

White, Theodore H. *The Making of the President 1964* (New York: Signet Books, 1966).

—. *The Making of the President 1968* (New York: Pocket Books, 1970).

Wilson, Andrew. *The Concorde Fiasco* (Baltimore: Penguin Books, 1973).

Zaloga, Steven. *Target America: The Soviet Union and the Strategic Arms Race, 1945-1964* (Novato, California: Presidio Press, 1993).

# REPORTS AND OTHER PUBLICATIONS: AIAA, CORPORATIONS, NASA

The title of a report is often followed by an accession number: CASI (Center for Aerospace Information) or SHHDC (Shuttle History Historical Documents Collection), located at NASA's Marshall Space Flight Center. Documents are grouped by source below. Except as noted, cited corporate documents are executive summaries.

**Aerospace Corp.:**

ATR-72 (7231)-1. August 1, 1971. *Integrated Operations/Payloads/Fleet Analysis Final Report*. CASI 72N-26790.

**AIAA (American Institute of Aeronautics and Astronautics):**

69-1064. September 1969. L. J. Walkover and A. J. Stefan. *Space Station Design for Flexibility*.

70-044. June 1970. F. M. Stewart and R. L. Wetherington. *Space Shuttle Main Propulsion*.

70-1249. October 1970. Philip P. Antonatos, Alfred C. Draper and Richard D. Neumann. *Aero Thermodynamic and Configuration Development*.

71-658. June 1971. Richard C. Mulready. *Pratt & Whitney's Space Shuttle Main Engine*.

71-804. July 1971. J. F. Yardley. *McDonnell Douglas Fully Reusable Shuttle*.

71-805. July 1971. Bastian Hello. *Fully Reusable Shuttle*.

71-806. July 1971. Robert N. Lindley. *The Economics of a New Space Transportation System*.

73-31. January 1973. Eugene S. Love. *Von Karman Lecture: Advanced Technology and the Space Shuttle*.

73-73. January 1973. M. W. Hunter II, R. M. Gray and W. F. Miller. *Design of Low-Cost, Refurbishable Spacecraft for Use with the Shuttle*.

89-2387. July 1989. D. Warren and C. Langer. *History in the Making—The Mighty F-1 Rocket Engine*.

2987. April-May 1991. B. L. Koff. *Spanning the Globe with Jet Propulsion*.

**Boeing:**

D2-114012-1. November 1, 1967. *Saturn V Single Launch Space Station and Observatory Facility*. CASI 70N-27668.

December 1971. *Ballistic Recoverable & Reusable LOX/RP Flyback Boosters*. SHHDC-2965.

**General Dynamics:**

GD/C-DCB-65-018. April 1965. *Reusable Orbital Transport Second Stage Detailed Technical Report*.

GDC-DCB-67-031. September 25, 1967. *Multipurpose Reusable Spacecraft Preliminary Design Effort*.

GDC-DCB-68-017. November 1968. *Reusable Launch Vehicle/Spacecraft Concept*. SHHDC-0023.

GDC-DCB-69-046. October 31, 1969. *Space Shuttle Final Technical Report*. CASI 70N-31536.

SAMSO-TR-68-171. April 1968. *Multipurpose Reusable Spacecraft Cruise Vehicle Preliminary Design Effort*. CASI 81X-77461.

November 17, 1971. *Fully Reusable Shuttle*.

**Grumman:**

B35-43 RP-5. December 31, 1970. *Alternate Space Shuttle Concepts Mid-Term Report*.

# Bibliography

B35-43 RP-11. July 6, 1971. *Alternate Space Shuttle Concepts Study.*

B35-43 RP-21. September 1, 1971. *Alternate Space Shuttle Concepts Study; Design Requirements and Phased Program Evaluation.* CASI 73N-17877.

B35-43 RP-28. November 15, 1971. *Definition of Mark I/II Orbiters & Ballistic & Flyback Boosters.* SHHDC-3004.

B35-43 RP-30. December 15, 1971. *Shuttle Systems Evaluation and Selection.* CASI 72N-18871.

B35-43 RP-33. March 15, 1972. *Space Shuttle System Program Definition; Phase B Extension Final Report.* CASI 74N-30308.

February 16, 1972. *Shuttle Systems Evaluation and Selection; Phase B Extension Final Briefing.*

**Lockheed:**

LMSC-A946632. March 1969. *Space Transport and Recovery System (Space Shuttle).* SHHDC-0048.

LMSC-A959837. December 22, 1969. *Final Report: Integral Launch and Re-entry Vehicle.* CASI 70N-31831.

LMSC-A989142. June 4, 1971. *Study of Alternate Space Shuttle Concepts.* CASI 74N-76538.

LMSC-A990556. June 30, 1971. *Payload Effects Analysis Study.* CASI 71N-37496.

LMSC-A990558. June 30, 1971. *Design Guide for Space Shuttle Low-Cost Payloads.* CASI 72N-31883.

LMSC-A990594. June 30, 1971. *Final Report Summary, Payload Effects Analysis.* CASI 71N-35028.

LMSC-A991394. September 1, 1971. *Alternate Space Shuttle Concepts Study; Interim Review Presentation.* CASI 74N-75495.

LMSC-A995887. November 3, 1971. *Alternate Space Shuttle Concepts Study.* CASI 74N-76542.

LMSC-A995931. November 15, 1971. *Alternate Concepts Study Extension.* CASI 73N-30844.

LMSC-D152947. February 22, 1972. *Space Shuttle Final Review.* CASI 75N-74948.

LMSC-D153024. March 15, 1972. *Space Shuttle Concepts.*

LMSC-D157302. February 15 1972. *Space Shuttle Final Review.*

LPC 629-6. March 15, 1972. *Study of Solid Rocket Motors for a Space Shuttle Booster.* CASI 72N-22772.

LR 18790. May 21, 1965. *Design Studies of a Reusable Orbital Transport, First Stage.* CASI 65X-19708.

**Martin Marietta:**

M-69-36. December 1969. *Spacemaster: A Two-Stage Fully Reusable Space Transportation System.* SHHDC-214.

MCR-71-309. October 1971. *Study of Titan IIIL Booster for Space Shuttle Application.*

**Mathematica, Inc.:**

See Bibliography under "Heiss" and "Morgenstern."

**McDonnell Douglas—Space Shuttle:**

G994. February 7, 1969. *Low-Cost Reusable Orbital Transport System.* CASI 72X-82639.

H321. August 21, 1969. *Big G: Logistic Spacecraft Evolving from Gemini.* CASI 69X-18044.

MDC E0049. November 1, 1969. *Integral Launch and Reentry Vehicle System.* CASI 70N-15212.

*443*

MDC E0056. December 15, 1969. *A Two-Stage Fixed Wing Space Transportation System.* CASI 70N-31597.

MDC E0308. June 30, 1971. *Space Shuttle System Phase B Study Final Report.* SHHDC-2825.

MDC E0376-1. June 30, 1971. *External LH2 Tank Study Final Report.*

MDC E0497. November 15, 1971. *Phase B System Study Extension Final Report.* CASI 76N-71552.

MDC E0558. March 15, 1972. *Space Shuttle Phase B System Study Extension Final Report.* CASI 77X-78418; SHHDC-2861.

September 1, 1971. *Interim Report to OMSF: Phase B System Study Extension.* SHHDC-2854.

February 22, 1972. *Space Shuttle Design Review: Phase B System Study Extension.*

**McDonnell Douglas—Space Station:**

MDC G2570. November 1971. *Space Station, Laboratory in Space.*

MDC G2727. April 1, 1972. *Space Station Executive Summary.* CASI 72N-27912.

**NASA:**

NAS 9-10960. June 1970. *Study Control Document: Space Shuttle System Program Definition (Phase B).* SHHDC-2916.

SP-196. 1969. *Proceedings of the Winter Study on Uses of Manned Space Flight, 1975-1985.* NASA Science and Technology Advisory Committee.

SP-440. 1981. Donald D. Baals and William R. Corliss. *Wind Tunnels of NASA.*

SP-468. 1985. Laurence K. Loftin. *Quest for Performance.*

SP-4011. 1977. Roland W. Newkirk, Ivan D. Ertel and Courtney G. Brooks. *Skylab: A Chronology.*

SP-4012. 1988. Linda Neuman Ezell. *NASA Historical Data Book: Programs and Projects.* Vol. II: 1958-1968. Vol. III: 1969-1978.

SP-4026. 1995. Hermann Noordung. *The Problem of Space Travel: The Rocket Motor* [English translation].

SP-4102. 1982. Arnold S. Levine. *Managing NASA in the Apollo Era.*

SP-4205. 1979. Courtney G. Brooks, James M. Grimwood and Loyd S. Swenson, Jr. *Chariots for Apollo: A History of Manned Lunar Spacecraft.*

SP-4208. 1983. W. David Compton and Charles D. Benson. *Living and Working in Space: A History of Skylab.*

SP-4210. 1977. R Cargill Hall. *Lunar Impact: A History of Project Ranger.*

SP-4303. 1984. Richard P. Hallion. *On the Frontier: Flight Research at Dryden, 1946-1981.*

SP-4307. 1993. Henry C. Dethloff. *'Suddenly Tomorrow Came...': A History of the Johnson Space Center.*

SP-4308. 1995. James R. Hansen. *Spaceflight Revolution: NASA Langley Research Center from Sputnik to Apollo.*

SP-4407. John M. Logsdon, gen. ed. *Exploring the Unknown: Selected Documents in the History of the U.S. Civil Space Program.* 1995: Vol. I. *Organizing for Exploration.* 1996. Vol. II. *Relations with Other Organizations.*

TM X-52876. July 1970. *Space Transportation System Technology Symposium.* NASA Lewis Research Center.

TT F-9227. December 1964. Hermann Oberth. *Rockets in Planetary Space* [English translation].

*Bibliography*

**North American Aviation:**

AL-1347. October 1951. *Development of a Strategic Missile and Associated Projects.*

**North American Rockwell—Space Shuttle:**

LE 71-7. 1971. *Shuttle - The Space Transporter of the 1980's.*

SD 69-573-1. December 1969. *Study of Integral Launch and Reentry Vehicle System.* CASI 70N-31832.

SD 71-114-1. June 25, 1971. *Space Shuttle Phase B Final Report.* CASI 74N-75830.

SD 71-342. November 12, 1971. *Space Shuttle Phase B Extension.* CASI 72N-13860.

SD 72-SH-0012-1. March 15, 1972. *Space Shuttle Phase B Final Report.* CASI 74N-31330.

SSV 72-2. October 1972. *Space Shuttle Summary Briefing.*

SV 71-28. July 19, 1971. *Fully Reusable Shuttle.*

SV 71-40. September 1, 1971. *Space Shuttle Phase B Extension Mid-Term Review.*
Vol. I: *Executive Summary.*
Vol. II: *Mid-Term Presentation.* CASI 90N-70121.

SV 71-50. November 3, 1971. *Space Shuttle Phase B Extension, 4th Month Review.* SHHDC-2961.

SV 71-59. December 15, 1971. *Space Shuttle Program Review.*

SV 72-14. February 22, 1972. *Space Shuttle Phase B Final Briefing.* SHHDC-203.

SV 72-19. July 8, 1972. *Space Shuttle System Summary Briefing.*

**North American Rockwell—Space Station:**

SD 70-153. July 1970. *Space Station Program Phase B Definition.* CASI 70X-18668.

SD 70-160. July 24, 1970. *Space Base Definition.* CASI 70X-17323.

SD 70-536. November 1970. *Space Station Mockup Brochure.* CASI 71X-10114.

SD 71-214. January 1, 1972. *Modular Space Station Phase B Extension Period.* CASI 72N-18889.

SD 71-217-1. January 1, 1972. *Modular Space Station Phase B Extension: Preliminary System Design.* CASI 72N-18881.

SD 71-576. November 4, 1971. *Modular Space Station Phase B Extension.* CASI 72N-22904.

**Pan Am (Pan American World Airways):**

December 8, 1970. Lehman Brothers. *Prospectus: $87,375,000: 11-1/8% Guaranteed Loan Certificates Due December 16, 1986.*

**Pratt & Whitney:**

GP 70-35. February 7, 1970. *Space Shuttle Engine.*

GP 70-271. August 12, 1970. *Space Shuttle Main Engine Orientation.*

PWA FP 71-50. April 28, 1971. *Proposal for Space Shuttle Main Engine.*

**Rocketdyne:**

RSS-8333-1. June 23, 1971. *Space Shuttle Main Engine Phase B Final Report.* Volume I: *Summary.*

*445*

# THE SPACE SHUTTLE DECISION

**Thiokol:**

TWR-5672. March 15, 1972. *Study of Solid Rocket Motors for a Space Shuttle Booster.* CASI 73N-29820.

**United Technology:**

UTC 4205-72-7. March 15, 1972. *Study of Solid Rocket Motors for a Space Shuttle Booster.* CASI 72N-22784.

# Index

Page references to illustrations or photos are in italics.

## A

Aerojet General Corp., 16, 24, 48, 80, 106
    SSME, 249, 255, 266
    solid rockets, 45, 46, 353
Aeronautics and Astronautics Coordinating Board (NASA-USAF), 83-84, 201
Aerophysics Laboratory (NAA), 24
Aerospace Corp., 54, 264, 267, 277-279, 357
aerospace industry, 291, *292*, 395
Aerospaceplane, 78-79, 82
Agenda rocket stage, 194, 197, 227
Agnew, Spiro, 125-126, 167, 170
    background, 144-145
    supports Mars flight, 126, 145-146, 148, 150, 165-166, 178
Air Force:
    and CIA, 58, 194, 195, 197, 205, 214, 216, 259
    high-speed flight, 25, 28, 30-32, 33, 36-39, 78-79
    lifting bodies, 40-41
    missiles, postwar, 17, 18, 19, 22, 24, 26, 45-46, 61
    and NASA, 83-84, 132-133, 199-202, 215, 224-225
    rocket engine research, 235-236, 239
    space shuttle studies, 83-84, 116, 117, 210-213, 331-332
    sonic boom research, 309-310
    *See also* Dyna-Soar; Manned Orbiting Laboratory
Air Force officials:
    Crotty, Maj. Patrick, 232
    Estes, Lieut. Gen. Howell, 54
    Hall, Lt. Col. Edward, 17, 20, 45
    Putt, Maj. Gen. Donald, 22
    Roth, Col. M. S., 19
    Schriever, Maj. Gen. Bernard, 45, 54, 115
    Steelman, Lt. Col. Donald, 66
    Yarymovych, Michael, 215; quoted, 216
    Zuckert, Eugene, 54
    *See also* Hansen; Seamans
Allen, H. Julian (NACA), 26, 29, 35-36, 206
Allen, William (Boeing), 300, 301, 302, 303, 307
American Airlines, Inc., 252-253, 319
    Kolk, Frank, 319
American Institute of Aeronautics and Astronautics, 142-144
    Newbauer, John, 98
Anderson, Sen. Clinton, 95, 178, 203, 423-424, 426, 427; quoted, 178
Apollo program, 25, 63, 64, 100, 101, 102, 112, 136, 154-155, 158-159, 174, 176, 186, 206, 224, 229, 253
Apollo 7, 73, 113
Apollo 8, 66, 113-114, 126, 147
Apollo 9, 125-126

447

Apollo 11, 66, 145-146, 147, 389-390
Apollo 13, 113n, 431
Apollo 16, 366
Apollo 17, 186, 270, 271, 366
    lunar module, 112, 113, 125-126, 227
Apollo Applications Program, 62-66, *64*, 67, 98, 99, 101, 102, 103, 138-139, 189, 434
    Apollo Telescope Mount, 63, 65
    wet vs. dry workshop, 65-66, 67
    *See also* Skylab
Armstrong, Neil, 35, 187
Army, 18, 23, 24, 43
    Lemnitzer, Gen. Lyman, 153-154
Army Air Forces, 14
ASSET (hypersonic research vehicle), 36-37, 41, 53, 210, 221
Astro launch vehicle concept, 80
Astronauts:
    Anders, William, 113, 382, 408, 422
    Armstrong, Neil, 35, 187
    Borman, Frank, 113, 126
    Chaffee, Roger, 100
    Grissom, Gus, 100
    Lovell, James, 113
    Schmitt, Harrison, 186
    White, Ed, 100
astronomy in space, 2, 8, 63, 71, 423
Astroplane launch vehicle concept, 80
Astrorocket launch vehicle concept, 80-81
Atlas rocket, 92
    for space flight, 25, 37, 197, 199, *399*
    as ICBM, 24, 45, 50, 196
Atomic Energy Commission, 94, 105, 108, 394, 423-424
    Seaborg, Glenn, 125
automated checkout, in airliners, 88-89, 252-253
aviation, commercial, 84-85, 260, 292-293, 295, 318, 328
    *See also* Boeing 727; Boeing 747; Concorde; Lockheed L-1011; Pan American; SST; Trippe
aviation in Europe, 305. *See also* Concorde

# B

B-52 bomber, 40-41, 193
B-58 bomber, 210, 310
Becker, John, 27, 28, 29, 30, 32, 35
Bell Aircraft Corp., 27, 30, 49, 50
    Woods, Robert, 27
Bellcomm, 136-137, 139, 146
Bell Telephone Laboratories, 56-57, 214
    Boyle, William, 214
    Pierce, John, 57
    Smith, George, 214
Bissell, Richard (CIA), 193, 194; quoted, 193, 284-285
Boeing Company, 50, 51, 70
    aviation, 295, 297, 300-301, 302-303, 311, 318, 327-328
    Connelly, Bruce, 298
    Haynes, Hal, 303

# Index

space shuttle studies, 224, 265, 267, 348-349, 351-352, 416-417, 418-419
    Sutter, Joseph, 295, 296-297
    *See also* Allen; Boeing 747; SST; Steiner
Boeing 747, 295-302, 304, 319, 328
    design of, 295-297, 301-302
    engines, 296, 298, 301-302
Bollay, William, 15, 16-17
Bonestell, Chesley, 3, 5, 56
    magazine covers, *4, 57*
Borman, Frank, 113, 126
Bossart, Karel, 26
Branscomb, Lewis, quoted, 131
Bristol Siddeley, 305, 306
British Aircraft Corp., Ltd., 305, 306
Bureau of the Budget, 63-65, 109-110, 114, 121-123, 203, 204, 269
    discounted dollars, in analyses, 170, 255-259
    as federal institution, 159, 177, 269
    Gordon, Kermit, 203
    Rhode, Earl, 263
    Schlesinger, James, 159, 161
    *See also* Mayo; Office of Management and Budget
Bussard, Robert, 105

# C

C-5A transport, 294-295, 297, 320-321, 323
C-47 aircraft ("Gooney Bird"), 40, 41
California Institute of Technology, 43, 119, 123
Cape Canaveral, 198, 200-201, *202,* 214, 215, 275-277, 425, 426-427
    Vehicle Assembly Building, 275, *276,* 297, 427
Caravelle airliner, 305
Castenholz, Paul, 24, 25, 240-243, 272; quoted, 242, 243, 248
Centaur rocket, 48-49, 135, 235
Central Intelligence Agency, 50, 58, 191, 192, 193-196, 198, 225
    Helms, Richard, 204
    Lundahl, Arthur, 194-195
    Wheelon, Albert, 273
    *See also* Bissell; Discoverer; reconnaissance
Charyk, Joseph, 51
China, 22, 153
    Chairman Mao, quoted, 81
Chrysler Corp., 266, 267
civil rights movement:
    Abernathy, Ralph, 151-152
    *Brown* v. *Board of Education,* 156
    Connor, Eugene "Bull," 157
    Hamer, Fannie Lou, quoted, 157-158
    King, Martin Luther, 151, 152, 157
    Warren, Earl, 156
Civil War, 152, 157, 159
Clarke, Arthur C., 4, 11, 57, 187; quoted, 55-56
Clauser, Francis, quoted, 119, 245
Cleaver, Arthur, quoted, 190
Cold War, 3-4, 153-155

449

Castro, Fidel, 155
    See also Reconnaissance; Soviet Union
Collier's magazine (series on space flight), 2-4, 5, 6, 11, 12, 56, 57, 127
    Manning, Gordon, 2, 3
    Ryan, Cornelius, 2-3
Comet airliner, 305
communications satellites, 8, 55-56, 57, 135, 226, 277, 286
    Pierce, John, 57
Concorde airliner, 305-307, 328-329
    background of, 304-306
    engine, 305
Congress, 158, 204, 330, 362
    and NASA budget, 65, 101, 177-186, 189-190, 230-231, 329
    and SST, 313-318
    and Lockheed L-1011, 326-327, 329
    See also House of Representatives; Senate
Conquest of Space, The (George Pal film), 5, 11
Conrad, Paul, 316, 317
Cosenza, Charles, 36, 210

# D

Dassault, Marcel, 305
David, Edward, 362, 382, 393, 400, 406, 409, 410, 422
Day, LeRoy, 131-132, 134-135, 139, 233; quoted, 368
de Gaulle, Charles, 306
de Havilland, Sir Geoffrey, 305
De Lauer, Richard, 272
Delta launch vehicle, 25, 85, 277, 399, 419
Democratic Party, 326, 413
    Westwood, Jean, quoted, 427-428
Die Rakete zu den Planetenraumen (Oberth): quoted, 7-9
discounted dollars, 255-256
Discoverer reconnaissance satellites, 58, 59, 194-196, 284-285
    resolution, 195, 197
    achievements, 197, 198
Disney, Walt, 5, 11
Domestic Council, 268-269
Donlan, Charles, 34, 233, 270, 361, 375-376, 401-402, 403; quoted, 357, 377, 410
Dornberger, Walter, 27, 28, 30; quoted, 92-93
Douglas Aircraft Co., 69, 80, 85, 204, 227
    Bono, Philip, 92
    Douglas Skyrocket, 25-26, 30-31
Drew, Russell, 124, 143, 148-149, 164, 167, 362
DuBridge, Lee, 123-124, 125, 126, 136, 142, 144, 148, 150, 164-165, 192, 362
Dyna-Soar, 49-54, 52, 80-81, 203, 205, 221

# E

Economists:
    Arrow, Kenneth, 315
    Burns, Arthur, 123, 176-177
    Friedman, Milton, 283, 315
    Galbraith, John Kenneth, 315

*Index*

Heller, Walter, 315
Leontief, Wassily, 315
Okun, Arthur, 315
Samuelson, Paul, 315
Wallich, Henry, 315
Edwards Air Force Base, 39-40, 41, 51, 201, 248, 310
    Bikle, Paul, 40
Eggers, Alfred, 27, 35-36, 206
Ehrlichman, John, 173, 176, 189, 269, 383, 393, 395, 404, 411-412, 414; quoted, 165-166, 389, 392, 414, 415
Eisenhower, Dwight, 153, 154, 156, 192, 193, 194
electronics, 55, 56-58, 68
    charge-coupled devices, 214, 225
engines, 6, 12, 19-20. *See also* rocket (several entries); ramjets; turbofans
environmental movement, 311-312, 318, 394
    Brower, David, 312
*Exploration of Space, The* (Clarke), 4

# F

F-100 fighter, 26, 305
FAA, 307, 308, 309, 310, 311, 318
Faget, Maxime, 34-35, 206-207, 221, 232
    space shuttle concept, 208-213, 217-223, 224, 233-234, 337, 401
    and external tanks, 341-344, *342-343*, 346, 351, 415-416
    foreshadows final shuttle design, *343*, 344
Fairey Delta FD-2, 305
Ferri, Antonio, 78, 79
Flanigan, Peter, 159, 165, 175-176, 177, 188-189, 273, 383, 385, 395, 400-401, 404, 406, 409, 411, 413, 422
Flax, Alexander, 363, 369-372, 373, 381, 382
Fletcher, James: quoted, 350, 402, 403, 404
    background, 272-273
    meets with Nixon, 408, 411-413
    and NERVA, 424
    and space shuttle, 350, 351, 363-364, 369, 380-381, 383, 384, 387, 400, 402-406, 408-411, 414, 422, 423
    shuttle contractor selection, 428, 429, 430-431, 432
Fubini, Eugene, 369, 382

# G

Gardner, Trevor, 32, 192
Garrett AiResearch, 77
Garwin, Richard, quoted, 314
Gemini program, 48, 53, 129-130, 203, 206, 253
    Big G, 129-130, 289, 370, 383-384, 385
General Accounting Office, 243
    Staats, Elmer, 244, 433-434
General Dynamics Corp. (Convair), 26, 75, 77, 82, 90, 92, 220, 222, 227
    Triamese, 89-90, *91*, 118, 218, 219-220
    shuttle studies of 1970-1971, 224
General Electric Co., 111, 188, 293, 296-305, 311
    Parker, Jack, 188

Gilpatric, Roswell, 195, 199
Gilruth, Robert, 28, 34-35, 72, 98, 187, 232; quoted, 128-129
Goddard, Robert, 6-7
Grand Tour (planetary mission), 392
Grumman Aerospace Corp., 227, 264-265, 267, 434
    Mead, Lawrence, 337
    space shuttle, contract competition, 428, 430-431
    space shuttle studies, 337-341, *339*, 344, 348, 354, 356, 374, *375*

# H

Haber, Heinz, 3
Halaby, Najeeb, 306, 308
Hansen, Grant, 133, 215, 216, 224, 226
Haughton, Daniel, 319-321, 323-326, 360
Hayden Planetarium, 2
    Coles, Robert, 2
heat sink (thermal protection), 30, 32, 349, 355, 356, 358
Heath, Prime Minister Edward, 322, 323-324
Heiss, Klaus, 281, 359, 373, 376-377, 410
Hilton hotels, 94, 246
historians, quoted:
    Benson, Charles, 65
    Catton, Bruce, 157
    Compton, David, 65
    Hunley, John, 11
    Horwitch, Mel, 310
    Koppes, Clayton, 285
    Logsdon, John, 103, 114
    Newhouse, John, 301
    Richelson, Jeffrey, 58, 204
    White, Theodore, 414
Holloman Air Force Base, 426
Hornig, Donald, 96, 203
hot structures, 30, 32, 34, 36-37, 38-39, 52, 53, 221, 248, 349
    for space shuttle, 221-222, 333-336, 340, 346, 348-349, 357
    *See also* heat sink; Inconel; thermal protection
House of Representatives, members:
    Clawson, Del, 432
    Fulton, James, 424
    Fuqua, Don, quoted, 182, 362
    Koch, Edward, quoted, 182
    Miller, George, 182; quoted, 178
    Patman, Wright, 326
    Roudebush, Richard, quoted, 182
    Yates, Sidney, 317
    *See also* Karth; Teague
Hunter, Max, 85-86, 88, 89, 207, 245-246, 247
hypersonic flight, 26-33, 34, 35-39, 50, 51, 53
    crossrange, 37-38, 213, 216-217

*Index*

# I

ICBMs, 26, 27, 45, 58, 194-197
Inconel metals, 30, 32, 34, 335
Institute for Defense Analyses, 363, 375
    Brady, George, 375-376
    Finke, Reinald, 375
Integral Launch and Reentry Vehicles, 92, 117

# J

Jet Propulsion Laboratory, 43, 123, 272, 285
    Dunn, Louis, 272
    Pickering, William, 285
    Ranger lunar program, 285
Johnson, Clarence "Kelly," 31, 192
Johnson, Lyndon, 100, 114, 156, 158, 194; quoted, 101, 191
    and space program, 64-65, 94-97, 101, 191, 203
    and SST, 306, 307-308

# K

Karth, Congressman Joseph, 180-182; quoted, 178, 181, 182
Kartveli, Alexander, 78
Kennedy, John F., 157; quoted, 154, 307, 390-391
    and Apollo, 59, 152, 390-391
    and civil rights, 156, 157
    and Cold War, 130-131, 152-156, 195
    and SST, 306, 307
Killian, James, 192
King, Martin Luther, 151, 152, 157
Kissinger, Henry, 173-174, 204-205, 269
Kiwi nuclear rocket, 105-108
Korean War, 22, 153
Kryter, Karl, quoted, 310
Kubrick, Stanley, 11

# L

Land, Edwin, 192
Lang, Fritz, 9
Lapp, Ralph, 259
Ley, Willy, 1-2, 3, 4, 5, 9
lifting bodies, 35-36, 37-42, 89
    Peterson, Bruce, 41
    Reed, R. Dale, 39-40
    as space shuttles, 80, 82, 88-89, 207-208, 209, 218, 220, 337
    Thompson, Milt, 40
    HL-10, 40, 41, 218
    M2-F1, *38*, 39-40, 41
    M2-F2, *38*, 40, 41
    X-24A, *39*, 40
    X-24B, 41, *42*
Lilly, William, 404-405, 422; quoted, 404, 409

Lincoln, Abraham, 152
Lindbergh, Charles, 315
Lindley, Robert, 259-264, 274-275, 376
Lockheed Aircraft Corp., 31, 82, 85, 218, 220, 221
  C-5A, 294, 320-321, 323, 360
  Johnson, Clarence "Kelly," 31, 192
  L-1011 airliner, 323, 324-327, 329-330
  Root, Eugene, 85
  Shuttle contract competition, 428
  shuttle studies, 1970-1972: 224, 264, 265-266, 267, 278, 354, 356, 360, 419
  solid-fuel rockets, 46, 353, 419, 422
  Space Shuttle, final design foreshadowed, 222, 331
  Star Clipper concept, 88-89, 118, 206, 207, 210, 218, 220, 265, 336-337
  *See also* Haughton; Hunter; SR-71
Lockheed L-1011, 319-327
  and DC-10, 319, 326, 327
  engines, 320, 321-323, 326
Lord Carrington, 324
Lord Cole, 322, 323
*Los Angeles Times*, 79
  Conrad, Paul, *316*, 317
  Miles, Marvin, 79
Low, George, 112, 267-268, 372; quoted, 382-383, 424
  and budgets, 270-271, 331
  and contractor selections, 428, 430, 431, 432, 434
  and space shuttle, 270-271, 289, 366-367, 383-384, 385-386, 387, 396, *397*, 400, 402-403, 405-406, 407-408, 409-411, 418, 422, 426-427

# M

MacLeish, Archibald, quoted, 147
Magruder, William, quoted, 313
  New Technology Opportunities Program, 393, 410
  SST, 313, 330
Manhattan Project, 3
*Man in Space* (Disney), 5, 11
Manned Orbiting Laboratory (USAF), 64, 203-205, 214
Manned Orbiting Module (NASA), 127
Marquardt Co., 77
Mars: automated missions, 98, 99, 101-102, 176, 181-182, 285
Mars: piloted missions, 3, 5, 60, 70-71, 120, 126, *138*, 140, 165
  and BoB, 171
  and Congress, 178, 181-182, 216, 424
  engines for, 94-95, 105. *See also* NERVA
  as NASA goal, 96, 98, 101, 137-138, 142, 146-147, 149-150, 164, 166, 167, 180, 425
Martin Marietta Corp. (also Martin Aircraft Co.):
  LeVine, David, 352
  lifting bodies, 37, 40, 41
  space shuttle, 80-81, 118, 224, 352, 367, 434
  Titan launch vehicles, 49, 86
  *See also* Titan
Mathematica, Inc.:
  recommends TAOS, 378-379
  *See also* Heiss; Morgenstern

*Index*

Mayo, Robert:
    director, Budget Bureau, 121, 159, 204-205, 259, 269
    and NASA budget, 121-123, 160, 161, 163-164, 170, 174-175, 177
    and space shuttle, 262-263, 274
    and Space Task Group, 125, 148, 164-165, 173
McCurdy, Richard, 412, 428-429; quoted, 430, 431, 432
McDonnell Aircraft Corp., 36, 89, 227
McDonnell Douglas Corp., 89, 218, 219, 220, 222, 319, 434
McDonnell, Sanford, 432
    shuttle contract competition, 428, 429-430
    shuttle studies, 1970-1972: 224, 226-227, 253, 255, 266, 332-336, *335*, 341, 348, 349-350, 354, 356, 374, 419
    space station studies, 227-228, 229, *230*
    Tip Tank concept, 89, *90*, 118, 207, 218
    Yardley, John, 332, 348, 416
McNamara, Robert, 53-54, 195, 203, 213, 286, 293, 308
    and SST, 308
Meany, George (AFL-CIO), 317; quoted, 269
Mercury piloted spacecraft, 36, 206
missiles:
    Jupiter, 24, 196
    Minuteman, *44*, 45-46, 61, 196
    MX-770, 16, 18-19
    Polaris, 44, 45, 196
    R-7 (Soviet ICBM), 195, 196-197
    Redstone, 22-24
    *See also* Atlas; Navaho; Thor; Titan; V-2
Mondale, Senator Walter, 183, 184, 231, 329; quoted, 183
Moon: automated missions, 99, 112-113, 285
Moon: piloted missions, 3, 5, 59-60, 70, 99, 140, 150, 155
    *See also* Apollo
Morgenstern, Oskar, 264, 377, 409
MORL space station concept, 69-70, 72, 126-127, 129
Morrow, Ian, 322
Mudd, Henry, 428
Mueller, George, 61-62, 98, 112, 119, 133, 179, 272; quoted, 92-94, 246
    and Apollo Applications, 62-66, 102, 200, 203, 204
    and post-Apollo future, 136-141, 146, 148, 149, 167, 170, 189, 423, 434
    and space shuttle, 86, 92-94, 116, 117, 131-132, 133-134, 215, 224, 246, 251
Myers, Dale, 25, 189, 223, 233, 270, 272; quoted, 179-180
    and space shuttle, 179-180, 224, 226, 229, 234, 351, 368, 387, 400, 405, 407, 421-422, 426, 428

# N

NACA (National Advisory Committee for Aeronautics), 25, 27, 28-29, 30, 31-32, 33
    Ames Aeronautical Laboratory, 35-36
    Dryden, Hugh, 32
    Langley Aeronautical Laboratory, 27-28, 34-35, 206
    Pilotless Aircraft Research Division (Langley), 28, 34, 206
    Thompson, Floyd, 27-28, 72
NASA (National Aeronautics and Space Administration), 6, 25, 59, 85, 102-103, 106, 110, 152, 159, 223, 233, 288-289, 358-359, 424-425
    Ames Research Center, 39, 40, 81, 115, 201
    BoB, OMB views of, 109-110, 171-172, 360-361, 364-366, 396, 398-400

budget and staff level, 64-65, 94, 101, 102, 110, 121-123, 159, 161-164, 167-170, 174-177, 185-186, 259-260, 270-272, 291, 331, 363-364, 366, 369, 418
    cost overruns, 253-254, 287
    Electronics Research Center, 176
    Langley Research Center, 40, 67, 68, 72, 74, 79, 201
    long-range plans, 6, 97-100, 137-141, 149-150, 162, 166-170, *168*, *169*
    Michoud Assembly Facility, 60, 425
    nuclear propulsion, 106, 423-424
    Office of Manned Space Flight, 63, 86, 116, 119, 129, 136, 141, 180, 421-422
    Office of Space Science and Applications, 63, 108, 136, 141, 180
    Planning Steering Group, 109, 136, 141
    procedures, planning, 108-109, 136-137
    program phases, 254-255
    space shuttle studies, 116-118, 218, 223-224, 226-227, 264-267, 332, 337-338, 357, 402-403, 405
    White House views of, 160-162, 165-166, 273-274
NASA Manned Spacecraft Center, 34-35, 70, 101, 112, 131, 358-359
    space shuttle, 82-83, 116, 223-224, 238, 359, 408, 417
    *See also* Faget; Gilruth
NASA Marshall Space Flight Center, 74, 358-360
    and Apollo, 60-61, 224, 358
    committed to liquid-fuel rockets, 359, 376, 416
    Mars missions, 96, 146-148
    Rees, Eberhard, 242, 422
    solid-propellant rockets, 46, 48, 359, 376, 417, 422
    space shuttle, 82-83, 84, 116, 131, 223-224, 338, 353, 416-417, 421, 422
    space stations, 71-72
    SSME, 236-238, 242-243
NASA officials:
    Akridge, Max, quoted, 82, 116, 220
    Culbertson, Philip, quoted, 286
    Hearth, Donald, 137
    Love, Eugene, 40, 416
    Mark, Hans, quoted, 115
    Mathews, Charles, 129, 131, 209
    McGolrick, Joseph, quoted, 263, 286
    Phillips, Sam, 112
    Schneider, William, 66
    Schnyer, Daniel, 116
    Silveira, Milton, 220, 222; quoted, 208
    Silverstein, Abe, 127, 128
    Wyatt, Dale, 360
    *See also* Day; Donlan; Faget; Fletcher; Gilruth; Lilly; Lindley; Low; McCurdy; Mueller; Myers; Newell; Paine; Shapley; Von Braun; Webb
National Aeronautics and Space Council, 125, 382
National Air and Space Museum, 66
National Science Foundation, 73, 395
    Bisplinghoff, Raymond, 395
National Security Council, 268, 269
Navaho missile, 20-25, 45, 179, 272
    engines, 19-24, *23*, 25, 26
Navy:
    and high-speed flight, 20, 30-31, 32
    and solid-propellant rockets, 43-44

*Index*

NERVA nuclear rocket, 94-95, 100, 106-108, *107*, 139, 239, 365-366, 423-425, 427
   termination of, 270, 424-425
Newell, Homer, 63, 108-109, 136-141; quoted, 109
New Technology Opportunities Program, 392-395, 414-415
Nexus launch vehicle concept, 75, *76*, 79, 92
Nicholson, Rupert, 323
Nixon, Richard, 115, 121, *122*, 123, 130-131, 144-145, 160, 161, 175, 177, 179, 184, 188, 189, 204-205, 243, 326, 389-390, 392-393, 424
   approves space shuttle, 386, 387, 400, 408, 411-413, 414-415
   management, interest in, 268-269, 414
   sets up Office of Management and Budget, 268-270
   space, statements on, 147, 391-392, 412-413
   space, views on, 115-116, 161-162, 170, 366, 368, 389, 435
   and Space Task Group, 123, 125, 166, 167, 170, 390
Nixon Administration officials:
   Burns, Arthur, 123, 176, 177
   Connally, John, 325
   David, Edward, 362, 382, 393
   Harper, Edwin, quoted, 395
   Helms, Richard, 204
   Kriegsman, William, 390
   Laird, Melvin, 125, 204-205
   Peterson, Peter, 383; quoted, 395
   Romney, George, 177
   Rose, Jonathan, 383
   Packard, David, 402
   Ruckelshaus, William, 317
   Schlesinger, James, 159, 161
   Seaborg, Glenn, 125
   Staats, Elmer, 244, 433-434
   Train, Russell, quoted, 314
   Ziegler, Ron, 167
   *See also* Flanigan; Kissinger; Magruder; Mayo; Office of Management and Budget; Rice; Rose; Seamans; Shultz, Weinberger; Whitehead
Noordung, Hermann (pen name of Herman Potočnik), 9-11, 67, 68
North American Aviation, 14-15, 24-25, 26, 32, 35, 67-68, 75, 82, 117-118
   Atwood, J. Leland, quoted, 16, 17
   Bollay, William, 15, 16-17
   corporate divisions, 24-25
   Kindelberger, James "Dutch," 14-15
   Moore, John R., 25
   and postwar missiles, 16, 17, 20, 22, 23
   Rice, Raymond, 15
   Storms, Harrison "Stormy," 35
   Tinnan, Leonard, 75
North American Rockwell, 117, 118, 218, 220, 222, 227, 431-432, 434
   Anderson, Robert, 240, 243
   Bergen, William, 434
   foreshadows final shuttle design, 344-346, *345*
   Hello, Bastian, 332
   space shuttle, contract competition and award, 427-428, 431-432
   space shuttle studies, 1970-1972: 224, 226-227, 253, 255, 266, 332-333, *334*, 335, 344-348, *345*, *347*, 354, 356, 419, *429*
   space station studies, 227-228, 229

457

Northrop Corp., 40, 41
nuclear weapons, 22, 44, 48, 50, 191-192, 193

## O

Oberth, Hermann, 6-9
Office of Management and Budget, 269, 270-271, 282-283, 288, 331, 363-364, 380-382, 384-386, 396, 399-401, 403-408, 414, 422, 424
    Loweth, Hugh, 393
    Niskanen, William, 283-284, 286, 404
    Sullivan, John, 275, 360, 372, 381, 384, 406, 420-421
    Taft, Daniel, 271, 288, 360, 362, 363, 384, 417
    Young, John, 364, 372
    *See also* Bureau of the Budget; Rice; Schulz; Weinberger
Office of Science and Technology, 98, 123-124, 143, 362, 393, 410
    Goldmuntz, Lawrence, quoted, 394

## P

Paine, Thomas, 66, *111*, 114-116, 124, 125, 126, 147, 150, 188-189, 229, 267-268; quoted, 121, 124, 127-128, 146-147, 177, 179, 183-184, 187-188, 389-390
    and Apollo, 112-114, 186, 229
    background, 111, 114-115, 188, 272
    boldness of, 114, 121-122, 127-128, 170, 187-188, 400
    and civil rights demonstrators, 151-152
    and critics, 179
    and federal budget, 160, 164, 174-177, 183-184, 229
    and post-Apollo future, 118, 125-126, 130-131, 136, 141, 144, 146-149, 165, 189
    and space shuttle, 132-133, 224-225, 425
Pal, George, 4-5, 11
Pan American World Airways, 295, 297, 328
    airliners, fault detection in, 252-253
    Boeing 747, 295, 297, 298-299, 300
    Borger, John, 295
    Kuter, Laurence, quoted, 299
payload effects, 260-264, *261*, 267, 275, 278, 279, 284, 286
Pearson, Sir Denning, 322
Penkovskiy, Oleg, 196-197
Penn Central Railroad, 326, 393-394
Phoebus nuclear rocket, 108
Pierce, John, 57
Potočnik, Herman, *see* Noordung
Pratt & Whitney, 235-236, 238-239, 255, 266
    appeals to GAO, 243-244, 432-434
    jet engines of, 296, 298-299
    Mulready, Richard, 243
    Torell, Bruce, 243
    RL-10 rocket engine, 80, 235, 244, 248
    XLR-130 rocket engine, 236, *237*, 238, 240, 241-242, 249, 433
President's Science Advisory Committee, 98-100, 119, 131, 142-144, 204, 362
PRIME (hypersonic research vehicle), 37-39, 40, 41
*Problem of Space Travel, The* (Noordung), 10
project management:
    cost overruns, 116-117

# Index

preliminary design, 381
program phases, 254-255
studies, 116-117
Proxmire, Senator William, 184, 186, 326, 329; quoted, 311, 313, 413
opposes SST, 311, 313-314

## Q

Q-ball instrument, 34
Quarles, Donald, 50

## R

*Raketenflugtechnik* (Sanger), 12
ramjets, 19-20, 22, 77-78
    supersonic combustion (scramjet), 78-79, 81, 82, *83*, 84
    Worth, Weldon, 20, 78-79
Rand Corp., 26, 159, 286
reaction controls, 33-34
Reaction Motors, Inc., 32
Reagan, Ronald, 230
reconnaissance, 2, 8, 50, 53, 191-198, 203, 216, 284
    Big Bird, 225, 232, 259, 273
    Corona, 58, *59*, 194-198, 213-214, 225, 227, 259, 284-285
    Gambit, 197, 213
    Kennan, 225
    real-time, 213-214
    significance of, 58, 191, 197, 198
    *See also* Central Intelligence Agency; Discoverer
reentry, 26, 29, 34, 35-36, 37-38, 41, 206, 209, *217*, 401-402
*See also* hypersonic flight; thermal protection
Rice, Donald, 270-271, 286-287, 288, 360-361, 363, 382, 383, 393, 400, 403-404, 406, 407, 409, 410; quoted, 361
Richelson, Jeffrey, quoted, 58, 204
riots, 100-101
Rockefeller, Nelson, 144
Rocketdyne, 24-25, 74, 235, 238, 239, 240-244, 255, 266, 432
    aerospike engine, 235, *236*, 238, 266
    Biggs, Robert, quoted, 240, 242, 250
    Brennan, William, 240, 243
    F-1 engine, 248, 249, 360, 424-425
    Hoffman, Sam, 25, 272
    J-2 engine, 25, 235, 240, 248, 249, 425
    J-2S engine, 265, 351, 356
rocket, liquid fuel:
    airbreathing (LACE), 77
    early concepts, 6-7, 8-9, *15*
    injector, 17-18, *19*, 21, 240
    in 1960s, 235-236
    postwar development, 16-18, 20-24, 25, 26
    principles of, 7, 21, 235-236, *397*
    reusable, 74, 80, 248
    test facilities, 16, 17, 21, 60
    turbojets, compared to, 12

for X-15, 32, 34
    *See also* Pratt & Whitney; Rocketdyne; Space Shuttle Main Engine
rocket, nuclear, 105-108, 139, 148
    Bussard, Robert, 105
    Dewar, James, quoted, 106
    Finger, Harold, 106
    *See also* NERVA
rocket propellants:
    liquid, 6, 7, 8, 45, 338
    solid, 43-44, 45
rocket, solid fuel, 43, 45, 46, 48, *49*, 84, 86, 88, 199, 421-422
    Atlantic Research Corp., 43
    Henderson, Charles, 43, 44
    Rumbel, Keith, 43, 44
    safety of, 419-420
    Sun Shipbuilding and Dry Dock Co., 46
    *See also* Aerojet; Lockheed; Thiokol; United Technology
rocket, winged, 12-14, *13*, 15-16, 18-20, 22, 25, 74-75, 79-81
    Bomi, 27, 49
*Rockets, Missiles and Space Travel* (Ley), 1-2
Rogallo, Francis, 74
Rolls-Royce, Ltd., 322-324
    Avon turbojet, 305
    Huddie, David, 320
    RB-211 turbofan, 320, 321-322, 324
ROMBUS launch vehicle concept, 92
Rose, Jonathan, 383, 406, 408, 414, 422
Ryan, Cornelius, 2-3

# S

Sänger, Eugen, 12-13, 27, 50
Saturn I, 25, 62
Saturn I-B, 25, 60, 74, *83*, 84, 98, 99-100, 199-200, 266, 370
    and Apollo Applications, 62, 63, 66, 67
    cost, 73, 199, 277
    S-IVB (second stage), 62, 63, 67, 70
    and space stations, 69, 70, 126-127, 129
Saturn V, 6, 25, 35, 59, 60, 62, 65-66, 98, 102, 112, 125, 229, *276*, 348
    and Apollo Applications, 63, 65, 67, 147
    cost, 48, 73-74, 246-247, 251-252
    production, 65, 99, 174, 176
    S-IC (first stage), 270, 351, *352*, 353, 416, 418-419, 425
    and space stations, 67-68, 70, 72, 129-130, 180, 228-229
Schlesinger, James, 159, 161
Schriever, Maj. Gen. Bernard, 45, 54, 115
Science and Technology Advisory Committee (NASA), 119, 137, 139
science fiction, 1, 4-5, 9, 11
Seamans, Robert, 119, 126; quoted, 142, 234
    Air Force Secretary, 125, 224-225
    background, 110-111
    endorses space shuttle, 142, 234
    and Space Task Group, 125, 132-133, 136, 142, 143, 144, 148, 150, 164
Seattle, 303-304

*Index*

Senate, members:
    Cannon, Howard, 424
    Cranston, Alan, 327
    Fulbright, William, 184, 413
    Goldwater, Barry, 184
    Jackson, Henry, 312, 315
    Kennedy, Edward, 179
    Magnuson, Warren, 315
    Mansfield, Mike, 178
    McCarthy, Joseph, 153
    McGovern, George, 413
    Metcalf, Lee, 327
    Nelson, Gaylord, quoted, 313
    Pastore, John, quoted, 231
    Scott, Hugh, 183-184, 424
    Smith, Margaret Chase, quoted, 178
    Symington, Stuart, 193, 194
    *See also* Anderson; Johnson, Lyndon; Mondale; Proxmire
Shapley, Willis, 109, 115, 286, 331; quoted, 109, 115-116, 410
Shultz, George, 269-270, 368, 390, 395, 404, 406, 409-410, 415, 418
Sierra Club, 312, 313
Silveira, Milton, 220, 222; quoted, 208
Six-Day War (1967), 213
Skylab, 12, 25, 60, 66, 67, 138-139, 147, 434
    *See also* Apollo Applications
SNECMA (France), 305, 306
solar power, 10-11, 62, 63, 412
Soviet Union, 9, 13, 22, 50, 130-131, 154-155, 195
    armaments, 13, 191-197, 198
    invasion of Czechoslovakia, 213-214
    Khrushchev, Nikita, 155
    Mir space station, 67
    Nedelin, Marshal Mitrofan, 197
    Plesetsk rocket center, 198
    Severodvinsk naval base, 198
    Stalin, 13, 155
    Tyuratam rocket center, 193-194, 196, 198
    Zond 5 (lunar mission), 112-113
    *See also* Cold War
Space Act (1958), 98, 124-125, 329
space base (crew of 50 to 100), 129-130, 137, *140*, 150
spacecraft, automated, 56-58, *59*, 139, 194
    design of, 260-261, 278, 284-285
    low cost, *see* payload effects
    Vanguard, 57-58
space flight:
    changing prospects, 56-59
    criticized by BoB and OMB, 109-110, 171-172
    early visions of, 6-13, 55-57
    long durations with crews, 67
    and national prestige, 154
    national significance, 414
    public opinion of, 178-179
    as wasteful, 74

*THE SPACE SHUTTLE DECISION*

Space Shuttle, 1, 22, 54, 73, 81, 84, 120, 293, *399*, 422-423, *429*
    and Air Force, 133, 214-217, 224-226, 231-234, 401, 402, 422, 427
    aluminum or titanium in construction, 222, 331-332, 416
    approved by Nixon, 386-387, 408-415
    and BoB, 170-171, 174, 255-259
    booster, *see* space shuttle, booster
    concept of 1952: 3, *4*, *5*, 6
    concepts of 1960-1969: 75-84, *81*, *83*, 86-91, 93-94, 116-118, 134-135, 217-223
    and Congress, 179-184, 230-231, 329
    contract award, 427-432
    cost-benefit tradeoff, 90-91, 337, 373, *374*, 384, *385*
    costs, projected for program, 258, 277, 279, *280*, 287, 340, 341, 350, 351, 356-357, 360, 370-372, 373, 378-380, 405, 420-421, 422
    economic studies, 86, 90-91, 255-260, 262-263, 267, 274-275, 279-283, *280*, 370-374, 410, 435
    with external tanks, 336, 337-346, *339*, *342*, *343*, *345*, 434
    fully reusable, 81, 82-84, *83*, 85, 89, 118, 135, 208-210, 217-221, *219*, 274-275, 289, 331, *333*, *335*, 337, 357-358, 361, 387
    funding and budget, 118, 131, 169, 180, 183-184, 185, 218, 223, 227, 255, 266-267, 369, 410, 418
    with jet engines, 232, 277, 346, 348
    launch and abort, 215, 275-277
    launch site selection, 425-427
    low-cost flight, technology for, 246-253, 254, 287, 435
    mission models, 263-264, 277-278, 281, 284
    missions and rationale, 135-136, 179-180, 205, 214-215, 218-219, 229
    name, origin of, 6, 92, 413
    as NASA goal, 137, 142-144, 149-150, 164, 189-190, 233, 267
    and OMB, 271-272, 282-288, 331, 337, 358, 360-361, 380-382, 383-387, 396, 399-401, 403-408, 417-418
    onboard checkout, 88-89, 117, 133-134, 251-253, 261-262, 278
    orbiter, *see* space shuttle, orbiter
    partially reusable, 85, 88-89, 91, 217-218, 265, 275, 278, 336-337
    payload capacity, 117, 133, 225, 231-233, 341, 405-407, 409, 415, 420
    phased development, 265, 267, 270, 350, 356, 359, 370, 415-416
    reviewed by Flax panel, 362-363, 369-372, 385
    and space station, 73, 93, 116, 117, 118, 130, 131, 133, 135-136, 179-180, 225-226, 228-230, 329, 434
    and Space Task Group (1969), 133-136, 137, 150
    TAOS, 373-380, *375*, *377*, 384, 396, *398*, 417
    technical background for, 33-34, 48, 49
    and White House officials, 382-383, 395. *See also* Nixon
    *See also* Air Force; NASA; Space Shuttle Main Engine
space shuttle, booster, 346-350, *347*, 351-358, 359-360, 416-423
    expendable, 117-118, 134-135, 218, 265, 351-354, *352*
    pressure-fed, 353, *354*, 355-356, 379, *398*, 417-418
    pump-fed, 416-417, 419, 421
    solid propellant, 118, 353, 354, 374-375, 406, 415, 417, 420-421
    and staging velocity, 338, 340, 341, 344, 348-349, 354-355, 421
    winged S-IC, 74-75, *355*, 356-357, 359-360, 361, 381, 416
space shuttle, orbiter, 331-346, *334*, *336*, *345*, 357
    final design, 331, 344-346, *345*, *399*
    glider, 367-368, 369-370, 384, 385-386, 411
    Phase B concepts, 332-336, *334*
Space Shuttle Main Engine:
    background, 235-239, 248-250, 255, 266

*Index*

    research and test, 240-242, *241*
    contract and protest, 239, 243-244, 432-434
    NASA role for, 34, 271, 351, 356, 401, 416
space station, 12, 59-60, 66, 120, 229-230
    and Air Force, 203, 205
    and BoB, 171, 174
    bypassed by technology, 55-59
    concepts, early, 2-3, 6-11, *10*, 55-56, 57
    concepts of 1960s: 67-73, *68*, *69*, *71*, 126-127, 225
    concepts of 1969-1970: 128-130, 131, 140, 180, 227-228
    funding and budget, 131, *169*, 223, 272
    Heberlig, Jack, 227, 228-229
    modular designs, 225-226, 229, *230*
    as NASA goal, 98, 99, 100, 137, 142-144, 149-150, 164
    rationales for, 68-69, 71-72, 93
    *See also* Apollo Applications; MORL; Skylab
Space Task Group, 125-126, 130-131, 132-133, 136, 137-138, 141-142, 144, 145-150, 160, 164-170, 174, 181, 189, 190, 224
space tug, 139, 226, 232, 377, 378, 379, 423
SR-71 aircraft, 210, 213, 221, 222
SST (supersonic transport), 304, 306-318, 319, 329-330, 392
    and Congress, 313-318
    and environment, 311, 312-314, 315-317, *316*
    financing of, 306, 307, 308, 313, 318
    sonic booms, 308-310
SST, supporters and opponents:
    Beranek, Leo, 317
    Black, Eugene, 307-308
    Greif, Kenneth, 312, 315
    Johnston, Harold, 316-317
    Kellogg, William, 317
    Leovy, Conway, 314
    Lindbergh, Charles, 315
    Lindsay, John, 314
    McDonald, James, 315-316
    Osborne, Stanley, 307-308
    Ruckelshaus, William, 317
    Shurcliff, William, 311, 312
    Train, Russell, quoted, 314
    *See also* Magruder; McNamara; Proxmire
State Department, 125, 268
    Acheson, Dean, quoted, 414
    Johnson, U. Alexis, 125
Steiner, John (Boeing), 295; quoted, 302, 303, 327-328
submarines, 6, 7, 44-45, 111, 198
Sud Aviation (France), 305, 306
Suharto (Indonesia), 155
supersonic flight, 19-20, 25-26, 27, 305
    sonic boom, 308-309

# T

Taylor, Charles, 193
Teague, Congressman Olin "Tiger," 178, 180-181, 362, 426; quoted, 180
telemetry, 56, 284-285
test pilots, 92, 247
    Knight, William, 33
    Peterson, Bruce, 41
    Thompson, Milt, 40
    Walker, Joseph, 32
Texas Instruments, Inc., 115, 245-246
    Haggerty, Patrick, 115
*Theory of Games and Economic Behavior* (Morgenstern, Von Neumann), 264
thermal protection, 36, *37*, 38
    materials, 36, 221-222, 251, 331, 334, 335, 346, 348
    for shuttle booster, 222, 346-350, *347*, 356, 358
    for shuttle orbiter, 212, 217, 222, 250-251, 333-336, *334*, 416
    *See also* heat sink; hot structures
Thiokol Chemical Corp., 42-43, 45, 46, 86, 353, 422, 434
    Crosby, Joseph, 43, 45
Thomas, Dylan, quoted, 110
Thor rocket:
    and space flight, 25, 36, 37, 85, 194, 197, 199
    as weapon, 24, 45, 50, 196
    *See also* Delta
Thorpe, Jeremy, 324
*Time* magazine, 4, 6, 114
Titan ICBM, 24, 45, *47*, 48, 50, 51, 196, 199
Titan II, *47*, 48-49, 51
Titan III, *47*, 48-49, 51, 88, 102, 199-200, 203, 225, 232, *399*, 400
    alternative to shuttle, 170, 216, 255-259, 288, 380, 406-407
Titan III-L, 352-353, 354, 367
Titan III-M, 49, 84, 86-88, *87*, 91, 99-100, 129, 199, 352, 370
Townes, Charles, 119
    committee report (1969), 119-121, 123, 125, 126, 141, 245
Treaty of Tordesillas, 198-199
Trippe, Juan, 295, 297, 306-307
Truman, Harry, 153
TRW, Inc., 115, 272-273
    Mettler, Ruben, 272-273
    Ramo, Simon, 115
    Solomon, George, 273
Tsiolkovsky, Konstantin, 9
turbofans, 293, *294*, 319, 323
    CF-6 (General Electric), 323
    JT-9D (Pratt & Whitney), 296, 298-299, 323
    ovalization problem, 301-302
    RB-211 (Rolls-Royce), *294*, 320, 321, 324
    TF-39 (General Electric), 293, 296
TWA (Trans World Airlines), 325, 328
Tyuratam (Soviet rocket center), 193-194, 196, 198

# Index

## U

U-2 aircraft, 192, 193-194, 195, 213
United Airlines, 305
United Technology Corp., 46, 352-353

## V

V-2 missile, 1-2, 3, 9, *13*, 14, 15-16, 56, 251
    engine, 16-18, 21, 23-24
V-2, project staff:
    Dannenberg, Konrad, 16-17
    Dornberger, Walter, 27, 28, 30; quoted, 92-93
    Ehricke, Krafft, 75
    Huter, Hans, 16-17
    Huzel, Dieter, 17
    Rees, Eberhard, 242, 243
    Riedel, Walter, 17
    Riedel, Walther, 16-17, 21
    Roth, Ludwig, 14
    Thiel, Walter, 17-18
    *See also* von Braun
Van Allen, James, 351
Vandenberg Air Force Base, 38, 194, 198, 214-215, 233, 427
Vietnam War, 100, 101, 114, 153-154, 155-156, 204, 291
    Ho Chi Minh, 153
von Braun, Wernher, 9, 56, 72, 117, 187, 242, 266; quoted, 60, 102, 129, 146, 152
    and Apollo, 60, 62, 102, 152
    colleagues of, 16-18. *See also* V-2, project staff
    and *Collier's* series, 2-3, 6, 11, 56, 57
    flight to Mars, 146, 148
    publicizes space flight, 4, 5
    and rocket engines, 22-24, 236-238, 359
    and space stations, 2-3, 11, 67, 68, 98, 129
von Harbou, Thea, 9
von Neumann, John, 264

## W

Webb, James, 65, 94-98, *95*, 102, 108, 110, 112-113, 114, 199, 203, 272, 283; quoted, 97
Weisner, Jerome, 95, 361
Weinberger, Caspar, 363-366, *365*, 380-382, 404, 406, 409, 414, 418, 422; quoted, 381-382
Wellington, Duke of, quoted, 190
Westinghouse Electric Corp., 106, 239
Whipple, Fred, 3, 11
Whitehead, Clay, 159, 160-161, 176, 273, 383, 390
White Sands Missile Range, 426-427
Whittle, Sir Frank, 305
wings, 30, 36, 209-213, *211*, 233, 401-402
    canards, 18, 212
    Draper, Alfred, 210, 212
Worth, Weldon, 20, 78-79
Wright-Patterson Air Force Base (Wright Field), 19-20, 32, 36, 78, 83, 210

# THE SPACE SHUTTLE DECISION

## X-Y-Z

X-1 aircraft, 25
X-1A aircraft, 27
X-2 aircraft, 27, 28, 29
X-7 missile, 77-78
X-15 aircraft, 32-35, 41, 42, 49-50, 75, 92, 201, 209, 221, 349
    ground operations, 247-248
X-15, personnel:
    Brown, Clinton, 27-28
    Knight, William, 33
    McLellan, Charles, 29
    Soulé, Hartley, 28, 32
    Walker, Joseph, 32
XB-70 aircraft, 201, 210, 308, 310
XF-103 fighter, 78

York, Herbert, 50-51

# The NASA History Series

*Reference Works, NASA SP-4000:*

Grimwood, James M. *Project Mercury: A Chronology*. (NASA SP-4001, 1963).

Grimwood, James M., and Hacker, Barton C., with Vorzimmer, Peter J. *Project Gemini Technology and Operations: A Chronology*. (NASA SP-4002, 1969).

Link, Mae Mills. *Space Medicine in Project Mercury*. (NASA SP-4003, 1965).

*Astronautics and Aeronautics, 1963: Chronology of Science, Technology, and Policy*. (NASA SP-4004, 1964).

*Astronautics and Aeronautics, 1964: Chronology of Science, Technology, and Policy*. (NASA SP-4005, 1965).

*Astronautics and Aeronautics, 1965: Chronology of Science, Technology, and Policy*. (NASA SP-4006, 1966).

*Astronautics and Aeronautics, 1966: Chronology of Science, Technology, and Policy*. (NASA SP-4007, 1967).

*Astronautics and Aeronautics, 1967: Chronology of Science, Technology, and Policy*. (NASA SP-4008, 1968).

Ertel, Ivan D., and Morse, Mary Louise. *The Apollo Spacecraft: A Chronology, Volume I, Through November 7, 1962*. (NASA SP-4009, 1969).

Morse, Mary Louise, and Bays, Jean Kernahan. *The Apollo Spacecraft: A Chronology, Volume II, November 8, 1962-September 30, 1964*. (NASA SP-4009, 1973).

Brooks, Courtney G., and Ertel, Ivan D. *The Apollo Spacecraft: A Chronology, Volume III, October 1, 1964-January 20, 1966*. (NASA SP-4009, 1973).

Ertel, Ivan D., and Newkirk, Roland W., with Brooks, Courtney G. *The Apollo Spacecraft: A Chronology, Volume IV, January 21, 1966-July 13, 1974*. (NASA SP-4009, 1978).

*Astronautics and Aeronautics, 1968: Chronology of Science, Technology, and Policy*. (NASA SP-4010, 1969).

Newkirk, Roland W., and Ertel, Ivan D., with Brooks, Courtney G. *Skylab: A Chronology*. (NASA SP-4011, 1977).

Van Nimmen, Jane, and Bruno, Leonard C., with Rosholt, Robert L. *NASA Historical Data Book, Volume I: NASA Resources, 1958-1968*. (NASA SP-4012, 1976, rep. ed. 1988).

Ezell, Linda Neuman. *NASA Historical Data Book, Volume II: Programs and Projects, 1958-1968*. (NASA SP-4012, 1988).

Ezell, Linda Neuman. *NASA Historical Data Book, Volume III: Programs and Projects, 1969-1978*. (NASA SP-4012, 1988).

Gawdiak, Ihor Y., with Fedor, Helen. Compilers. *NASA Historical Data Book, Volume IV: NASA Resources, 1969-1978*. (NASA SP-4012, 1994).

*Astronautics and Aeronautics, 1969: Chronology of Science, Technology, and Policy*. (NASA SP-4014, 1970).

*Astronautics and Aeronautics, 1970: Chronology of Science, Technology, and Policy*. (NASA SP-4015, 1972).

*Astronautics and Aeronautics, 1971: Chronology of Science, Technology, and Policy*. (NASA SP-4016, 1972).

*Astronautics and Aeronautics, 1972: Chronology of Science, Technology, and Policy*. (NASA SP-4017, 1974).

*Astronautics and Aeronautics, 1973: Chronology of Science, Technology, and Policy*. (NASA SP-4018, 1975).

# THE SPACE SHUTTLE DECISION

Astronautics and Aeronautics, 1974: Chronology of Science, Technology, and Policy. (NASA SP-4019, 1977).

Astronautics and Aeronautics, 1975: Chronology of Science, Technology, and Policy. (NASA SP-4020, 1979).

Astronautics and Aeronautics, 1976: Chronology of Science, Technology, and Policy. (NASA SP-4021, 1984).

Astronautics and Aeronautics, 1977: Chronology of Science, Technology, and Policy. (NASA SP-4022, 1986).

Astronautics and Aeronautics, 1978: Chronology of Science, Technology, and Policy. (NASA SP-4023, 1986).

Astronautics and Aeronautics, 1979-1984: Chronology of Science, Technology, and Policy. (NASA SP-4024, 1988).

Astronautics and Aeronautics, 1985: Chronology of Science, Technology, and Policy. (NASA SP-4025, 1990).

Noordung, Hermann. *The Problem of Space Travel: The Rocket Motor.* Stuhlinger, Ernst, and Hunley, J.D., with Garland, Jennifer. Editor. (NASA SP-4026, 1995).

Astronautics and Aeronautics, 1986-1990: A Chronology. (NASA SP-4027, 1997).

## Management Histories, NASA SP-4100:

Rosholt, Robert L. *An Administrative History of NASA, 1958-1963.* (NASA SP-4101, 1966).

Levine, Arnold S. *Managing NASA in the Apollo Era.* (NASA SP-4102, 1982).

Roland, Alex. *Model Research: The National Advisory Committee for Aeronautics, 1915-1958.* (NASA SP-4103, 1985).

Fries, Sylvia D. *NASA Engineers and the Age of Apollo.* (NASA SP-4104, 1992).

Glennan, T. Keith. *The Birth of NASA: The Diary of T. Keith Glennan.* Hunley, J.D. Editor. (NASA SP-4105, 1993).

Seamans, Robert C., Jr. *Aiming at Targets: The Autobiography of Robert C. Seamans, Jr.* (NASA SP-4106, 1996)

## Project Histories, NASA SP-4200:

Swenson, Loyd S., Jr., Grimwood, James M., and Alexander, Charles C. *This New Ocean: A History of Project Mercury.* (NASA SP-4201, 1966).

Green, Constance McL., and Lomask, Milton. *Vanguard: A History.* (NASA SP-4202, 1970; rep. ed. Smithsonian Institution Press, 1971).

Hacker, Barton C., and Grimwood, James M. *On Shoulders of Titans: A History of Project Gemini.* (NASA SP-4203, 1977).

Benson, Charles D. and Faherty, William Barnaby. *Moonport: A History of Apollo Launch Facilities and Operations.* (NASA SP-4204, 1978).

Brooks, Courtney G., Grimwood, James M., and Swenson, Loyd S., Jr. *Chariots for Apollo: A History of Manned Lunar Spacecraft.* (NASA SP-4205, 1979).

Bilstein, Roger E. *Stages to Saturn: A Technological History of the Apollo/Saturn Launch Vehicles.* (NASA SP-4206, 1980).

SP-4207 not published.

Compton, W. David, and Benson, Charles D. *Living and Working in Space: A History of Skylab.* (NASA SP-4208, 1983).

Ezell, Edward Clinton, and Ezell, Linda Neuman. *The Partnership: A History of the Apollo-Soyuz Test Project.* (NASA SP-4209, 1978).

Hall, R. Cargill. *Lunar Impact: A History of Project Ranger.* (NASA SP-4210, 1977).

Newell, Homer E. *Beyond the Atmosphere: Early Years of Space Science.* (NASA SP-4211, 1980).

# The NASA History Series

Ezell, Edward Clinton, and Ezell, Linda Neuman. *On Mars: Exploration of the Red Planet, 1958-1978.* (NASA SP-4212, 1984).

Pitts, John A. *The Human Factor: Biomedicine in the Manned Space Program to 1980.* (NASA SP-4213, 1985).

Compton, W. David. *Where No Man Has Gone Before: A History of Apollo Lunar Exploration Missions.* (NASA SP-4214, 1989).

Naugle, John E. *First Among Equals: The Selection of NASA Space Science Experiments.* (NASA SP-4215, 1991).

Wallace, Lane E. *Airborne Trailblazer: Two Decades with NASA Langley's Boeing 737 Flying Laboratory.* (NASA SP-4216, 1994).

Butrica, Andrew J. Editor. *Beyond the Ionosphere: Fifty Years of Satellite Communication* (NASA SP-4217, 1997).

Butrica, Andrews J. *To See the Unseen: A History of Planetary Radar Astronomy.* (NASA SP-4218, 1996).

Mack, Pamela E. Editor. *From Engineering Science to Big Science: The NACA and NASA Collier Trophy Research Project Winners.* (NASA SP-4219, 1998).

Reed, R. Dale. With Lister, Darlene. *Wingless Flight: The Lifting Body Story.* (NASA SP-4220, 1997).

## Center Histories, NASA SP-4300:

Rosenthal, Alfred. *Venture into Space: Early Years of Goddard Space Flight Center.* (NASA SP-4301, 1985).

Hartman, Edwin, P. *Adventures in Research: A History of Ames Research Center, 1940-1965.* (NASA SP-4302, 1970).

Hallion, Richard P. *On the Frontier: Flight Research at Dryden, 1946-1981.* (NASA SP-4303, 1984).

Muenger, Elizabeth A. *Searching the Horizon: A History of Ames Research Center, 1940-1976.* (NASA SP-4304, 1985).

Hansen, James R. *Engineer in Charge: A History of the Langley Aeronautical Laboratory, 1917-1958.* (NASA SP-4305, 1987).

Dawson, Virginia P. *Engines and Innovation: Lewis Laboratory and American Propulsion Technology.* (NASA SP-4306, 1991).

Dethloff, Henry C. *"Suddenly Tomorrow Came...": A History of the Johnson Space Center.* (NASA SP-4307, 1993).

Hansen, James R. *Spaceflight Revolution: NASA Langley Research Center from Sputnik to Apollo.* (NASA SP-4308, 1995).

Wallace, Lane E. *Flights of Discovery: 50 Years at the NASA Dryden Flight Research Center.* (NASA SP-4309, 1996).

Herring, Mack R. *Way Station to Space: A History of the John C. Stennis Space Center.* (NASA SP-4310, 1997).

Wallace, Harold D., Jr. *Wallops Station and the Creation of the American Space Program.* (NASA SP-4311, 1997).

## General Histories, NASA SP-4400:

Corliss, William R. *NASA Sounding Rockets, 1958-1968: A Historical Summary.* (NASA SP-4401, 1971).

Wells, Helen T., Whiteley, Susan H., and Karegeannes, Carrie. *Origins of NASA Names.* (NASA SP-4402, 1976).

Anderson, Frank W., Jr. *Orders of Magnitude: A History of NACA and NASA, 1915-1980.* (NASA SP-4403, 1981).

Sloop, John L. *Liquid Hydrogen as a Propulsion Fuel, 1945-1959.* (NASA SP-4404, 1978).

Roland, Alex. *A Spacefaring People: Perspectives on Early Spaceflight.* (NASA SP-4405, 1985).

Bilstein, Roger E. *Orders of Magnitude: A History of the NACA and NASA, 1915-1990.* (NASA SP-4406, 1989).

Logsdon, John M. Editor. With Lear, Linda J., Warren-Findley, Jannelle, Williamson, Ray A., and Day, Dwayne A. *Exploring the Unknown: Selected Documents in the History of the U.S. Civil Space Program, Volume I, Organizing for Exploration.* (NASA SP-4407, 1995).

Logsdon, John M. Editor. With Day, Dwayne A., and Launius, Roger D. *Exploring the Unknown: Selected Documents in the History of the U.S. Civil Space Program, Volume II, Relations with Other Organizations.* (NASA SP-4407, 1996).

Logsdon, John M. Editor. With Launius, Roger D., Onkst, David H., and Garber, Stephen E. *Exploring the Unknown: Selected Documents in the History of the U.S. Civil Space Program, Volume III, Using Space.* (NASA SP-4407, 1998).

ACY2456

# LIBRARY
## LYNDON STATE COLLEGE
## LYNDONVILLE, VT 05851